ORGANIC ANALYSIS USING
ATOMIC ABSORPTION SPECTROMETRY

ELLIS HORWOOD SERIES IN ANALYTICAL CHEMISTRY
Series Editors: Dr. R. A. CHALMERS and Dr. MARY MASSON
University of Aberdeen

ORGANIC ANALYSIS USING ATOMIC ABSORPTION SPECTROMETRY

SAAD S. M. HASSAN

Professor of Chemistry
Ain Shams University
Cairo, Egypt

ELLIS HORWOOD LIMITED
Publishers · Chichester

Halsted Press: a division of
JOHN WILEY & SONS
New York · Chichester · Brisbane · Toronto

First published in 1984 by
ELLIS HORWOOD LIMITED
Market Cross House, Cooper Street, Chichester, West Sussex, PO19 1EB, England

The publisher's colophon is reproduced from James Gillison's drawing of the ancient Market Cross, Chichester.

Distributors:

Australia, New Zealand, South-east Asia:
Jacaranda-Wiley Ltd., Jacaranda Press,
JOHN WILEY & SONS INC.,
G.P.O. Box 859, Brisbane, Queensland 40001, Australia

Canada:
JOHN WILEY & SONS CANADA LIMITED
22 Worcester Road, Rexdale, Ontario, Canada.

Europe, Africa:
JOHN WILEY & SONS LIMITED
Baffins Lane, Chichester, West Sussex, England.

North and South America and the rest of the world:
Halsted Press: a division of
JOHN WILEY & SONS
605 Third Avenue, New York, N.Y. 10016, U.S.A.

QD
272
S6
H37
1984

© 1984 S.S.M Hassan/Ellis Horwood Limited

British Library Cataloguing in Publication Data
Hassan, S.S.M.
Organic analysis using atomic absorption spectrometry.
(Ellis Horwood series in analytical chemistry)
1. Atomic absorption spectroscopy
I. Title
543'.0858   QD96.A8

Library of Congress Card No. 83-22648

ISBN 0-85312-559-7 (Ellis Horwood Limited)
ISBN 0-470-27498-0 (Halsted Press)

Typeset in Press Roman by Ellis Horwood Ltd.
Printed in Great Britain by R.J. Acford, Chichester.

Table of Contents

Chapter 12 – Biological Materials

Table of Contents

Foreword

From the inception of analytical atomic absorption spectrometry some three decades ago to the present time, many analysts must have wondered (in both senses) about the apparently limitless supply of innovation and ingenuity from academics, manufacturers and users alike that has characterized its evolution. Even today the progress continues unabated. The technique has undeniably had a tremendous impact upon virtually every branch of physical and biological science, because of its sensitivity, and the speed and simplicity of analysis that stem in turn from its high degree of selectivity.

To the user, keeping abreast of developments can become a major headache. Computerized information retrieval helps up to a point but the outcome of computer searches for solutions to some types of problem may be limited by the quality of the abstracts on which they are based. Books on atomic absorption abound, but few provide comprehensive, in-depth treatment of specific problem areas.

Close scrutiny of this contribution from Hassan soon swept aside any cynical doubts I might have had about the need for yet another book on A.A.S. This book is undoubtedly the end-product of an extraordinary amount of work, bringing together as it does, not just information from the major analytical journals, but also the more useful contributions from those obscure specialist journals from other disciplines which are often so hard to obtain in a hurry. The work will be specially valuable to anyone involved in the determination of species other than the metal elements in a diverse range of matrices. It is impressive in a work of this kind that the author has obviously delved deeply into the original literature, and not relied on abstracts, an approach which all too often leads to little more than a sketchy and vague overview.

This book provides more than just a high-quality review of the determination of non-metallic species. It also contains a first-class introduction to both practical and theoretical aspects of atomic absorption. The coverage is very much state-of-the-art, and for this too the author is to be congratulated. And as if that is not enough, he goes on to provide, in the final chapter, a range of interesting experiments for students, complete with practical details.

The whole book is written in a clear, concise and very readable style, and can be recommended to novice and expert alike.

Malcolm S. Cresser
Aberdeen University

1

Basic principles and theoretical aspects

Atomic-absorption spectrometry (AAS) has now pervaded every field of chemical analysis more than any other technique in spectroscopy. The technique is used for the determination of at least 70 elements in amounts as low as 10^{-14} g with reasonable selectivity, little manipulation and minimum sample size. The sensitivity of the method favourably matches that of mass spectrometry or neutron-activation analysis and is much better than that of X-ray fluorescence and spectrophotometry. It is fortunate that the analysis of organic compounds has also benefited from this technique, which was originally developed only for metallic species, although it was several years after AAS became a viable analytical tool before papers began to appear describing methods for the analysis of various organic compounds.

Quantitative analysis by AAS is based on the measurement of the radiant energy absorbed by free atoms in the gaseous state. The technique owes its high selectivity to the fact that the spectra of gaseous atomic species consist of well-defined narrow lines at wavelengths characteristic of the element involved. In practice, the samples (solution or solid) are vaporized and on further heating the vapour dissociates into free atoms. The atomic vapour is then allowed to absorb radiant energy of a characteristic wavelength. Flames, electrical heating, inductively coupled plasmas and lasers are commonly used to convert the sample into atomic vapour [1].

Atomic absorption, emission and fluorescence spectrometric methods are similar in many respects and all use essentially the same technology. Atomic absorption and emission methods are about equally common for direct determination of metals and indirect measurement of non-metals and anions. Both techniques can be readily accommodated in a single instrument and all major atomic-absorption spectrometers now have an atomic-emission mode as well. However, AAS is exceptionally well suited for the analysis of samples containing organic materials, or proteins, or analytes dissolved in non-aqueous solvents. The presence of such substances in the analyte matrix affects and impairs the detection limit of the emission methods. Moreover, some metals (e.g., zinc) do not show emission lines but display strong absorption lines. Since practically all the atoms in the vapour phase are in the ground-state it might be expected that atomic absorption would always be more sensitive than atomic emission, but this is not always true, because at low analyte concentrations it is much easier to measure a very small emission signal with good precision than the very small difference between the two large signals from the sample and the reference beam in AAS. As a rough general rule, with a flame as the atom cell, elements with principle resonance lines at wavelengths shorter than 300 nm are measured with much greater sensitivity by AAS, whereas atomic-emission spectrometry is more sensitive for elements with resonance lines at wavelengths longer than 350 nm.

1.1 ATOMIC-ABSORPTION SPECTRA

The atomic-absorption spectra of most metals originate from the transitions of electrons from the ground state to excited states, referred to as resonance lines (Fig. 1.1). Some elements, however, have either complicated electronic structure

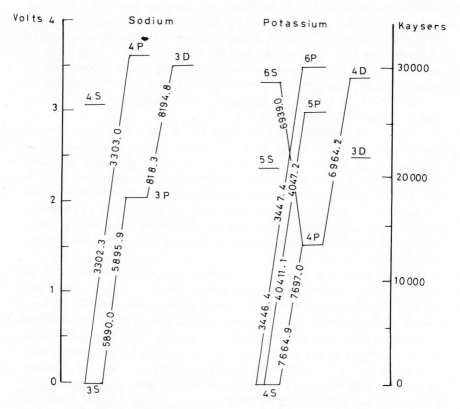

Fig. 1.1. Energy-level diagrams for atomic sodium and potassium.

(e.g. iron and manganese) or a low-lying multiplet component in the ground state (e.g. aluminium and tin). These elements frequently display several resonance lines close together. It should also be noted that the width of most atomic lines is inherently very small ($\sim 10^{-5}$ nm), and even when broadened in various ways the line width never exceeds 10^{-2} nm [2,3]. This means that if a continuous source of radiation were used for excitation, as is the case for most spectroscopic methods, atomic-absorption measurements would require a spectral band width of about 5×10^{-4} nm. Such high resolution is beyond the capacity of most monochromators.

The measurement of atomic-absorption spectra as a function of concentration was described in 1914 by Malinowski [4] who measured mercury vapour at 254 nm by a technique limited to the easily vaporized elements. It was not until 1955, when Walsh [1] and Alkemade and Milatz [5,6] independently but almost simultaneously used the flame for generating the atomic vapour, and a hollow-cathode lamp or gas discharge tube as excitation source, that analytical AAS was used for more difficultly vaporized elements. Instruments based on this principle became commercially available in 1959, and since that time the technique has flourished and now plays an ever-increasing role in the analysis of various substances. At least 8000 papers and many books [e.g. 7-17] have been published so far, reflecting the steady progress in various aspects of the field.

AAS in its present state involves the use of a source (hollow-cathode or electrodeless discharge lamp) to give strong sharp lines of appropriate wavelength. This radiation is allowed to pass through the atomic vapour of the analyte, thermally generated by flame, furnace, plasma or laser [18-22] and then through the monochromator to separate the resonance line from other radiation emitted by the source. Thermally excited emission by the analyte atoms is discriminated by modulating the source of radiation and using an a.c. amplifier, so that the unmodulated emission is not detected. The detector circuit may also be designed to reject the d.c. output from emission and to measure the a.c. absorption signal from the source and sample. Although 28 years have now elapsed since the publication of the landmark paper of Walsh, the principle of the instrumentation now in use is virtually unchanged from that described by this author. The current status and state of the art have been reviewed [23-29].

The magnitude of the atomic absorption is a complex function of the initial concentration of the analyte, and generally depends on the number of atoms capable of being excited. If the emission line from the source is very narrow, compared to that of the absorption line in the atomic vapour, and the sample is vaporized to the free atomic species with minimum loss, a linear relationship between line intensity and atomic concentration can be established. However, many factors affect the actual concentration of atoms. Expressions relating the atomic concentration to the analyte concentration in solutions, and the line intensity, have been reported [30-34].

1.2 INTENSITY OF ATOMIC LINES

1.2.1 Atomic-Absorption Coefficient

The total energy absorbed per unit time and atom-cell volume is proportional to the number of free atoms per unit volume in the ground state (N), the energy of the photon absorbed (E), and the probability of electronic transition [16]:

$$E = B_{ij} N I_\nu h\nu \qquad (1.1)$$

where B_{ij} is the probability factor or Einstein probability coefficient of absorption [35] from the ground state (i) to the excited state (j), I_ν is the intensity of the spectral radiation of frequency ν, and h is Planck's constant. Since the atoms can also be considered as an oscillating electric dipole, the total energy absorbed by such a harmonic oscillator per unit time can be expressed by the electrodynamic laws. Thus:

$$E = f \frac{\pi e^2}{m} I_\nu \qquad (1.2)$$

where e is the charge of the electron, m the electron mass, f a dimensionless factor (the so-called absorption oscillator strength), which is the average number of electrons per atom which can be excited by the incident radiation. This definition lost its classical meaning after the introduction of the quantum theory of radiation, but is still in use and may be derived by combining Eqs. (1.1) and (1.2) and considering that Eq. (1.2) is valid for one atom:

$$f = \frac{mh\nu}{\pi e^2} B_{ij} \qquad (1.3)$$

The atoms in the ground state absorb a number of photons that is proportional to the total number of free atoms and to the effective photon-capture cross-section of the atom [the so-called absorption coefficient (X)].

$$E = X N I_\nu c \qquad (1.4)$$

where c is the speed of light and hence $I_\nu c$ is the amount of radiation energy passing through unit volume in unit time. By combining Eqs. (1.1) and (1.4):

$$X = \frac{h\nu}{c} B_{ij} \qquad (1.5)$$

The absorption coefficient (X) is a measure of the quantity of light of frequency ν which can be absorbed by an atom, and has the dimension of an area, as expected from the effective cross-section. In practice it is more convenient to use an absorption coefficient (K) related to unit volume

$$K = XN \qquad (1.6)$$

where N is the number of atoms per unit volume capable of absorbing energy; K has the dimension length^{-1}.

$$K = \frac{\pi e^2}{mc} fN = 2.653 \times 10^{-2} fN \tag{1.7}$$

The intensity of the spectral line is usually expressed in terms of oscillator strength. The atomic concentration can be calculated from the absorption coefficient (K) when the oscillator strength is known. The theoretical calculation and experimental evaluation of the f values for many metals have been reported [1,19,36-38].

For derivation of Eq. (1.5), the equilibrium between atoms in the excited and ground states has been neglected. This means that the number of atoms not available for absorption is neglected. Furthermore, spontaneous emission (also referred to as negative absorption) brings about a reduction in the absorption coefficient, since this process adds photons to the transmitted beam and compensates in part for the absorption. Taking these phenomena into account the expression for the absorption coefficient becomes:

$$X = \frac{h\nu}{c} \left(1 - \frac{N_i g_i}{N_j g_j}\right) B_{ij} \tag{1.8}$$

where N_i and N_j indicate the number of excited and ground-state atoms respectively in unit volume, and g_i and g_j are the statistical weights of the excited and ground states, respectively. Since the ratio N_i/N_j is small under the usual conditions, (1.8) approximates to (1.5). The 'statistical weight' (g) is the number of atomic levels (of whatever kind) constituting the atomic state.

1.2.2 Number of Excited Atoms

The fundamental formula relating line intensity to the number of absorbing free atoms (N) at an absolute temperature T, in thermal equilibrium and without self-absorption, is expressed by Eq. (1.9)

$$I_\nu = B_{ij} N h \nu \frac{g_i}{F(T)} [\exp(-F_i/kT)] \tag{1.9}$$

where $F(T)$ is the partition function [39] and k is the Boltzmann constant. N can be expressed in terms of the partial pressure of the free element P_M. Since $P_M = NkT$, then

$$I_\nu = B_{ij} P_M \frac{h\nu}{kT} \frac{g_i}{F(T)} [\exp(-E_i/kT)] \tag{1.10}$$

This equation means that any factor influencing P_M will change the intensity of the absorption line. When self-absorption is not negligible, the ratio between I_ν and P_M becomes more complicated but the intensity I_ν will still be influenced by the factors governing P_M.

The integrated radiant flux of the signal (I_m) of an isolated emission line (in watts) is given by Eq. (1.11), assuming that a thin flame in thermal and chemical equilibrium is used [8]. This assumption is valid for a dilute analyte solution and the central region of the outer cone of most flames:

$$I_m = \frac{N_m A L h \nu_0 B_{ij} g_i \exp[-E_i/kT]}{4\pi\Omega F(T)} \tag{1.11}$$

where N_m is the minimum detectable number of atoms per unit volume (cm^3) of the flame, A is the area of the flame surface viewed by the optics (cm^2), L is the flame thickness along the optical path (cm), Ω is the solid angle (steradians) of the emission, h is Planck's constant (6.62×10^{-34} J.sec), ν_0 is the frequency of the line centre (sec^{-1}), B_{ij} is the transition probability (sec), $F(T)$ is the partition function, k is the Boltzmann constant (8.62×10^{-5} eV/deg), E_i is the energy of the excited state, g_i is the statistical weight of the excited state.

1.2.3 Effect of Temperature on the Number of Excited Atoms

The relation between the average number of particles n_i and n_j of a chemical species with energies E_i and E_j, respectively, in thermal equilibrium at absolute temperature T is given by the Boltzmann distribution:

$$\frac{n_i}{n_j} = \exp[-(E_i-E_j)/kT] \tag{1.12}$$

This equation holds for ionization as well as excitation of atoms and molecules, provided that the energy levels considered are non-degenerate. The equation is applicable to atomic excitation since an electron with an inner quantum number J has $2J + 1$ non-degenerate sub-levels of the same energy. Consequently the sub-levels of a given energy level are equally populated (i.e. $E_i = E_j$ and $n_i = n_j$). If there are g_i sub-levels for the level E_i, the total number of particles (N_i) with energy E_i is therefore $g_i n_i$. The ratio of N_i and N_j for the levels E_i and E_j is given by Eq. (1.13) [1,39]:

$$N_i/N_j = \frac{g_i n_i}{g_j n_j} = \frac{g_i}{g_j} \exp[-(E_i-E_j)/kT] \tag{1.13}$$

where g_i and g_j are the statistical weights of the excited and ground-state levels, respectively, and k is the Boltzmann constant. When the ground-state level is part of a multiplet, as in the case of vanadium, iron and cobalt, Eq. (1.13) becomes slightly more complicated. Table 1.1 shows the ratio between the number of atoms in the excited and ground states, based on Eq. (1.13). It is shown that at 2500 K, about $10^{-2}\%$ of sodium, $10^{-5}\%$ of iron, $5.2 \times 10^{-7}\%$ of magnesium and $6.2 \times 10^{-10}\%$ of zinc atoms are excited at 589, 371.99, 285 and 213.86 nm, respectively. It is evident from Eq. (1.13) that the fraction of atoms thermally excited increases with increase of temperature and is negligible compared to the

Table 1.1. Effect of temperature on the number of excited atoms

Element	Wavelength, (nm)	g_i/g_j	Excitation energy (eV)	N_i/N_j 2000 K	2500 K	3000 K
Cobalt	338.29	1	3.664	5.85×10^{-10}	4.11×10^{-8}	6.99×10^{-7}
Gold	267.59	1	4.632	2.12×10^{-12}	4.6×10^{-10}	1.65×10^{-8}
Iron	371.99	–	3.332	2.29×10^{-9}	1.04×10^{-9}	1.31×10^{-6}
Magnesium	285.21	3	4.346	3.35×10^{-11}	5.20×10^{-9}	1.50×10^{-7}
Silver	328.07	2	3.778	6.03×10^{-10}	4.84×10^{-8}	8.99×10^{-7}
Sodium	589.00	2	2.104	0.99×10^{-5}	1.14×10^{-4}	5.83×10^{-4}
Vanadium	437.92	–	3.131	6.87×10^{-9}	2.5×10^{-7}	2.73×10^{-6}
Zinc	213.86	3	5.795	7.45×10^{-15}	6.22×10^{-12}	5.50×10^{-10}

total number of atoms. If the temperature is below 4000 K, which is usually the case for flames and furnaces, the N_i/N_j ratio is very low for most elements and the total number of atoms (N) will be essentially equal to N_j. Atomic-absorption measurements are hence theoretically less dependent upon temperature because they are based on the relative number of unexcited atoms. Temperature fluctuation, however, will exert an indirect influence on absorption. For example, increasing the temperature increases ionization and thus causes broadening of the absorption line and decrease in its height. Thus, reasonable control of the atomization temperature is required for quantitative measurements.

1.3 WIDTH OF ATOMIC LINES

Spectral lines have a very small but finite width, as they usually spread over a range of wavelengths or frequencies, exhibiting breadth and shape, with a maximum at a certain wavelength or frequency. Different types of effect are known to contribute to this broadening [40–43]. However, the integral of the absorption coefficient, taken over the domain where it is non-zero, is proportional to the number of absorbing atoms N.

1.3.1 Natural Broadening

The mean lifetime of the atom in the excited state is defined by the radiative lifetime τ, which for the resonance level varies inversely with the transition probability. The larger the transition probability (B_{ij}), the shorter the radiative time

$$\tau = 1/B_{ij} \tag{1.14}$$

From Heisenberg's uncertainty principle, the natural breadth of a resonance line can be expressed by the equation

$$\Delta \nu_N = B_{ij}/2\pi \tag{1.15}$$

The transition probability is of the order of 10^8 sec^{-1}, so τ is about 10^{-8} sec. This leads to an absorption line with a width of about 10^{-5} nm. This width is negligible compared to that of lines broadened by various effects, which are in the range 5×10^{-4}-5×10^{-3} nm [44-46].

1.3.2 Doppler Broadening

This results from random thermal motion of the atoms [47,48]. When an atom that emits or absorbs a photon of frequency ν_0 moves relative to the observer with a radial velocity μ (i.e. μ is the velocity component on the line joining the atom to the observer), the frequency ν measured by the observer is no longer ν_0 but is shifted from it by a quantity $\nu-\nu_0$ related to ν_0 and μ by the relation:

$$(\nu-\nu_0)/\nu_0 = \mu/c \tag{1.16}$$

where c is the velocity of light. Since there are many atoms with various radial velocities emitting at the same time, the observer receives a continuous range of frequencies centred on ν_0, described by a Gaussian distribution (Fig. 1.2) and having a breadth ($\Delta\nu_0$) at half peak-height given by

$$\Delta\nu_D = \frac{2\nu_0}{c} \frac{\sqrt{(2 \ln 2)RT}}{M} = 7.15 \, \nu_0\sqrt{I/M} \tag{1.17}$$

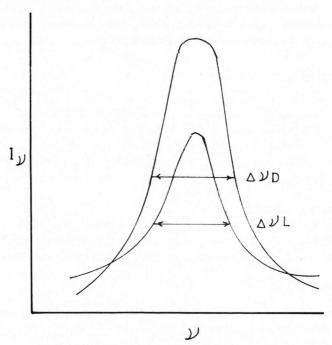

Fig. 1.2. Doppler ($\Delta\nu_D$) and Lorentz ($\Delta\nu_L$) broadening.

where T is the absolute temperature of the absorbing species, M is its atomic weight, R is the gas constant, c is the velocity of light, and $\Delta\nu_D$ is in cm^{-1}. The Doppler line width is therefore greater for elements of low atomic weight and becomes larger with increase in temperature. In common flame cells, the wavelength broadening is in the range 5×10^{-4}-5×10^{-3} nm (Table 1.2).

Table 1.2. Doppler ($\Delta\lambda_D$) and Lorentz ($\Delta\lambda_L$) widths of some spectral lines

Element	Wavelength	$\Delta\lambda_D$ (pm)		$\Delta\lambda_L$ (pm)	
	(nm)	2000 K	3000 K	2000 K	3000 K
Barium	553.56	1.5	1.8	3.2	2.6
Calcium	422.67	2.1	2.6	1.5	1.2
Cobalt	338.29	1.0	1.3	1.5	1.2
Copper	324.75	1.3	1.6	0.9	0.7
Gold	267.59	0.6	0.7	–	–
Iron	371.99	1.6	1.9	1.3	1.0
Magnesium	285.21	1.8	2.3	–	–
Silver	328.07	1.0	1.2	1.5	1.3
Sodium	589.00	3.9	4.8	3.2	2.7
Vanadium	437.92	2.0	2.4	–	–
Zinc	213.86	0.8	1.0	–	–

1.3.3 Lorentz Broadening

This results from collision of the absorbing or emitting atoms with atoms of other species [34,49-51]. Since the energy level of both the ground and excited states of an atom will be influenced by interaction with surrounding particles, such interaction causes a variation in the frequency of the radiation absorbed or emitted, and a broadening of the spectral line results (Fig. 1.2).

$$\Delta\nu_L = \sigma_L^2 N \sqrt{2RT(M_1 + M_2)/M_1 M_2 \pi} \qquad (1.18)$$

where σ_L^2 is the collisional cross-section for Lorentz broadening, N is the number of foreign gas atoms or molecules per unit volume, M_1 is the atomic or molecular weight of the foreign species and M_2 that of the absorbing species, R is the gas constant and T the absolute temperature. The collisional cross-section is best determined experimentally—but can be taken as the square of the mean of the diameters of the colliding species. The Lorentz half-breadth is of the same order of magnitude as the Doppler half-breadth (Table 1.2). The classical Lorentzian distribution predicts a symmetric line profile, but in practice there is an asymmetric profile and a red-shift of the maximum. The theory of the effect is still incomplete [52].

1.3.4 Holtsmark Broadening

This results from collision between atoms of the same element in the ground state and is commonly called resonance broadening. It results in an intensity distribution similar to that obtained by Lorentz broadening but without line asymmetry or shift. The effect depends on the concentration, but a half-width of only about 10^{-5} nm is obtained with $1M$ analyte solution, so the effect is negligible.

1.3.5 Quenching Broadening

When a foreign gas molecule, with vibrational levels very close to those of the excited state of the resonance line of the analyte atomic species, is present in the atomization cell, collisions can occur in which vibrational excitation energy is transferred, with a consequent broadening of the line. The effect is generally small in comparison with the Lorentz and Doppler broadening.

1.3.6 Field Broadening

This results from either an electrical field (the Stark effect) or a magnetic field (the Zeeman effect). The former results from splitting of the electronic levels of an atom by the presence of a strong non-uniform electric field or moving ions or electrons. The Zeeman effect arises from splitting similarly caused by the magnetic field, and is mainly of interest as a means of background correction (see Section 3.2.5.2, p. 83).

The Lorentz and Holtsmark broadenings are often referred to as "pressure" or "impact" or "collisional" broadening. With the concentrations of analyte used in work with flames, Holtsmark broadening is quite negligible compared to Lorentz broadening [40,53-56] or even absent. Field broadening manifests itself only when a strong field is applied to the emitting or absorbing medium, or when the medium is strongly ionized, such as in a plasma. Thus, the Doppler and Lorentz broadenings are the main factors affecting the total line width. When Doppler and Lorentz broadening occur simultaneously the line profile can be expressed by the Voigt function [57].

In AAS, a radiation source that emits lines with very small half-width is used. If the width of this emission line is neglected in comparison to that of the absorption line in the atomic vapour of the analyte, and assuming that the shape of the absorption line is determined entirely by Doppler broadening, then the relation between the absorption coefficient (K) of the measured line and the concentration of atoms in the vapour state is given by the relation:

$$K = \frac{2\sqrt{\ln 2}\ e^2 NF\sqrt{\pi}}{mc\,\Delta\nu_D} \tag{1.19}$$

1.3.7 Self-Absorption and Self-Reversal Broadening

Photons produced by atomic emission in the inner hot part of the flame or the hollow-cathode lamp can be absorbed again by ground-state atoms of the same metal present in the outer cooler part of the source [58,59]. As a result, the latter atoms become excited and may lose energy by emission of radiation or through radiationless collisional transitions or internal conversion. Because the atoms are moving more rapidly in the inner hot zone, the Doppler broadening of the emission line is greater than that of the absorption line, so the centre of the source line is more strongly absorbed than the edges. This effect is called self-absorption [7,8,60,61]. The effect increases with increase in the atomic concentration of the analyte in the flame or increase in the flame thickness or the current of the hollow-cathode lamp.

If the effect becomes large enough, the centre of the source line becomes less intense than the edges and the line looks like a doublet; it is then said to be self-reversed (Fig. 1.3) [62]. Under these circumstances, the total radiation intensity is not proportional to the number of analyte atoms present and Eq. (1.13) no longer holds.

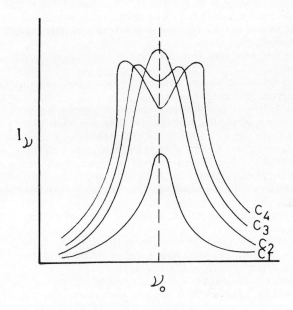

Fig. 1.3. Change in line profile due to self-absorption. C_1, C_2, C_3 and C_4 are increasing analyte concentrations.

The validity of the Beer–Lambert law for AAS, and the dependence of the absorption measurement on the widths of the source emission line and analyte absorption line have been investigated [63-67]. The dependence of flame atomic-absorption measurements on geometrical and temporal experimental parameters has also been discussed [68]. In AAS, the Beer–Lambert law is usually applicable to only a narrow range of concentration (one or two orders of magnitude), probably because of non-uniformity of the analyte distribution in the atomization cell. At higher concentrations of the analyte, the calibration graph becomes convex instead of linear, and the slope progressively decreases or even changes in sign. These effects can be avoided by using: (i) a non-resonance line for measurement; (ii) low lamp current; (iii) a dilute solution of the analyte; (iv) a shielded burner with uniform flame temperature.

REFERENCES

[1] A. Walsh, *Spectrochim. Acta,* 7, 108 (1955).
[2] M. L. Parsons, W. J. McCarthy and J. D. Winefordner, *Appl. Spectrosc.,* 20, 223 (1966); 22, 385 (1968).
[3] K. Yasuda, *Anal. Chem.,* 38, 592 (1966).
[4] A. V. Malinowski, *Ann. Phys.,* 44, 935 (1914).
[5] C. T. J. Alkemade and J. M. W. Milatz, *Appl. Sci. Res.,* B4, 289 (1955).
[6] C. T. J. Alkemade and J. M. W. Milatz, *J. Opt. Soc. Am.,* 45, 583 (1955).
[7] R. Mavrodineanu (ed.), *Analytical Flame Spectroscopy,* Macmillan, London, 1970.
[8] C. T. J. Alkemade and R. Herrmann, *Fundamentals of Analytical Flame Spectroscopy,* Hilger, Bristol, 1979.
[9] J. C. VanLoon, *Analytical Atomic Absorption Spectroscopy,* Academic Press, New York, 1980.
[10] G. G. Kirkbright and M. Sargent, *Atomic Absorption and Fluorescence Spectroscopy,* Academic Press, London, 1974.
[11] W. J. Price, *Spectrochemical Analysis by Atomic Absorption,* Heyden, London, 1979.
[12] M. Slavin, *Atomic Absorption Spectroscopy,* Wiley, New York, 1978.
[13] K. C. Thompson and R. J. Reynolds, *Atomic Absorption, Fluorescence and Flame Emission Spectroscopy, A Practical Approach,* Wiley, New York, 1978.
[14] G. D. Christian and F. J. Feldman, *Atomic Absorption Spectroscopy, Applications in Agriculture, Biology and Medicine,* Wiley–Interscience, New York, 1970.
[15] J. D. Winefordner (ed.), *Trace Analysis Spectroscopic Methods for Elements,* Wiley, New York, 1976.
[16] I. Rubeska and B. Moldan, *Atomic Absorption Spectrophotometry,* Iliffe, London; SNTL, Prague, 1967.
[17] B. Welz, *Atomic Absorption Spectroscopy,* Verlag Chemie, Weinheim, 1976.
[18] B. V. L'vov, *Inzh.-Fiz. Zh., Akad. Nauk Belorussk. SSR,* 2, 44 (1959).
[19] B. V. L'vov, *Spectrochim. Acta,* 17, 761 (1961).
[20] S. Greenfield, I. L. Jones and C. T. Berry, *Analyst,* 89, 713 (1964).
[21] V. G. Mossotti, K. Laqua and W. D. Hagenah, *Spectrochim. Acta,* 23B, 197 (1967).
[22] C. Veillon and M. Margoshes, *Spectrochim. Acta,* 23B, 503 (1968).
[23] S. R. Koirtyohann, *Anal. Chem.,* 52, 736A (1980).
[24] S. L. Ali, *Pharm. Ztg.,* 125, 450 (1980).
[25] K. C. Thompson, *Methods for the Examination of Waters and Associated Materials,* H.M. Stationery Office, U.K., 1980.
[26] P. L. Larkins and J. B. Willis, *Chem. Aust.,* 47, 130 (1980).
[27] G. Horlick, *Anal. Chem.,* 52, 290R (1980).
[28] G. Horlick, *Anal. Chem.,* 54, 276R (1982).
[29] W. Slavin, *Anal. Chem.,* 54, 685A (1982).
[30] J. D. Winefordner and T. J. Vickers, *Anal. Chem.,* 36, 1947 (1964).
[31] J. D. Winefordner, M. L. Parsons, J. Mansfield and W. J. McCarthy, *Anal. Chem.,* 39, 439 (1967).

[32] L. de Galan and J. D. Winefordner, *J. Quant. Spectrosc. Radiat. Transfer*, 7, 251 (1967).

[33] C. G. James and T. M. Sugden, *Proc. Roy. Soc.*, 47, 156 (1957).

[34] E. Hinnov and H. Kohn, *J. Opt. Soc. Am.*, 47, 156 (1957).

[35] A. Einstein, *Physik. Z.*, 18, 121 (1917).

[36] B. V. L'vov, *Spectrochim. Acta*, 24B, 53 (1969).

[37] B. J. Russell, J. P. Shelton and A. Walsh, *Spectrochim. Acta*, 8, 317 (1957).

[38] D. R. Bates and A. Damgaard, *Phil. Trans. Roy. Soc. London, Ser. A*, 242, 101 (1949).

[39] R. Mavrodineanu and H. Boiteux, *Flame Spectroscopy*, Wiley, New York, 1965.

[40] M. L. Parsons, W. J. McCarthy and J. D. Winefordner, *Appl. Spectrosc.*, 20, 223 (1966).

[41] M. Danos and S. Geschwind, *Phys. Rev.*, 91, 1159 (1963).

[42] W. Behmenburg, *J. Quant. Spectrosc. Radiat. Transfer*, 4, 177 (1964).

[43] R. B. King and D. C. Stockbarger, *Astrophys. J.*, 91, 488 (1940).

[44] G. F. Kirkbright and O. E. Troccoli, *Spectrochim. Acta*, 28B, 33 (1973).

[45] D. W. Posener, *Aust. J. Physics*, 12, 184 (1959).

[46] N. N. Sobolev, *Spectrochim. Acta*, 11, 310 (1956).

[47] H. Margenau and W. W. Watson, *Rev. Mod. Phys.*, 8, 22 (1936).

[48] B. P. Straughan and S. Walker, *Spectroscopy*, Vol. 3, Wiley, New York, 1976.

[49] E. Hinnov, *J. Opt. Soc. Am.*, 47, 151 (1957).

[50] T. Hollander, B. J. Jansen, J. J. Plaat and C. T. J. Alkemade, *J. Quant. Spectrosc. Radiat. Transfer*, 10, 1301 (1970).

[51] H. P. Hooymayers and C. T. J. Alkemade, *J. Quant. Spectrosc. Radiat. Transfer*, 6, 501 (1966).

[52] R. G. Breene, *The Shift and Shape of Spectral Lines*, Pergamon Press, Oxford, 1961.

[53] W. Behmenburg and H. Kohn, *J. Quant. Spectrosc. Radiat. Transfer*, 4, 163 (1964).

[54] H. Prugger, *Optik*, 21, 320 (1964).

[55] P. A. Rice and D. V. Ragone, *J. Chem. Phys.*, 42, 701 (1965).

[56] C. Van Trigt, T. Hollander and C. T. J. Alkemade, *J. Quant. Spectrosc. Radiat. Transfer*, 5, 813 (1965).

[57] A. C. G. Mitchell and M. W. Zemansky, *Resonance Radiation and Excited Atoms*, Cambridge University Press, London, 1961.

[58] A. I. Bodretsova, B. V. L'vov and V. I. Mosichev, *J. Appl. Spectrosc.*, 4, 149 (1966).

[59] G. F. Kirkbright and M. Sargent, *Spectrochim. Acta*, 25B, 577 (1970).

[60] P. W. J. M. Boumans, *The Theory of Spectrochemical Excitation*, Hilger and Watts, London, 1966.

[61] T. Tako, *J. Phys. Soc. Japan*, 16, 2016 (1961).

[62] H. M. Crosswhite, G. H. Dieke and C. S. Legagneur, *J. Opt. Soc. Am.*, 45, 270 (1955).

[63] J. D. Winefordner, *Appl. Spectrosc.*, 17, 109 (1963).

[64] K. Yasuda, *Anal. Chem.*, 38, 592 (1966).

[65] I. Rubeska and V. Svoboda, *Anal. Chim. Acta*, 32, 253 (1965).

[66] M. Schimazu and A. Hashimoto, *Sci. Light (Tokyo)*, 11, 131 (1962).

[67] L. de Galan and G. F. Samaey, *Spectrochim. Acta*, 24B, 679 (1969).

[68] J. M. Ramsey, *Anal. Chem.*, 52, 2141 (1980).

2

Instrumentation: theoretical and practical considerations

The basic principle of the AAS instrumentation now in use is virtually the same as that described in the first publication by Walsh. However, major technological improvements have been made in the past ten years [1]. The more sophisticated instruments now commercially available use integrated microcomputers for control, measurement, data-display and data-handling. Instant access to various modes of operation, the ability to change any parameter with only a few key-strokes, and the matching of each mode to the task to be performed are the main features of these modern instruments.

The basic components of atomic-absorption spectrometers are as follows:

(1) Radiation source (e.g. hollow-cathode or electrodeless discharge lamp) to emit the spectral line of the element of interest.
(2) Atomization system (e.g. flame or furnace) to provide sufficient energy for analyte dissociation and vaporization as free atoms.
(3) Monochromator for spectral dispersion and isolation of the spectral line to be measured.
(4) Detector and data-logging device to measure, amplify and display the results.

There are basically two types of atomic-absorption spectrometer, single- and double-beam (Fig. 2.1). In the single-beam system, the change in intensity of the beam from the radiation source, due to interaction with the atomic vapour of the element, is measured. In the double-beam system, the source beam is divided by a rotating mirror chopper, into a sample beam that is passed through the atomic vapour of the element to be measured, and a reference beam, the ratio of the two beam-intensities being measured. This system is more stable than the single-beam system because it compensates for any drift in the radiation-source intensity, the detector and the electronic system [2]. Liddell and Wildy [3] compared the noise levels of single- and double-beam instruments. They found that at low absorbance, where photon noise, lamp flicker and flame transmission noise predominate, the detection limits are better for the single-beam instrument.

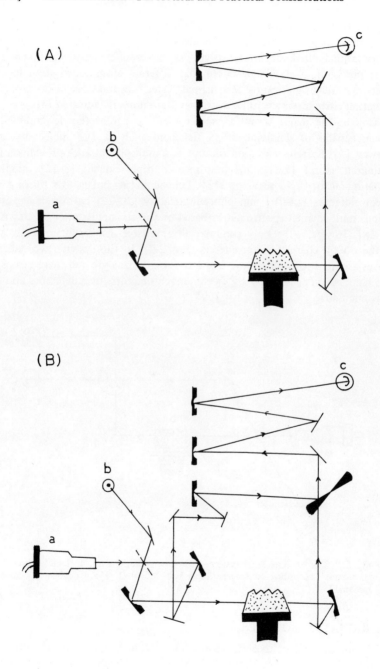

Fig. 2.1. Single- (A) and double-beam (B) optical systems of atomic-absorption spectrometers. a, hollow-cathode lamp; b, deuterium lamp; c, photomultiplier.

Provided that a blank correction is made before and after each sample measurement, the single-beam instrument also gives slightly better precision. However, if only occasional blank measurements are made, the double-beam instrument gives better precision. It should be noted that different atomic-absorption spectrometers may vary significantly in sensitivity, precision and slope of the analytical calibration curve, even when the instruments are properly adjusted [4].

One of the major trends in spectrometer design is the development of systems capable of simultaneous multielement analysis. This subject has been reviewed [5]. Systems based on the use of a multiple entrance-slit vidicon [6], continuum source [7,8], time multiplex multiple exit-slit [9–12], modified oscillating-mirror rapid scanning [13], Echelle grating linked to a silicon-target vidicon detector [14–19] and photodiode array [20–22] have been suggested. Various multichannel spectrometers based on these systems have been described [23–28]. However, the only commercially available instruments of such types are the double-channel spectrometers (Fig. 2.2). In this system, two radiation sources are used to irradiate the free atoms of the analyte and these beams are then handled independently by two monochromators, two detectors and two electronic units.

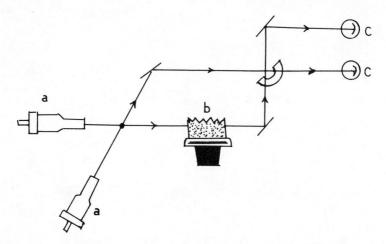

Fig. 2.2. Two-channel AAS system. a, hollow-cathode lamp; b, flame; c, monochromator [according to International Lab. Model IL 951. Courtesy of International Lab.].

2.1. RADIATION SOURCES

These sources are used to emit resonance radiation of the element, with a half-width less than the width of the absorption line. Since the early development of atomic spectrometry, hollow-cathode lamps have been used almost exclusively as the radiation source for most elements.

2.1.1 Hollow-Cathode Lamps (HCLs)

These sources of excitation were first described by Paschen in 1916 [29] and further simplified and modified by Walsh and co-workers [30,31]. They have been critically reviewed [32]. These lamps are low-pressure gaseous discharge tubes characterized by a unique design of the cathode. The lamps are vessels into which the electrodes are sealed, followed by evacuation and filling with an inert gas (e.g. argon, helium or neon). The cathode is a hollow cylinder (10 mm inner diameter or less) most often made of the (highly pure) metal to be determined, and an anode wire made of zirconium or some other gettering metal (Fig. 2.3) [32]. When a sufficient voltage is applied between the two electrodes, the inert gas ions produced at the anode are accelerated towards the cathode, where they arrive with sufficient energy to sputter out excited atoms of the cathode material. If the lamp is run at too high a current, the ground-state atoms will be present in high enough concentration to cause self-reversal of the resonance line by absorption of part of the emitted resonance radiation.

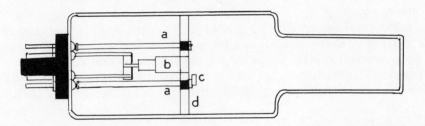

Fig. 2.3. Hollow-cathode lamp with simple disc electrode shields. a, insulated support; b, hollow cathode; c, anode; d, shield.

The nature of the filler gas depends on the nature of the metal used in the lamp. Neon is most commonly used. Argon is not used in chromium or lead lamps since the strongest chromium and lead lines at 357.9 nm and 217.0 nm may suffer interference by the argon lines at 357.7 nm and 217.14 nm, respectively. Furthermore, an increase in the lamp current of an argon-filled lamp often produces a lower sensitivity than that for lamps filled with helium or neon. However, the arsenic resonance line at 193.76 nm is subject to spectral interference from the neon lines at 193.89 and 193.01 nm in neon-filled lamps [34]. On the other hand, neon at a pressure of 3 mmHg secures a greater brightness than argon and is useful for elements such as iron, molybdenum, nickel, cobalt, chromium, copper and titanium. Argon at a pressure of 1–1.5 mmHg is used for aluminium, calcium, and magnesium lamps. With volatile elements (e.g. bismuth, cadmium, lead, tin and zinc), the brightness of radiation increases with decrease in filler-gas pressure.

A potential difference of 350–500 V is commonly applied, with a current ranging from 1 to 30 mA and exceptionally up to 50 mA. Each hollow-cathode lamp has a particular current for optimum performance. The stability of the spectral output depends on the stability of the power supply.

2.1.1.1 High-intensity hollow-cathode lamps. In these lamps, a cathode similar in design to that used in the conventional hollow-cathode lamp, and two auxiliary electrodes coated with electron-emissive materials, are used. This design results in the formation of an area of high electron density that interacts with the atomic vapour cloud, with consequent increase in the radiation intensity and decrease in self-absorption [35–39]. Enhancement of the radiation intensity by factors up to 100 is commonly obtained in these lamps. Enhancements between 50- and 800-fold relative to ordinary lamps have also been reported as obtained by use of a pulsed current with a frequency of 300 Hz, a pulse width of 15 μsec and a peak amplitude of about 600 mA [40]. However, pulsed lamps are used less extensively, probably because higher pulsed currents cause severe self-reversal of the neutral resonance lines, as well as increased depletion of the cathode material [41].

The radiofrequency-boosted pulsed hollow-cathode lamps are promising as a means of producing high-intensity radiation without the attendant reversal of the analytical line [42]. Design criteria, characterization data and sample performance data for these lamps have been discussed [43]. An output intensity 10–20 times that of commercial hollow-cathode lamps has been obtained [44].

2.1.1.2 Multielement hollow-cathode lamps. These are conventional hollow-cathode lamps in which the cathode is made of several metals. Mixed metal powders pressed in a die and sintered at an appropriate temperature [23,30, 45–49], rings of various metals pressed into a single supporting cathode [23,45] and true intermetallic compounds [50] have been employed. A single lamp containing two or more hollow cathodes in conjunction with a common anode has also been suggested [30]. However, multielement lamps made of alloys preferentially lose the more volatile metal, so the spectrum emitted by the lamp gradually weakens, and the lamp degenerates into a one-element lamp. Cathodes made from alloys of metals having similar melting points and volatilities (e.g. calcium–magnesium, silver–gold, copper–iron, zinc–cadmium) are more useful. Many multielement lamps are now commercially available, with cathodes containing up to seven elements. It is worth mentioning that some single-element hollow-cathode lamps, such as that for iron, give a spectrum containing lines lying close enough to the resonance lines of barium, copper, magnesium, manganese and nickel for these lamps to be usable for measuring these elements [51].

2.1.1.3 Demountable hollow-cathode lamps. These are hollow-cathode lamps with exchangeable cathodes for different elements, and are purged continuously with an inert gas or temporarily sealed (Fig. 2.4) [52-54]. These lamps may be cooled and the atomic vapour cloud removed by means of a flow of argon [55-57]. A modified version of the Grimm glow-discharge lamp has been developed for use as a demountable hollow-cathode emission source. The effect of anode distance from the hollow cathode on the intensity of the spectral lines has been investigated for three cathode materials, viz. aluminium, copper and graphite, with use of argon as carrier gas [58]. Cathodes prepared by electrolytic deposition rather than by sputtering have been used in demountable hollow-cathode lamps to produce slightly more intense and clear spectra [59]. A procedure for regeneration and modification of old hollow-cathode lamps (that would otherwise be discarded) by converting them into demountable form has also been described [60].

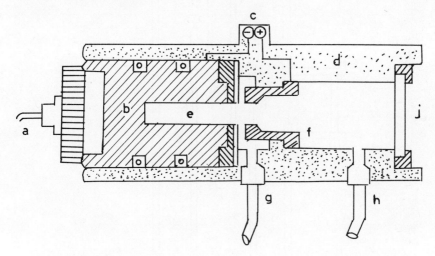

Fig. 2.4. Demountable hollow-cathode lamp. a, heat exchanger; b, cathode holder; c, power connector; d, main body; e, cathode; f, anode; g, vacuum port; h, ionizing gas; j, window.

2.1.2 Electrodeless-Discharge Lamps (EDLs)

These efficient high-intensity radiation sources were first used by Jackson in 1928 [61] and are widely used in both atomic-absorption and atomic-fluorescence spectroscopy. The lamps consist of a sealed transparent quartz or Pyrex glass bulb (3-8 cm long, 5-10 mm in diameter), and contain a few milligrams of the element of interest (Fig. 2.5). The pure metal, metal halide or a mixture of the metal with elemental iodine is used, and the tubes are filled with an inert gas at a pressure of a few mmHg. The optimum size of the tube for each element depends on the vapour pressure of the material used and the nature of the cavity. Tubes have been prepared for about fifty elements, the most successful being those for arsenic, bismuth, antimony, selenium, tellurium and phosphorus

[62,63]. A multielement source has also been prepared [64]. The elements in these tubes are excited at either radiofrequency (100 kHz–100 MHz) or microwave frequency (> 100 MHz).

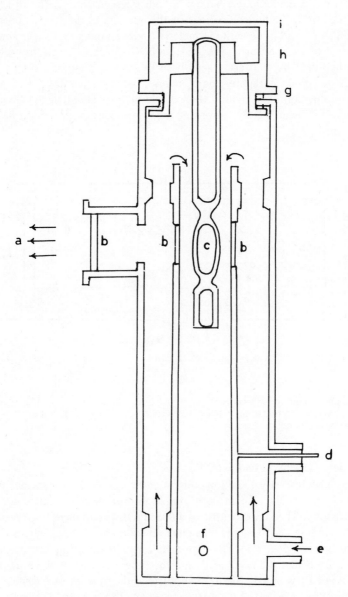

Fig. 2.5. Electrodeless-discharge lamp (¾ cavity). a, light output; b, silica windows; c, lamp; d, power input (2450 MHz); e, gas input; f, gas outlet; g, tuning ring; h, insulation; i antiradiation cap. (Courtesy of EMI Electronics Ltd.)

Radiofrequency EDLs give lower light intensity than the microwave lamps, but they are more stable and do not require temperature control. They are recommended for arsenic, selenium, tellurium, antimony and phosphorus and are also available for about ten other elements. Microwave EDLs have the advantages of useful operational lifetime and efficient radiation [65-67]. These lamps are usually operated with their walls kept at a controlled temperature sufficiently high to prevent condensation of the metal and to provide a stable output intensity. The EDLs are usually mounted within the coil of a high-frequency generator (~ 2400 MHz) and excited by a low power output of up to 200 W [68-70]. The excited electrons produced by ionization of the inert gas collide with the metal atoms and produce a pure intense spectrum with extremely narrow lines. The radiation intensity of the EDLs is several orders of magnitude greater than that of conventional hollow-cathode lamps.

Barnett *et al.* [71] have made a comparison of the analytical performance of the atomic-absorption instruments equipped with EDLs and HCLs. They found that the EDLs give 5-100 times more intense radiation than the HCLs. The sensitivity values are almost equal for the two sources but the detection limit offered by use of EDLs is lower by a factor of 1.7-2.5 than that given by HCLs (Table 2.1). A comparison has also been made of the radiant output of radiofrequency EDLs excited by pulsed and continuous radiofrequency power. Each mode of operation affects the output intensity of some metals [72]. The

Table 2.1 Comparison of sensitivity and detection limits for some elements [air–acetylene flames; hollow-cathode (HCL) and electrodeless-discharge (EDL) lamps as sources] [71]
(Perkin-Elmer 406 instrument)

(reproduced by permission of the copyright holders, Perkin-Elmer, Inc.)

Element	Wavelength (nm)	Band-width (nm)	Sensitivity (μg/ml/1%)			Detection limit (μg/ml)		
			HCL	EDL	Ratio	HCL	EDL	Ratio
Antimony	217.6	0.2	0.46	0.44	1.0	0.076	0.040	1.9
Arsenic	193.7	0.7	1.10	0.93	1.2	0.98	0.340	2.9
Bismuth	223.1	0.2	0.45	0.53	0.8	0.071	0.028	2.5
Cadmium	228.8	0.7	0.024	0.025	1.0	0.0026	0.0007	3.7
Germanium*	265.1	0.7	1.90	2.00	1.0	0.16	0.078	2.1
Lead	283.2	0.7	0.43	0.46	0.9	0.019	0.014	1.4
Mercury	253.6	0.7	6.4	4.5	1.4	0.39	0.15	2.6
Phosphorus*	213.6	0.2	260	300	0.9	93	33	2.8
Selenium	196.0	2.0	0.99	0.63	1.6	0.54	0.21	2.6
Tellurium	214.3	0.2	0.35	0.36	1.0	0.077	0.038	2.0
Thallium	276.8	0.7	0.43	0.45	1.0	0.050	0.013	3.8
Tin*	286.3	0.7	2.5	2.6	1.0	0.19	0.083	2.3
Zinc	213.9	0.7	0.014	0.016	0.9	0.0012	0.00094	1.3

* Elements measured with a nitrous oxide–acetylene flame.

high-frequency barium EDL, when operated at 25 W, gives an emission intensity about 12 times that of commercial hollow-cathode lamps and a detection limit three times better than that obtained with the HCLs [73]. A comparative study of the hollow-cathode, electrodeless-discharge and metal-vapour discharge lamps has been reported for atomic-absorption spectrometry [74]. Thermostatically controlled lamps [75] and a system using a twin-port power divider that allows operation of two EDLs simultaneously from one microwave generator [76] have been described.

However, the disadvantages of the EDLs are the short lifetime, long warm-up time, a tendency for the output intensity to drift and the need for a separate power supply. The construction and performance of an inexpensive radio-frequency generator for exciting the EDLs have been reported [77]. The generator operates at 149 MHz, which is much higher than the frequencies commonly used ($\sim$ 27 MHz). The maximum power level of the generator is 192 W and the typical power output to the lamps is about 60 W. Construction of an inexpensive microwave source for electrodeless-discharge lamps has also been described [78]. Applications [79–81] and rejuvenation [82] of the electrodeless-discharge lamps have been reported.

2.1.3 Controlled Temperature-Gradient Lamps (CTGL)

Gough and Sullivan [83] described a controlled temperature-gradient lamp as an excitation source. The lamp emits radiation by excitation of thermally produced vapour in a low-pressure gas discharge. An improved design was described later by the same authors and applied to the determination of arsenic, cadmium, phosphorus, potassium, rubidium, selenium, sodium, sulphur and zinc [84]. A schematic diagram of this lamp is shown in Fig. 2.6. The lamp consists essentially of three glass tubes, one each for the anode, the metal to be excited and the oxide-coated filament cathode. These tubes are attached to a bulb housing and a recessed silica exit-window. The anode compartment is connected to the central tube in such a way as to ensure that all the atomic vapour produced by heating the electrode will diffuse into the excited discharge. The vapour of the element in the discharge region diffuses and condenses on the cool portion of the lamp envelope. This lamp gives more intense radiation than the electrodeless-discharge lamps, at the same line-width, and is suitable for use in AAS. It warms up more quickly than the EDLs and gives resonance lines that are free from self-reversal. The line-widths are narrower than with the EDLs at the same power, except for arsenic [84].

2.1.4 Glow-Discharge Lamp (GDL)

This lamp consists of a common anode between a thermionic cathode and a cylindrical modulating cathode. The discharge from the first cathode is modulated by tuning the second cathode to the required resonance frequency. This system significantly reduces the relative intensity of the background and enhances the sensitivity of the measurements [85].

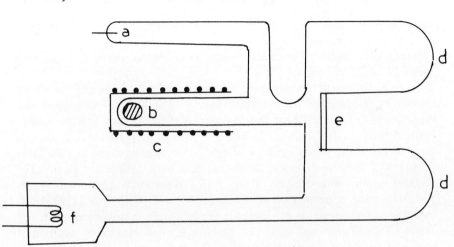

Fig. 2.6. Schematic diagram of controlled temperature-gradient lamp. a, anode; b, central tube (6 mm i.d.); c, furnace; d, lamp envelope; e, silica exit window (20 mm diameter); f, oxide-coated cathode [according to D. S. Gough and J. V. Sullivan, *Anal. Chim. Acta,* **124**, 259 (1981); by permission of the copyright holder, Elsevier Scientific Publishing Co.] .

2.2 FLAME ATOMIZATION

Historically, the flame has played a dominant role in AAS for generating atoms from both solids and soltuions, and still enjoys wide use as an atom cell because of its simplicity, the low cost of the equipment, and its versatility in excitation of various elements of different nature. The generation of atoms from a sample solution by the flame takes place through a series of processes: nebulization to form very fine aerosol droplets, evaporation of the solvent to form clotlets, volatilization of these solid particles, and dissociation of the vaporized analyte into free metal atoms.

2.2.1 Flames

A variety of flames have been utilized, depending on the nature of the element to be determined. Several fundamental and practical studies of various flame systems have been reported [86-90] . However, the first gas flame used for AAS measurements was air-coal gas. The temperature of this flame is low and insufficient to atomize many elements. The air–acetylene flame ($\sim 2300^\circ$C) is generally used for effective atomization of a large number of elements. This flame is completely transparent over a wide spectral range, with little self-absorption at wavelengths longer than 230 nm, and very low emission. The flame can be operated under stoichiometric or weakly oxidizing conditions [91,92] . The temperature of the flame, however, is insufficient to dissociate many of the oxides or to prevent their formation in the flame, which can only be achieved by using a hotter flame such as oxygen–acetylene or oxygen–cyanogen, or employing organic solvents, modified burners and complexing agents [93-108].

The most crucial development in work with flames was the use of the nitrous oxide–acetylene flame in 1965 [109]. The high temperature of this flame ($\sim$ 2900°C) is sufficient to atomize many elements, including those forming refractory oxides, without significant interferences [110-115]. It has also been reported that large numbers of free atoms are produced in this flame [116-118]. The flame velocity is low, so flash-back dangers are absent. The shielded nitrous oxide–acetylene flame is used in the vacuum ultraviolet region for determining elements such as iodine, sulphur and phosphorus [119]. However, the high temperature of this flame causes ionization of some elements and thereby reduces the sensitivity. This problem is often avoided by addition of excess of another easily ionized element. The strong self-emission of the flame may also be considered a drawback [120].

The analytical applications of the nitrous oxide–hydrogen flame have also been demonstrated [88,121-123]. This flame has a relatively low background, high temperature and reasonably good sensitivity for easily atomized elements. However, it shows poor atomization efficiency for refractory elements. This limits its usefulness as a general purpose flame. The advantages of using this flame with non-aqueous solvents have been reported [124-126]. The analytical utility of the substoichiometric nitrous oxide–hydrogen flame has been extended to allow use of a broad range of organic solvents which could not be used with hydrocarbon flames [127]. This permits direct analysis of petrochemicals and petroleum products without dilution. A study of the properties and possible uses of the toluene–nitrous oxide flame in AAS has been published; the flame has a maximum temperature of 2600 K [128].

The air–hydrogen flame ($\sim$ 2000°C) has been used for alkali metals, but it suffers from flash-back and possible explosion problems. The sensitivity of the flame is high and the self-absorption in the region of 200-230 nm is low [129-132]. The argon or nitrogen entrained-air flame, also known as the diffusion flame, where hydrogen serves as the fuel gas and argon or nitrogen for nebulizing the sample into the burner chamber, has also been used [133-136]. When the flame is ignited, the hydrogen is diluted with the inert gas and burns in the surrounding air. A three-slot burner is commonly used [133]. The flame provides a temperature that depends on the position in the flame. The boundary temperature is about 850°C, while that in the middle of the flame reaches only 300-500°C. In this flame, considerable chemical interference and high non-specific light-losses take place. The advantages of this flame are, however, its possible use in the vacuum ultraviolet range (owing to its high transparency compared with other flames) and its suitability for determining elements such as arsenic at 194 nm and selenium at 196 nm.

The physical characteristics of the helium–oxygen–acetylene flame as an atom cell have been investigated. Conditions were established for safe operation and for prevention of flash-back. The maximum temperature that could be attained was 2760 K [137]. Air-organic solvent (e.g. toluene, benzene, octane or hexane) and air–propane flames [139,140] have also been used for AAS.

All these types of flames are produced in suitable burners. The burner type is commonly referred to in terms of the type of gas flow, i.e. as laminar or turbulent. Commercially available burners of both types operate with pneumatic nebulizer systems and have a comparable uptake rate (2-10 ml/min) and temperature.

2.2.2 Burners

2.2.2.1 Laminar (premix) burner. In this type of burner system, the sample solution is sprayed into a mixing chamber by a pneumatic nebulizer to form an aerosol with the fuel–oxidant mixture [141,142]. The support gas (air or nitrous oxide) may be added either before or after the spray-chamber section. The liquid droplets emerge from the nabulizer and hit a glass impact-bead (Fig. 2.7) [143] or a flow spoiler (Fig. 2.8) [144] which serves to break up the drops. The large drops settle out and the aerosol (fuel gas-support gas-fine droplets mixture) is carried into the flame.

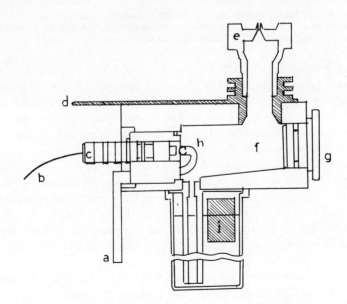

Fig. 2.7. Premix burner system with a glass impact-bead nebulizer. a, nebulizer interlock arm; b, capillary; c, nebulizer; d, burner rotation adjuster; e, burner head; f spray chamber; g spray chamber bung; h, glass impact bead; i, float (Courtesy of Varian).

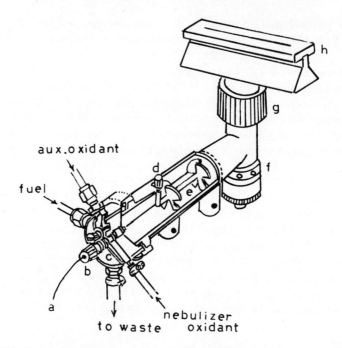

Fig. 2.8. Premix burner system with a spoiler. a, sample capillary; b, nebulizer adjusting knob; c, nebulizer; d, flow-spoiler retaining screw; e, flow spoiler (penton plastic); f, pressure relief vents; g, burner-head locking ring; h, burner head.

Pneumatic nebulizers typically produce an aerosol with droplets ranging in diameter from < 5 to 25 μm or greater. Optimum signal-to-noise ratio is achieved by carrying only the smallest droplets (< 10 μm diameter) to the flame. This fraction of the droplets probably constitutes no more than 10-15% of the total sample nebulized. Large droplets, which are incompletely vaporized, reduce the flame temperature and increase the noise.

For the analysis of very concentrated solutions or for use of the nitrous oxide-acetylene flame, the flow spoiler is recommended, for minimizing burner memory and clogging problems. A burner system has recently been designed to allow switching between a flow spoiler and an impact bead [145]. The mixing chamber of this burner is internally coated with polypropylene, which is chemically etched to ensure proper drainage and to resist chemical attack by corrosive solutions.

The body of most burners is made from stainless steel or other chemically resistant metal, or from glass, plastic and titanium to avoid contamination caused by corrosion [146,147]. The burner head has an array of holes, grooves [148-150] or slots [110,151,152]. A metal burner head with three parallel slots [133] and burners with adjustable slot width [153] have been described. Cooling of the burner head with a water jacket [154] both eliminates slow drift of the signals and slows the rate of carbon deposition when a hydrocarbon fuel is used [155]. However, excessive cooling may result in both deposition of solid material in the burner parts and condensation of sample spray in the body of the burner. Flame sheathing and/or separation can be used for separation of part of the interconal zone from the effect of the surrounding atmosphere or of the secondary reaction zone [156]. This results in an increase of the transparency of the flame at wavelengths shorter than 200 nm, owing to the removal of the oxidizing flame mantle and to the lack of oxygen species in the interconal zone. Such a flame is useful for the determination of a number of elements forming refractory oxides and elements which absorb in the ultraviolet region (e.g. arsenic, selenium, iodine, sulphur and phosphorus) [62,156-159]. The effect of gas composition and sheathing on the velocity profile of laminar-burner flames has been reported [160]. A theoretical model has been used to examine the dependence of AAS measurement on several geometrical parameters which have to be considered in burner design [161]. In general, the laminar burner can be used with flame gases of relatively low combustion speed (e.g. acetylene-air but not acetylene-oxygen). The burner provides a stable and reproducible absorption cell, low noise, sufficient sensitivity, little interference, a long rectangular transparent flame cell and reasonable safety [162-164].

2.2.2.2 Turbulent (total consumption) burner. The turbulent burner is a combination of the nebulizer burner with a specially shaped outlet jet. The burner consists of two concentric tubes through which fuel gas and oxidant are passed separately, and an extended central capillary tube which protrudes from the base of the burner and dips into a container of test sample solution (Fig. 2.9). The sample is aspirated and nebulized in the same way as for a laminar burner, but the whole of the aerosol passes into the flame. These burners have the advantages of simplicity and of supplying all the nebulized sample solution to the flame without loss. Combustible liquids and oxygen-hydrogen or oxygen-acetylene flames can be used. However, the introduction of a large volume of liquid into the flame causes an appreciable reduction of the flame temperature. Moreover, a relatively large fraction of the sample enters the flame as large droplets [165]. The noise and background radiation of the flame obtained with this burner are high.

2.2.3 Analyte Nebulization
Nebulization is the dispersion or breakdown of the sample solution into fine droplets below a certain diameter. The solution emerging from a capillary is shredded and contracted by the action of surface tension into small droplets which are dispersed into still smaller droplets by the action of the flame support gas or by striking an impact surface. Strong turbulence, shock waves, or ultrasonic vibration in the oxidant jet enhance the distribution of these fine particles within the flame.

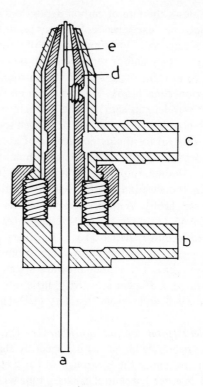

Fig. 2.9. Total-consumption nebulizer burner. a, capillary tube for sample solution; b, oxidant inlet; c, fuel inlet; d, fuel nozzle; e, oxidant nozzle.

Pneumatic nebulizers are the most frequently used for introducing analyte solutions into the flame, owing to their ease of operation and great reliability. The analyte solution is aspirated from the sample container through a capillary tube by suction (venturi action) caused by the rapid flow of support gas past the capillary tip. However, these nebulizers suffer from short-term noise, long-term drift and inefficient sample transport, and aspiration of samples with either a high dissolved-solid or suspended-particle content causes clogging [166]. Furthermore these nebulizers produce a wide range of droplet sizes.

The Babington nebulizer produces a droplet size-distribution similar to that of the pneumatic nebulizer but successfully reduces the problems due to nebulizer clogging [167]. Ultrasonic and electrostatic nebulizers [168-177] produce a very narrow droplet size distribution, and therefore exhibit higher transport efficiency [178]. These nebulizers are more complex and expensive than the pneumatic nebulizers. The use of a fritted glass disk as a nebulizer has been reported [179]. This nebulizer produces a smaller droplet size distribution and has a higher sample transport efficiency. The most significant limitation, however, is the slow sample equilibration and the requirement for wash cycles between samples

The droplet size is an important parameter in AAS [165,170,180-195]. The drop sizes found with the most commonly used nebulizers (pneumatic) are mostly between 1 and 50 μm, but mainly below 2 μm in diameter [170]. However, recombination of the drops to form larger droplets has been reported and is dependent on the initial droplet size distribution. The degree of recombination depends upon the travel time of the solution in the nebulizer [196], the type of nebulizer, and the flow-rate of the flame gases and the sample solution [197]. An automated method (based on laser diffractometry) for determining the droplet and particle size distribution in aerosol production, and for transport studies related to AAS, has been suggested [198].

The efficiency of nebulization is directly related to the subsequent atomization efficiency and depends mainly on the properties of the nebulizer, a significant fraction of the spray usually being lost in the expansion chamber and in the flame. The nebulization efficiency (ϵ) never attains 100%. Values of ϵ in the range 1-15% have been obtained for aqueous solvents [181,199,200]. For solvents containing 90% methanol, the ϵ values seem to be about three times that for water [201]. The physical properties of organic solvents (e.g. surface tension, density, viscosity, boiling point, vapour pressure and combustibility) obviously affect the sensitivity of the AAS measurement [202-204]. Detergents and surfactants alter the drop size and improve the nebulization efficiency.

It must be emphasized that ϵ is a property of the nebulizer and is considerably enhanced when the chamber wall is heated or the spray is irradiated by infrared radiation or subjected to a large electric field [205]. The values of ϵ can be increased by increasing the support gas flow while keeping the aspiration rate constant. It should also be noted that the concentration of the analyte in the flame is not necessarily proportional to the aspiration rate [206-209]. The fraction of sample lost is not constant, but increases with increasing aspiration rate, owing to deterioration of the nebulizing action [210]. High aspiration rate is accompanied by a decrease in the atomization efficiency, which in turn affects the desolvation, dissociation and excitation of the atomic species [181].

The aspiration efficiency is directly related to the atomization efficiency ϵ', and both govern the concentration of free atoms available for excitation in the flame, and hence the absorption intensity [187,211]:

$$N = n_R C \phi \epsilon' \beta (2.98 \times 10^{21})/n_T QT \qquad (2.1)$$

where n_R is the number of moles of flame-gas products produced at 298 K, n_T is the number of moles of burnt gases produced at T K per mole of unburnt fuel-oxidant mixture, C is the molar concentration of the analyte in the aspiration solution, ϕ is the flow-rate (ml/min) of the solution in the aspiration, ϵ' is the atomization efficiency, β is a factor to account for incomplete compound dissociation and atomic losses by ionization, Q is the flow-rate (ml/sec) of the unburnt gases supplied to the burner at 298 K, N is the concentration of analyte atoms in the flow gases (atoms/ml) and T is the flame temperature (K).

One of the limitations of pneumatic nebulization is the sample uptake rate, which depends on the nebulizer gas flow-rate and the sample viscosity [212]. This renders a constant rate of sample delivery to the flame difficult to achieve. Improved concentric pneumatic nebulizers that have a fixed and adjustable sample uptake have been designed [213,214]. Several methods of automating sampling have also been suggested [215-218]. A multichannel peristaltic pump that permits on-line dilution or reagent addition by pumping the sample through one channel and the reagent (ionization buffer or chemical releasing agent) through another channel and mixing of the solution before nebulization, has been used [215]. Automated flow-injection with aqueous and non-aqueous solvents [216,217,219] and nebulizer parameters for on-line flame atomic-absorption detectors in liquid chromatography [220] have been described. A microsampling nebulizer has also been proposed, to reduce analysis time, sample size and memory effects [221]. The sample ($\sim$ 100 μl) is pipetted into a conical PTFE cup and is drawn from the bottom of the cup to a standard nebulizer through a tube. Analysis of $\sim$ 100-μl samples by use of a PTFE cup and nebulization through an ordinary capillary tube has also been reported [222].

2.2.4 Analyte Desolvation

Evaporation of the solvent and volatilization of the solid particles determine the number of free atoms available for excitation in the flame. Partial desolvation of the spray affects the size of the droplets reaching the flame [223-227]. When the analyte solution is sprayed into the flame, the spray droplets are rapidly heated to the boiling point of the solvent. The solvent then starts evaporating at a rate depending on the rate of heat transfer from the ambient flame gas to the boiling droplet. This heat is consumed by the processes of solvent evaporation and vapour heating. On the other hand, the vapour pressure varies from one atmosphere at the surface of the droplet to almost zero at distances far from the droplet. The vapour pressure varies in such a way that the transportation of the vapour from the droplet surface by diffusion balances the rate of evaporation.

The rate of droplet desolvation can be simply expressed by Eq. (2.2) [228] provided that the heat is transferred to the droplet materials by conduction through the ambient gas:

$$-\mathrm{d}m/\mathrm{d}t = (4\pi r\lambda/C_\mathrm{p}) \ln \left[1 + C_\mathrm{p}(T - T_\mathrm{b}/L\right] \qquad (2.2)$$

where r is the droplet radius, λ is the average thermal conductivity of the ambient gas, C_p is the average specific heat of the solvent vapour at constant pressure, T is the ambient flame gas temperature, T_b is the boiling point of the solvent, and L is the specific heat of vaporization of the solvent.

Under ideal conditions, all drops entering the flame are desolvated immediately. In practice, desolvation of the spray is enhanced by using organic solvents

[229,230] , a narrow capillary for aspiration [231] , preheated air for nebulization [232, 233] and hot aqueous analyte solution. The effect of organic solvents is due to their ease of dissociation, low specific heat of vaporization and the heat released in the combustion of their vapour [103,229,230,234-238] . When combustible organic solvents are used, the rate of desolvation is given by the equation:

$$-dm/dt = (4\pi\lambda/C_p)\ln\left[1 + C_p(T - T_b)/L + Qf/L\right] \qquad (2.3)$$

where Q is the heat of combustion per unit mass of reaction mixture, and f is the ratio of actual oxygen concentration in the flame to the oxygen concentration required for complete combustion of the solvent.

The low surface tension and viscosity of many organic solvents lead to an increase of the flow-rate and favour spray dispersion, and consequently increase the concentration of atoms in the flame [239] . This is especially true for the total consumption burner, where the flow-rate is extremely critical. The effect of aspiration temperatures of 20-90°C for aqueous solutions, on the absorption of several metals, reveals signal enhancement ranging up to fourfold, depending on the temperature [240,241] . However, this enhancement is smaller by a factor of two than the enhancement with organic solvents.

2.2.5 Analyte Vaporization

Desolvation of the analyte solution gives an aerosol consisting of solid or molten particles which change into vapour on heating to the boiling point. Since the intensity of the absorption signals depends on the number of free atoms formed during the limited time of residence in the flame, incomplete vaporization means loss of analyte species capable of absorbing radiation and impairs the detection limit. The completeness of conversion of the aerosol into vapour depends on the chemical composition of the analyte, the radius of the particles atomized in the flame, and the flame temperature and gases.

Low-boiling and sublimable salts are easily vaporized whereas salts that decompose in the flame to give stable metal oxides (e.g. magnesium, calcium, aluminium) [86,242,243] are not vaporized completely at the usual flame temperatures. Addition of fluoride to aluminium or zirconium solutions, however, enhances the vaporization, because of formation of AlF_3 and $ZrOF_2$, respectively. Similarly, addition of EDTA, oxine or calcium promotes quick vaporization of magnesium and enhances the signal intensity [242,244,245] . Dispersion or entrapment of the analyte in a less volatile matrix (e.g. of sodium in presence of excess of calcium), and chemical changes (before vaporization) involving formation of covalent compounds or complexes with the matrix (e.g. calcium in the presence of aluminium phosphate and silicate) hinder the liberation of the analyte atom in the flame [242,246-249] . The temperature and composition of the flame gases also seem to have an influence on the vaporization of some elements. Non-volatilized metal oxides are formed in much higher concentration in the oxidizing or stoichiometric hydrogen flame than in the fuel-rich or reducing flame [250]. In the reducing flame, these refractory oxide particles are apparently readily reduced to the free atoms or more volatile species.

The rate at which the analyte particles vaporize depends on many different factors such as the particle radius and mass, the saturation vapour pressure at the surface of the solid or molten particles, the molecular weight of the analyte, and the flame temperature. The larger the particle size, the more time needed for complete vaporization. Larger droplets generally produce larger particles which are more difficult to vaporize. The rate of vaporization of molten particles ($-dm/dt$) at temperatures below their boiling point (i.e. when the flame temperature is between the melting and boiling points) is expressed by Eq. (2.4) and the total time (t) required for complete vaporization is given by Eq. (2.5) [251].

$$-dm/dt = 4\pi rMDP_s/RT \qquad (2.4)$$

$$t = QRTr_0^2/(2MDP_s) \qquad (2.5)$$

where m is the mass of the molten drop, r is the radius of the drop, r_0 is the initial radius of the molten particle, M is the molecular weight of the analyte, D is the diffusion coefficient of the vapour in the ambient flame gas, P_s is the saturation pressure, Q is the density of the molten particle, R is the gas constant and T is the absolute temperature of the molten drop.

The rate of vaporization can be increased by (a) decreasing the concentration of the analyte or matrix or both; (b) formation of a volatile derivative of the analyte; (c) formation of fine particles of aerosol by using a chamber nebulizer or decreasing the aspiration rate; (d) the use of higher temperature or a reducing flame.

2.2.6 Analyte Dissociation

It is well known that flames possess electrical conductivity and contain molecular, radical and atomic species. These species take part in a number of equilibria with the analyte, leading to the formation of metal oxides, hydroxides and hydrides [252,253]. Thus, only part of the element to be determined appears in the flame in the form of free neutral atoms, and the remaining part in molecular form does not contribute to the atomic absorption. The ratio of atomic to molecular species depends on the nature of the metal, the flame temperature and the gas composition. Some metals, particularly copper, silver, thallium and zinc exist in the usual flames entirely as free atoms [252-255]. Alkaline-earth elements, however, are present in the flame predominantly in the molecular form. The concentration of their monoxides and hydroxides often exceeds that of the free atoms even in stoichiometric and fuel-lean flames [256]. Similarly, titanium, silicon, tungsten, boron, zirconium, vanadium, scandium, tantalum, uranium and the lanthanides show a marked tendency to form refractory oxides even in strongly reducing flames.

Formation of metal hydroxides in the flame is also known, especially for alkali metals [257,258]. At the temperature of the hydrogen flame the free-atom:hydroxide ratio ranges from 0.1 in the case of lithium to an insignificant value for potassium, compared with a value of about 100 for magnesium, copper and manganese. Reactions involving hydride formation have also been reported, but so far only CuH, AuH, PtH, TlH and to a lesser extent AgH and PdH species are known to be present in noticeable amounts in the flame gases. In stoichio-metric flames, the proportion of stable hydrides relative to free atoms of the metal never exceeds 1% [259].

Formation of the molecular species is a reversible process and the atomic and molecular concentrations are thus mutually related through the law of mass action [260,261]:

$$K_d(T) = [M][X]/[MX] \qquad (2.6)$$

where M denotes a metal atom and X the other moiety of the MX molecule. The dissociation constant $K_d(T)$ depends on the temperature and the nature of the molecule. The higher the flame temperature and the larger the value of K_d, the more complete is dissociation of the molecule. The intensity of a given line originating from the neutral atom varies with the reducing or oxidizing character of the flame. A reducing flame contains a low oxygen concentration and thus displaces the equilibrium towards liberation of free atoms.

$$M + O \rightleftharpoons MO, \ P_M = K_1 \frac{P_{MO}}{P_O} \qquad (2.7)$$

$$I = BK_1 \frac{P_{MO}}{P_O} \frac{h\nu}{kT} \frac{g_i}{F(T)} [\exp(-E_i/kT)] \qquad (2.8)$$

Under the usual flame conditions, dissociation seems to be practically com-plete if the dissociation energy is less than about 3 eV. The atomic concentration is greatly enhanced by strongly increasing the fuel gas supply (i.e. hydrogen or acetylene) or by using organic solvents [100,262-269].

The degree of dissociation of barium oxide in an oxygen–acetylene or fuel-rich oxygen–hydrogen flame is about 1%, compared to 30% for magnesium oxide and about 90% for iron oxide [270]. However, the atomic concentration is generally still adequate for atomic-absorption measurements. In a fuel-rich cool hydrogen-air flame, SnO, LiOH and BOH dissociate appreciably. The non-premixed luminous oxygen–acetylene flames with excess of fuel and with alcoholic solvents are frequently utilized for the dissociation of metal oxides. The fuel-rich nitrous oxide–acetylene flame is also useful, and is more advantage-ous than the fuel-rich nitrous oxide-hydrogen, oxygen–hydrogen and air-hydrogen flames [88,122,271,272]. In strongly reducing acetylene flames, long-lived CN, NH, CH and C_2 radicals are formed in the reaction zone and can also play a significant role in the reduction of some stable metal oxides [271, 273-275].

2.2.7 Analyte Ionization

There is little ionization of the analyte in combustion mixtures that involve air as the oxidant, and it can generally be neglected. In high-temperature flames (e.g. with nitrous oxide or oxygen as support gas), neutral atoms or molecules of the analyte may produce singly ionized ions and free electrons during the thermal vaporization. Doubly ionized ions have not yet been reported in the commonly used flames. Ionization in the flame can be brought about by (a) collision of the metal atoms with thermally excited flame molecules [276-280]; (b) chemical association or dissociation associated with charge transfer [281-283]; (c) interaction with radicals [258,282-285].

At a certain distance from the flame reaction zone (i.e. in the outer zone) a state of thermal equilibrium is reached, and the absorption of light is dependent on the position of the equilibrium, but in the inner cone the processes are not balanced. The equilibria concerned involve neutral atoms, their ionized species and electrons [286,287].

$$M \rightleftharpoons M^+ + e^-; \quad K_i = P_{M^+}P_e/P_M \tag{2.9}$$

The change in degree of ionization (K_i) with temperature is given by the Saha equation [288]:

$$\log K_i = -5050(E_i/T) + \tfrac{5}{2} \log T - 6.49 + \log[(P_{M^+}P_e)/P_M]$$

where K_i is the equilibrium constant, E_i the ionization potential (eV), T the absolute temperature, and P_M, P_{M^+} and P_e are the statistical weighting factors for the metal atoms, metal ions and electrons, respectively.

The degree of ionization depends not only on the vaporization temperature and the ionization potential but also on the total element concentration. Elements with $E_i \geqslant 7.5$ eV (e.g. boron and gallium) are ionized only slightly or not at all, whereas those with $E_i < 7$ eV (e.g. aluminium, sodium and lanthanum) are appreciably ionized [120]. The alkali metals show a strong tendency to form ions in the flame, and the observed atomic line intensity corresponds exactly to the concentration of free atoms remaining [86,289,290]. This tendency decreases from caesium (E_i 2.89 eV) to lithium (E_i 5.39 eV). The alkaline-earth elements (E_i 5.21-6.11 eV) are less easily ionized, but molecular ions of the type MOH^+ can occur in considerable concentration [285,291].

In general, the higher the vaporization temperature and the lower the ionization energy, the larger will be the number of ions produced, so the fraction of free atoms is reduced and the limit of detection deteriorates. It should also be noted that hydrochloric acid enhances ionization. The electron affinity of the chlorine atom (3.8 eV) is sufficiently high to produce chloride ions and reduce the concentration of electrons in the flame, which promotes further ionization of the analyte metal and results in depression of the atomic line intensity.

However, the ionization is diminished to a great extent by (a) addition of an ionization suppressor (i.e. an electron donor); (b) using a cool flame [131] whenever possible; (c) avoiding the use of halogen acids. Addition of a second ionizable metal to the analyte increases the concentration of electrons and shifts the ionization equilibrium towards the formation of free atomic species of the analyte. For the determination of sodium or potassium, caesium, which fairly readily gives off electrons, can be added to the test solution.

2.3. ELECTROTHERMAL ATOMIZATION

The efficiency of atomization in flames is very poor, as the atomic concentration attainable is limited by the dilution effect of the high flow-rate of unburnt gas used to support the flame and transport the sample. Absorption takes place by one atom in *ca.* every 10^8 atoms that are fed into the flame. These drawbacks stimulated many workers to develop other types of atom cell. One of the first successful approaches was the use of the electrothermal or 'flameless' technique. The advantages inherent in the application of these methods in AAS are now well recognized [292-294]. The long residence time of the atoms in these devices ($\sim 10^5$ times that in the flame) increases the sensitivity of the method and reduces the volume of sample required for analysis. Moreover, the viscosity, surface tension and density of the sample solution have little influence on the atomization process. The use of an inert gas in most of these devices (to exclude oxygen) is a further advantage, because this enhances the reducing properties of carbon and so permits decomposition of many metal oxides, and also allows atomic-absorption measurement in the vacuum ultraviolet region.

However, flame and electrothermal atomization each have certain advantages which complement each other. Generally speaking, samples which can be analysed by either technique are best analysed in the flame, because it is faster and more convenient, with few interferences. Electrothermal atomization is used when (a) the analyte concentration begins to approach the detection limit offered by the flame; (b) the analyte concentration in the test solution is extremely low; (c) only a small volume of analyte solution is available. The most important advantage of electrothermal atomization is the limit of detection, which for most elements lies in the low picogram region. Thus many elements can be determined at concentration levels lower by a factor of 1000 than those that can be detected in the flame. Table 2.2 shows a comparison of flame and electrothermal atomization systems [294a].

2.3.1 Graphite Tube Furnace

The history of the use of a graphite furnace for generation of metal atoms dates back to King in 1908 [295]. Five decades later, on the basis of this work, L'vov developed a graphite tube furnace for use in AAS [296,297]. The tube (10 cm long, 10 mm external diameter and 3 mm internal diameter) is internally lined with tantalum or tungsten foil to eliminate diffusion of vapour through the porous graphite walls. The sample is placed on a carbon electrode and dried,

Table 2.2 Comparison of flames and graphite-tube furnaces as atomization cells [294a]
[Reprinted from *American Laboratory,* **14**, No. 6, 80 (1982). Copyright 1982 by
International Scientific Communications, Inc.]

Parameter	Flame	Furnace
Minimum sample volume	> 200 μl	~ 1 μl
Sensitivity	~ 1000 times lower	~ 1000 times higher
Nature of atomization	continuous	transient
Atomization (%)	10–20	total
Application to solids	low compatibility	compatible with solids
Background effect	low interference	high interference
Temperature	static for a given gas type and mixture	programmed from ambient to 3000°C at variable heating rates and in various steps
High-temperature effect	significant ionization	insignificant ionization

then introduced through a hole into the graphite tube. An auxiliary electrode is
used for arcing. In an improved version a shorter tube (3-5 cm long, 2.5-5 mm
internal diameter), lined on the inside and covered on the outside with a layer of
pyrolytic graphite, is used as the counter-electrode (Fig. 2.10). In this way the
arc which atomizes the sample is formed the moment the electrode is introduced
into the tube. The furnace is heated electrically by an alternating current from a
4-kW 220/10-V transformer. The whole arrangement is mounted in a sealed
chamber containing argon and having a quartz window to allow the light to pass
through. This arrangement permits determination of 10^{-10}-10^{-14} g of many
metals. L'vov and his co-workers continued their extensive studies with various
models of this original cuvette [298-302].

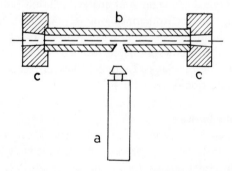

Fig. 2.10. L'vov atomization graphite furnace. a, electrode with the sample;
b, graphite tube; c graphite contacts placed inside cooling jackets [according to
B. V. L'vov, *Spectrochim. Acta,* **24B**, 53 (1969); by permission of the copyright
holder, Pergamon Press].

Woodriff [303,304] designed a graphite tube furnace similar in some respects to that of L'vov (25 cm long, 6 mm internal diameter and 8 mm external diameter), and held in a stainless-steel housing. Although the residence time of the atoms in the furnace is directly proportional to the square of the tube length, the large size of such a design militates against its use in a commercial spectrometer. A length of 3–5 cm has, however, been shown to be adequate [305,306]. A radiofrequency-heated T-shaped carbon furnace similar in design to a combination of the Woodriff and L'vov furnaces has been suggested [307]. The design (Fig. 2.11) consists of a horizontal graphite tube (75 mm long, 17.5 mm external diameter and 4.8 mm internal diameter) and a vertical tube (64 mm long, 7.5 mm external diameter and 6.2 mm internal diameter). The internal diameter of the bottom section is 1.6 mm. The furnace is preheated to 2600°C, the sample is introduced into the vertical tube, and the atomic vapour is swept into the horizontal tube by a stream of inert gas.

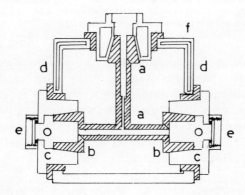

Fig. 2.11. T-shaped carbon furnace. a, graphite T-cell; b, graphite contacts; c, water-cooled electrodes; d, insulators; e, quartz windows; f, water-cooled housing [according to J. W. Robinson and D. K. Wolcott, *Anal. Chim. Acta,* 74, 43 (1975); by permission of the copyright holder, Elsevier Scientific Publishing Co.].

A compact and simplified version of King's original furnace has been described by Massmann [308,309]. It consists of a straight graphite tube (5.5 cm long, 6.5 mm internal diameter, 8 mm external diameter) with a hole (~ 2 mm) for sample introduction (Fig. 2.12). The tube is supported between two water-cooled steel cones. A high current (~ 500 A) at low voltage (~ 10 V) is applied to the end of the tube to cause heating to about 2600°C. The tube is continuously flushed with an inert gas stream, a difference from the closed argon chamber of the L'vov design. Solid (*ca.* 10 mg) and liquid (5–200 μl) samples can be analysed with this furnace. Although the detection limit is about ten times that given by the L'vov furnace, probably because of the use of a forced

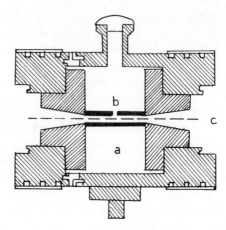

Fig. 2.12. Massmann graphite atomizer. a, end cones; b, graphite tube; c, light path [according to H. Massmann, *Spectrochim. Acta*, **23B**, 215 (1968); by permission of the copyright holder, Pergamon Press].

inert-gas flow through the tube, as little as 10^{-13} g of many elements can easily be determined. The Massmann design is currently one of the most popular models commercially available. Automation of the sample delivery step [310] or evaporation temperature [311] gives higher reproducibility than manual introduction of the sample does. Atomization under a pressure up to 4 kg/cm^3 above atmospheric [312], the effect of various inert sheath gases [313] and the use of a contoured tube [314] have also been described.

2.3.1.1 Pyrolytic carbon and metal carbide coating materials. The effect of variations in the quality of different batches of graphite tubes on the precision of AAS measurements has been demonstrated [315,316]. Since graphite is a porous material and is easily penetrated by both liquids and gases, especially when it is hot, coating with a thin homogeneous layer of pyrolytic graphite has been recommended for reducing the porosity [317-321]. This treatment increases the signal intensity for several important metals by a factor of four. Manning and Ediger [322] showed that the sensitivity improvement is, however, short-lived when the tube operation temperature exceeds 2700°C. These authors recommended a coating technique involving passing a stream of methane gas (10% methane and 90% nitrogen) for 10 min into a tube furnace preheated to 2200°C. The tube may also be heated in a stream of pure argon at 2300°C before passage of methane [323]. The results of *in situ* pyrolytic coating have also been discussed [324].

It has been reported that nitrate [325], perchlorate [326,327], chromate [328], chloride [329] and sulphate [330] ions have a destructive effect on the pyrolytic coating, probably due to thermal decomposition of these ions into oxidizing species. Unfortunately, these ions are usually present in high concentration in the digestion solutions of organic compounds. This necessitates complete evaporation of such solutions before injection into the graphite furnace [331].

Coating the internal surface of the graphite tube with a hydrophobic film of polystyrene [332] and carbide-forming elements (e.g. boron, barium, lanthanum, molybdenum, niobium, silicon, tantalum, vanadium and zirconium) has also been shown to enhance the analytical response of some elements significantly [333–336]. Zirconyl salts have been recommended as efficient surface modifiers [337–340]. The graphite tube is treated with an aqueous solution of zirconyl salts, followed by heating at 2000°C [341]. Similarly, a tube coated with tantalum carbide can be prepared and used for up to 400 atomization runs at 2700°C [342]. General comments on the carbide-coating technique and application of tubes coated with tantalum, zirconium and tungsten carbides in AAS have been published [343,344]. Silanation of the graphite tube by spraying the tube with dichlorodimethylsilane in toluene, followed by heating at 120°C, has also been described [345].

2.3.1.2 L'vov platform. For good selectivity and complete atomization in the graphite tube furnace, it is necessary to ensure good contact between the sample and the hot graphite surface for as long as possible, and to avoid non-uniformity of temperature. In the flame, the sample is always surrounded by the hot flame gases and therefore has the most intimate contact with the flame. With all commercial graphite tube furnaces, however, the sample is directly applied onto the inside wall of the cold graphite tube, followed by stepwise heating to volatilize the analyte at temperatures dependent on the matrix composition of the sample. Thus, the tube wall is heated considerably faster than the inert gas environment, causing a temperature gradient at a given time. This results in condensation or recombination of the atoms. Several workers have pointed out the advantages of introducing the sample into a hot environment that is isothermal in both time and space [298,303,346–349].

L'vov suggested a small platform made of solid pyrolytic or carbide-coated graphite. The platform is placed inside the furnace (Fig. 2.13) and heated mainly

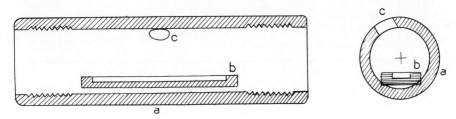

Fig. 2.13. L'vov platform. a, graphite-tube furnace; b, L'vov platform; c, sample hole.

by radiation from the red-hot tube walls and by collision with the hot gas molecules. The analyte vaporization and atomization are delayed until the tube has
more nearly attained constant temperature. Thus, the analyte is vaporized from
this surface into a gas that is hotter than the surface [346,347]. Consequently,
little chance for condensation or recombination is expected, and even the molecular species volatilized from the platform have a good chance for dissociation.
Although matrix modification can also be used with the platform [350], tremendous improvements in the determination of many elements are obtained without
the need for matrix modification. Figure 2.14 shows the results obtained for
measurement of 1 ppm of arsenic by the graphite tube furnace with and without
the L'vov platform. Graphite cups may also be used [351]. However, the L'vov
platform technique is beginning to be fairly widely utilized, owing to its effectiveness in reducing matrix interferences [352,353] without alteration of the
peak area. The platform can be simply fabricated from pyrolytically coated
graphite tubes that have been treated with tantalum solution [354].

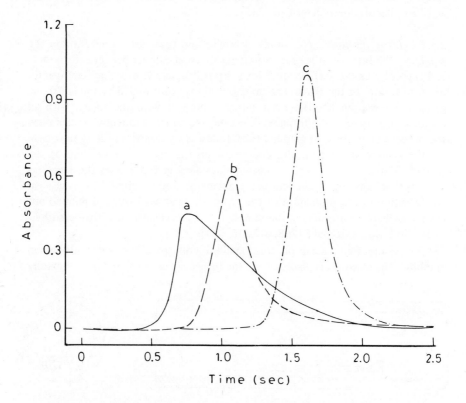

Fig. 2.14. Atomic-absorption signals for 1-ppm arsenic, obtained with a, pyrolytic carbon tube; b, standard graphite tube; c, L'vov platform.

Holcombe *et al.* [355], however, showed that the gas-phase interaction between interfering species and analyte can still occur when the L'vov platform is used, and some physical interferences as well as scattering from particulate matter may produce analytical error. In an attempt to minimize these problems while maintaining the platform's advantages, these authors suggested a simple modification involving the use of a loose-fitting graphite plug to be inserted into a slot milled in the top of the graphite tube furnace (Fig. 2.15). During the heating cycle, the plug temperature lags behind the wall temperature by as much as 900°C. This allows an extremely high temperature to be used for vaporizing the analyte.

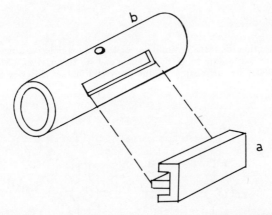

Fig. 2.15. Graphite plug. a, slotted graphite plug; b, graphite-tube furnace (according to J. A. Holcombe, M. T. Sheehan and T. Rettburg, Pittsburgh Conference on Analytical Chemistry and Applied Spectroscopy, Atlantic City, 1982).

2.3.1.3 Metal inserts. Wall [356] found that the sensitivity of the graphite tube atomizer could be improved by inserting a tungsten tube inside the graphite tube, and atomizing the analyte from the metal surface rather than the graphite surface. A tantalum cylinder [357] and foil [358] have similarly been used. The limit of detection for 23 elements has been lowered 3–100-fold with a reduction of 200–900°C in the atomization temperature. It seems that such metal surfaces act like the L'vov platform.

Several workers have begun to use wire filaments with graphite tube furnaces (Fig. 2.16) [358–361]. The technique is useful for preconcentration of many metal ions by repeated soaking of the wire in the test solution and evaporation. This permits drying and charring the sample outside the furnace, to avoid contamination, after which the wire is introduced into the preheated furnace at the required atomization temperature [358,359,362].

Finally, powdered biological materials can be analysed by introducing the samples into the furnace with a metal insertion tool [363]. This technique has

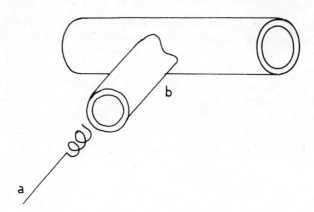

Fig. 2.16. Graphite-tube furnace modified for the wire sampler. a, tungsten wire; b, graphite tube [according to D. C. Manning and W. Slavin, *Anal. Chim. Acta*, **118**, 301 (1980); by permission of the copyright holder, Elsevier Scientific Publishing Co.].

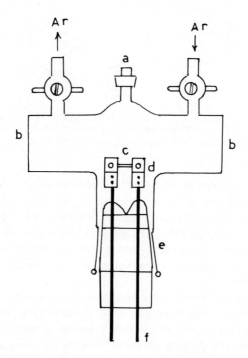

Fig. 2.17. Carbon-filament atom reservoir for atomic absorption. a, sample inlet; b, silica windows; c, carbon filament; d, stainless-steel electrodes; e, B55 Pyrex cone; f, tungsten rods [according to T. S. West and X. K. Williams, *Anal. Chim. Acta*, **45**, 27 (1969); by permission of the copyright holder, Elsevier Scientific Publishing Co.].

been utilized for electrochemical preconcentration and AAS determination of many metals, with flame and electrothermal atomization in graphite or quartz tubes [364-371].

2.3.2 Graphite Filament

Generation of atomic vapour with graphite filament atomizers has also been reported. The filament (1-4 cm long and 1-2 mm diameter) has a depression for sample application and is mounted on water-cooled stainless-steel electrodes. It is heated electrically to 2000-2500°C by passage of a current of 100 A at 5 V. The assembly is housed in an argon-filled chamber fitted with a quartz window (Fig. 2.17) [372]. Modified designs of this atomizer, including an atom reservoir for use in an open atmosphere (Fig. 2.18) have been suggested [373-379].

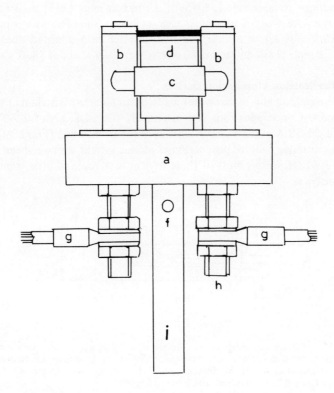

Fig. 2.18. Filament atomizer for use in open atmosphere. a, base; b, water-cooled electrodes; c, water link between electrodes; d, laminar-flow box; e, filament; f, inlet for shield gas; g, transformer terminals; h, water inlet; i, support stem for reservoir [according to D. J. Johnson, T. S. West and R. M. Dagnall, *Anal. Chim. Acta*, **67**, 79 (1973); by permission of the copyright holder, Elsevier Scientific Publishing Co.].

2.3.3 Metal Tube and Metal Filament

Determination of some elements in the graphite tube furnace is impaired by carbide formation. Metal tube atomizers made of molybdenum or tungsten (~ 25 mm long, 1.5 mm diameter) have been suggested. The design and performance of these tubes have been described, together with the atom formation process [380-383]. Püschel et al. [384] showed that isothermal atomization is achieved more simply and cheaply for most elements with a tungsten tube heated at the rate of 0.5-20°C/msec than by capacitative discharge heating of anisotropic graphite tubes. Problems arising from samples forming graphite-metal compounds of low volatility are also avoided. Stainless-steel [385] and platinum tube furnaces have also been used for AAS. The sensitivity of most of these atomizers can be greatly improved by using a microcomputer to measure and enhance the signal [386].

Electrically heated metal filament atomizers made from platinum or tungsten wire loops are also used [387-390]. Cedergren et al. [341] used a tungsten wire for the AAS analysis of sample sizes as low as 5 µl. The wire is preheated by application of a low voltage and then inserted into a heated silica tube in which the sample is rapidly vaporized electrothermally at about 1800°C.

2.3.4 Miscellaneous Atomization Devices

Boats (50 mm long and 6 mm wide) made of graphite, or tantalum or tungsten foil, supported on copper rods and electrically heated at about 2200°C by a current of 30-50 A at 12 V, have been used for atomization (Fig. 2.19) [392]. Cathodic sputtering cells for production of atomic vapour have also been suggested [393-399]. However, most of these devices have received little attention for routine analysis.

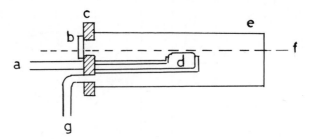

Fig. 2.19. Electrically heated boat atomizer. a, to power; b, quartz window; c, O-ring seal; d, boat; e, quartz envelope; f, light-path; g, vacuum gas [reprinted with permission from H. M. Donega and T. E. Burgess, *Anal. Chem.*, **42**, 1521 (1970). Copyright 1970 American Chemical Society].

2.3.5 Temperature Programming

All the electrothermal atomizers are resistively heated devices that heat the analyte in a programmable sequence controlling the time and temperature. The most commonly used graphite tube atomizers are heated by application of constant electrical power. Three main stages of heating are used: (a) drying for solvent evaporation; (c) charring for matrix elimination; (c) atomization for free atom formation from the analyte. A further stage may also be used, in which the tube is cleaned by heating at a high temperature to remove the last traces of sample before the next run. The drying step controls the analytical precision, whereas the objective of the ashing step is to separate selectively, by distillation, as much of the unwanted species as possible. With many samples, the analyte salt may volatilize before the unwanted matrix components. In such cases, a combination of precise and accurate temperature control and matrix modification is required. Microcomputer-controlled autosamplers allow matrix modification [400]. During the atomization cycle, a sufficiently high temperature is applied to volatilize the analyte salt, to dissociate the salt to its atomic form and to prevent condensation within the atomizer. The exact temperature and duration of each stage depend on the nature of the analyte. They can be adjusted and automatically controlled by a preselected program in the instrument.

The temperature of the graphite tube furnace may be increased either stepwise or continuously (in a slow controlled-rate 'ramp') (Fig. 2.20). In general,

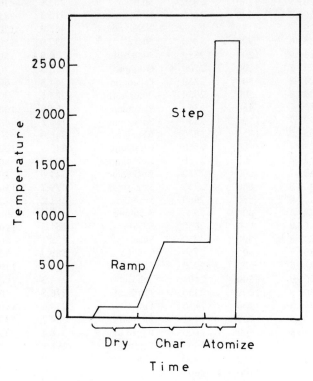

Fig. 2.20. Diagram of temperature programming of the furnace.

slow ramp heating is used for the drying and charring processes, and a rapid stepwise increase in temperature is used for atomization. The atomization temperature required for optimum sensitivity is lower when rapid heating is used for the atomization. However, the actual temperature of the atomizer is a complicated function of time and various other factors such as the input power, the mass of the atomizer and the heat losses by convection, conduction and radiation. The temperature and time must be rigidly controlled if reasonable precision is to be obtained. The temperature and time used depend on the type of sample and the element involved (Table 2.3).

The shape and width of the absorption signals are reported to be improved if the rate of heating is increased by heating the tube rapidly to the desired temperature and then maintaining this by application of power pulses [401].

Table 2.3 Recommended temperatures for electrothermal atomization of some elements

Element	Wavelength (nm)	Temperature (°C) Ashing	Temperature (°C) Atomization	Element	Wavelength (nm)	Temperature (°C) Ashing	Temperature (°C) Atomization
Aluminium	309.3	1100	2850	Mercury	253.6	<100	1500
Antimony	206.8	500	2800	Molybdenum	313.3	850	2900
Arsenic	193.7	350	2200	Neodymium	463.4	1200	2800
Barium	553.6	900	2800	Nickel	232.0	1100	2850
Beryllium	234.9	1200	2850	Niobium	334.4	1500	2900
Bismuth	223.1	400	2100	Osmium	290.9	1700	2500
Boron	249.8	500	2850	Palladium	244.8	600	2750
Cadmium	228.8	500	2050	Phosphorus	213.6	300	2750
Caesium	852.1	450	2350	Platinum	265.9	1600	2800
Calcium	422.7	1100	2850	Potassium	766.5	400	2250
Chromium	357.9	1350	2850	Rhenium	346.0	200	2250
Cobalt	240.7	1100	2850	Rhodium	343.5	1000	2750
Copper	324.8	800	2750	Rubidium	780.0	800	2100
Dysprosium	421.2	1000	2750	Ruthenium	349.9	500	2700
Erbium	400.8	800	2750	Samarium	429.7	1500	2550
Europium	459.4	850	2600	Scandium	391.2	700	2450
Gallium	287.4	800	2500	Selenium	196.0	900	2800
Germanium	265.2	700	2650	Silicon	251.6	1100	2850
Gold	242.8	500	2750	Silver	328.1	500	2750
Hafnium	307.3	1000	2750	Sodium	589.0	300	2250
Holmium	410.4	600	2900	Strontium	460.7	1100	2500
Indium	303.9	500	2400	Tellurium	214.3	600	2400
Iridium	208.9	1000	2750	Thallium	276.8	500	1600
Iron	248.3	1100	2450	Tin	224.6	1100	2850
Lead	217.0	700	2450	Titanium	364.3	1500	2900
Lithium	670.8	700	2400	Vanadium	318.5	800	2900
Lutetium	336.0	1100	2900	Ytterbium	398.8	1400	2800
Magnesium	285.2	1100	2800	Zinc	213.9	900	2750
Manganese	279.5	1100	2800				

Rapid heating in the atomization step allows the use of lower atomization temperatures for many elements, and increases the sensitivity [402]. It is well known that with pulse-heated electrothermal atomizers, the analytical peak is generally observed while the furnace wall temperature is still increasing [403]. Modifications to reduce the instrumental response time to 10 msec in order to monitor the absorption pulses accurately have been described [404]. The temperature variation along the graphite tube furnace can be reduced from 1000 to 100°C by using a contoured tube, tapering toward the ends [314].

2.3.6 Nature of Absorption Signals and Methods of Measurement

In contrast to the steady-state signals obtained with the flame AAS technique, the analytical signals generated by electrothermal atomization are transient pulses of, at most, a few seconds duration. The exact shape of these signals (signal–time relation) for all elements is determined by the physical and chemical properties of the sample matrix, physicochemical nature of the atomizers (i.e. construction material, porosity, reactivity, permeability and density) and the rate of heating.

The difference between signals obtained by flame and electrothermal atomization can be understood from the nature of the atomization process. In the flame, so long as the sample is being aspirated, the atom population in the flame is maintained and the absorption is constant. In the graphite tube atomizer, however, the sample aliquot is introduced into the furnace and totally atomized in a few seconds or even milliseconds. Consequently the absorption in the flame is directly related to the analyte concentration in the test solution. In the furnace, however, the sample is vaporized at a rate independent of the initial volume of the sample used, and thus the absorption depends on the total amount of analyte contained in the test solution, and not on the concentration.

The most convenient method for measuring the signals given by electrothermal atomizers is measurement of the peak height, although measuring the area under the signal pulse over a period of time is less affected by interference [405]. A simple method has been suggested for improving control of the time constant in recording transient signals in AAS [406]. A digital integrator that provides a read-out of absorbance has also been designed [407]. The relative advantages of signal height and area for AAS measurement have been discussed [408-410]. The strip-chart recorder provides valuable data to accompany the electronic peak height or peak area readings.

On the other hand, double peaks for relatively volatile metals (e.g. lead, cadmium and zinc) have been found to occur with a variety of sample types and atomizer designs [411-417a]. The non-reproducible appearance of these double peaks introduces significant error if the height of the peak is measured. Salmon *et al.* [418] showed that the presence of oxygen in the graphite atomizer has a profound effect on the atomizer surface and causes the appearance of double peak signals for volatile metals. It has also been noticed that the vaporization temperature of these metals lies between the temperature for optimum oxygen adsorption ($500°C$) and that for total desorption ($950°C$).

A number of workers [419–422] have used a storage oscilloscope and special high-speed electronic and computer systems to study the peak shape obtained with the graphite tube furnace in analytical applications. It has been reported [420] that the peak absorbance is a complex function of both the time taken to atomize the sample and the residence time of the atomic vapour within the analytical volume [Eq. (2.10)], whereas the integrated absorbance is dependent solely on the residence time [Eq. (2.11)] :

$$A_p = \frac{2K_A N_0 J_2^2}{J_2^2} [\frac{J_1}{J_2} - 1 + \exp(-J_1/J_2)] \tag{2.10}$$

$$\int_0^\infty A(t)dt = K_A N_0 J_2 = A_0 J_2 \tag{2.11}$$

where A_p is the peak absorbance, K_A a constant relating the absorbance to atom concentration, N_0 the number of atoms of the element to be determined in the sample, J_1 the atomization time, J_2 the residence time of the atomic vapour, and A_0 the absorbance that would result if all the analytical atoms were confined within the analytical volume at a single instant of time.

If a sample can be completely atomized within a time $< J_2$, Eq. (2.11) reduces to

$$A_p \approx K_A N_0 = A_0 \tag{2.12}$$

and the peak absorption corresponds to the mean of the total number of atoms in the sample, provided that $J_2 > J_1$. If, however, $J_1 > J_2$, the peak absorbance depends on both J_1 and J_2 so that a simple relationship between A_p and N_0 is possible only if J_1/J_2 is constant. On the other hand, the integrated absorbance is dependent on neither the rate nor the time. This is extremely significant since measurement of $\int_0^\infty A(t)dt$ prevents varying sample composition from affecting the analytical results.

2.3.7 Mechanism of Atomization

A number of studies employing mathematical modelling, theoretical mechanisms and kinetics for understanding atom formation and losses have been reported [423–436]. As in the flame, desolvation, decomposition, thermal dissociation and reduction of oxides all take place. Free atoms are produced from the metal oxides either by dissociation or reduction.

$$M_x O_{y(S,L)} \rightleftharpoons M_x O_{y(g)} \rightarrow x M_{(g)} + y O_{(g)} \tag{2.13}$$

$$M_x O_{y(S,L)} + y C \rightarrow x M_{(S,L)} + y CO_{(g)} \tag{2.14}$$

$$\Updownarrow$$

$$\tfrac{1}{2} x M_{2(g)} \rightarrow x M_{(g)} \tag{2.15}$$

Although the maximum temperature attained by the atomic vapour in electrothermal atomizers is lower than that attained in the nitrous oxide-acetylene flame, sufficient energy is available through collisional excitation processes to cause ionization of some elements [437–439]. Microwave attenuation measurements of the concentration of the free electrons generated in graphite and tantalum tube atomizers, and application of the Saha equation, indicate that analyte ionization is negligible for those elements having ionization potentials > 4.6 eV [440]. Sturgeon and Berman [441] demonstrated that the analyte ionization in graphite tube atomizers is very small compared to that obtained for the same elements in the flame. This is probably due to the different time and temperature dependences of the atom and ion populations as well as the high background concentration of free electrons generated in the furnace.

2.4 MONOCHROMATORS

The sole purpose of the monochromator is to separate the resonance line of the element under study from other emission lines. Both prism and grating monochromators can be used. The Ebert, Czerny-Turner and Littrow grating mountings are in common use. A dual grating to cover the entire range of both the ultraviolet and visible regions has also been used. Resonance detectors can serve a dual function, acting also as monochromators in AAS [442–446]. The radiation emitted from a high-intensity hollow-cathode lamp falls on the detector, which contains a cloud of ground-state atoms obtained by cathodic sputtering, and the resonance radiation emitted is measured at a convenient take-off angle.

The slits of the monochromator determine the fraction of the spectrum that falls on the photomultiplier. The slits should be as narrow as possible to reduce the amount of flame emission reaching the photomultiplier. Some commercially available atomic-absorption spectrometers contain two sets of slits, for use with the flame and electrothermal atomization, respectively. Both mirrors and lenses are used to reflect or transmit the radiation. The lens system is used for focusing ultraviolet radiation and is not perfect for visible radiation.

2.5. DETECTORS

Photomultipliers with 10 dynode stages are frequently used in atomic-absorption spectrometers to convert the light signal into an electrical signal. However, to eliminate unwanted radiation from the flame or the furnace from reaching the detector, the light-source emission is modulated. In all modern instruments the light-beam is modulated either mechanically with choppers (rotating sectors or a vibrating shield), or electrically by supplying the radiation source with a pulsed current. In mechanical modulation, the radiation source is run at constant current and the recorder reading indicates the actual current flow. In electronic modulation, however, the current is switched on and off at a rapid rate. The recorder will not follow these rapid fluctuations in current but will indicate a lower current that is actually flowing during the time the lamp is on (Fig. 2.21). As a result of this phenomenon, the lamp current settings are lower for instruments which modulate the source electronically.

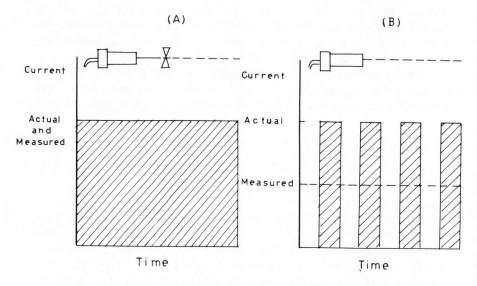

Fig. 2.21. Mechanical (A) and electrical (B) modulation of light.

A new approach to wavelength-specific photoelectric detection has been described by Stephens [447]. Modulation of the beam at a fixed frequency and setting of the amplifier electronics to the same frequency allow amplification of only the light from the source that has the modulator frequency. In this case the emission from the flame or furnace is not modulated and will be eliminated. Selective modulation may also be used for multielement analysis [448-451].

Whatever the method used for light detection, the beam intensity after absorption by the free analyte atoms can be displayed on a suitably calibrated voltmeter or a digital display screen. However, most modern instruments use microprocessors not only to replace the conventional electronic systems, but also for (a) controlling the sequence of processes performed by the instrument; (b) optimization of the analytical conditions for analysis; (c) processing the results to produce the results more quickly and reliably [452-466]. A fully automated system has been developed for processing the AAS output both numerically and graphically [467]. In general, automation minimizes the analysis time. For example, the Perkin-Elmer model 5000, coupled with its automated sample handling system, can determine six elements (e.g. calcium, magnesium, manganese, iron, copper and zinc) in 50 samples in less than 25 min [468].

REFERENCES

[1] J. A. C. Broekaert, *Spectrochim. Acta,* **36B**, 931 (1981).
[2] H. L. Kahn, *Atom. Absorp. Newslett.* **2**, 35 (1963).
[3] P. R. Liddell and P. C. Wildy, *Spectrochim. Acta,* **35B**, 193 (1980).
[4] H. P. J. Van Dalen and L. de Galan, *Analyst,* **106**, 695 (1981).
[5] K. W. Busch and G. H. Morrison, *Anal. Chem.,* **45**, 712 (1973).
[6] K. W. Busch and B. Malloy, *Anal. Chem.,* **51**, 670 (1979).
[7] J. M. Harnly and T. C. O'Haver, *Anal. Chem.,* **49**, 2187 (1977).
[8] J. M. Harnly, T. C. O'Haver, B. Golden and W. R. Wolf, *Anal. Chem.,* **51**, 2007 (1979).
[9] E. D. Salin and J. D. Ingle, Jr., *Anal. Chem.,* **50**, 1737 (1976).
[10] E. D. Salin and J. D. Ingle, Jr., *Anal. Chem.,* **50**, 1745 (1976).
[11] E. D. Salin and J. D. Ingle, Jr., *Appl. Spectrosc.,* **32**, 579 (1978).
[12] E. D. Salin and J. D. Ingle, Jr., *Anal. Chim. Acta,* **104**, 267 (1976).
[13] O. Rose, Jr., W. R. Heineman, J. A. Caruso and F. L. Fricke, *Analyst,* **103**, 113 (1978).
[14] H. L. Pardue and H. L. Felkel, *Dev. Toxicol. Environ. Sci.,* 375 (1977).
[15] H. L. Felkel and H. L. Pardue, *Clin. Chem.,* **24**, 602 (1978).
[16] K. W. Jackson, K. M. Aldous and D. G. Mitchell, *Spectrosc. Lett.,* **6**, 315 (1973).
[17] D. G. Mitchell, K. W. Jackson and K. M. Aldous, *Anal. Chem.,* **45**, 1215A (1973).
[18] K. W. Jackson, K. M. Aldous and D. G. Mitchell, *Appl. Spectrosc.,* **28**, 569 (1974).
[19] K. M. Aldous, D. G. Mitchell and K. W. Jackson, *Anal. Chem.,* **47**, 1034 (1975).
[20] E. G. Codding, J. D. Ingle, Jr. and A. J. Stratton, *Anal. Chem.,* **52**, 2133 (1980).
[21] G. Horlick and E. G. Codding, *Appl. Spectrosc.,* **29**, 167 (1975).
[22] F. S. Chuang, D. F. Natusch and K. R. O'Keefe, *Anal. Chem.,* **50**, 525 (1978).
[23] L. R. P. Butler and A. Strasheim, *Spectrochim. Acta,* **21**, 1207 (1965).
[24] A. Strasheim and H. G. C. Human, *Spectrochim. Acta,* **23B**, 265 (1967).
[25] R. Mavrodineanu and R. C. Hughes, *Appl. Opt.,* **7**, 1281 (1968).
[26] L. C. J. Pickford and G. Rossi, *Analyst,* **98**, 329 (1973).
[27] J. F. Alder, D. Alger, A. J. Samuel and T. S. West, *Anal. Chim. Acta,* **87**, 301 (1976).
[28] A. M. Ure, G. J. Ewen and M. C. Mitchell, *Anal. Chim. Acta,* **118**, 1 (1981).
[29] A. Paschen, *Physik,* **50**, 901 (1916).
[30] W. G. Jones and A. Walsh, *Spectrochim. Acta,* **16**, 249 (1960).
[31] B. J. Russell and A. Walsh, *Spectrochim. Acta,* **10**, 883 (1959).
[32] M. E. Pillow, *Spectrochim. Acta,* **36B**, 821 (1981).
[33] B. J. Russell, J. P. Shelton and A. Walsh, *Spectrochim. Acta,* **8**, 317 (1957).
[34] G. H. C. Freeman, M. Outred and L. R. Morris, *Spectrochim. Acta,* **35B**, 687 (1980).
[35] J. V. Sullivan and A. Walsh, *Spectrochim. Acta,* **21**, 721 (1965).
[36] J. V. Sullivan and A. Walsh, *Spectrochim. Acta,* **21**, 727 (1965).
[37] J. S. Cartwright and D. C. Manning, *Atom. Absorp. Newslett.,* **5**, 114 (1966).
[38] J. S. Cartwright, C. Sebens and D. C. Manning, *Atom. Absorp. Newslett.,* **5**, 91 (1966).
[39] J. S. Cartwright, C. Sebens and W. Slavin, *Atom. Absorp. Newslett.,* **5**, 22 (1966).
[40] J. B. Dawson and D. J. Ellis, *Spectrochim. Acta,* **23A**, 565 (1967).
[41] G. J. Dejong and E. H. Piepmeir, *Spectrochim. Acta,* **29B**, 159 (1974).
[42] T. Araki, J. P. Walters and S. Minami, *Appl. Spectrosc.,* **34**, 33 (1980).
[43] P. B. Farnsworth and J. P. Walters, *Anal. Chem.,* **54**, 885 (1982).
[44] S. A. Myers, *Spectrochim. Acta,* **34B**, 213 (1979).
[45] H. Massmann, *Z. Instrumentenk.,* **71**, 225 (1963); *Appl. Spectrosc.,* **16**, 56 (1962).
[46] C. Sebens, J. Vollmer and W. Slavin, *Atom. Absorp. Newslett.,* **3**, 165 (1964).
[47] W. Slavin, *Atom. Absorp. Newslett.,* **3**, 93 (1964).
[48] F. J. Fernandez, D. C. Manning and J. Vollmer, *Atom. Absorp. Newslett.,* **8**, 117 (1969).
[49] P. Heneage, *Atom. Absorp. Newslett.,* **5**, 67 (1966).

[50] D. C. Manning, D. J. Trent and J. Vollmer, *Atom. Absorp. Newslett.*, **4**, 234 (1965).
[51] C. W. Frank, W. G. Schrenk and C. E. Meloan, *Anal. Chem.*, **38**, 1005 (1966).
[52] K. B. Mitchell, *J. Opt. Soc. Am.*, **51**, 846 (1961).
[53] A. Strasheim and L. R. P. Butler, *Appl. Spectrosc.*, **16**, 109 (1962).
[54] G. I. Goodfellow, *Appl. Spectrosc.*, **21**, 39 (1967).
[55] R. E. Popham and W. G. Schrenk, *Appl. Spectrosc.*, **22**, 192 (1968).
[56] S. R. Koirtyohann and E. E. Pickett, *Anal. Chem.*, **37**, 601 (1965).
[57] G. Rossi and N. Omenetto, *Appl. Spectrosc.*, **21**, 329 (1967).
[58] S. Caroli, O. Senofonte, A. Alimonti and N. Violante, *Spectrosc. Lett.*, **13**, 905 (1980).
[59] T. M. Niemczyk and J. Erspamer, *Appl. Spectrosc.*, **33**, 61 (1979).
[60] W. Parker, R. Lozada and J. J. Labrecque, *Appl. Spectrosc.*, **34**, 94 (1980).
[61] D. A. Jackson, *Proc. Roy. Soc.*, **A121**, 432 (1928).
[62] G. F. Kirkbright, T. S. West and P. J. Wilson, *Atom. Absorp. Newslett.*, **11**, 53 (1972).
[63] G. F. Kirkbright, T. S. West and P. J. Wilson, *Atom. Absorp. Newslett.*, **11**, 113 (1972).
[64] R. M. Dagnall and T. S. West, *Appl. Opt.*, **7**, 1287 (1968).
[65] E. Jacobsen and G. R. Harrison, *J. Opt. Soc. Am.*, **39**, 1054 (1949).
[66] A. T. Forrester, R. A. Gudmundsen and P. O. Johnson, *J. Opt. Soc. Am.*, **46**, 339 (1956).
[67] W. F. Meggers and F. O. Westfall, *J. Res. Natl. Bur. Stds.*, **44**, 447 (1950).
[68] R. M. Dagnall, K. C. Thompson and T. S. West, *Atom. Absorp. Newslett.*, **6**, 117 (1967).
[69] R. M. Dagnall, K. C. Thompson and T. S. West, *Talanta*, **14**, 551 (1967).
[70] J. M. Mansfield, M. P. Bratzel, H. O. Norgordon, D. C. Knopp, K. E. Zacha and J. D. Winefordner, *Spectrochim. Acta*, **23B**, 389 (1968).
[71] W. B. Barnett, J. W. Vollmer and S. M. DeNuzzo, *Atom. Absorp. Newslett.*, **15**, 33 (1976).
[72] J. W. Novak and R. F. Browner, *Anal. Chem.*, **50**, 407 (1978); **50**, 1453 (1978).
[73] K. Kuga, K. Tsujii, S. Murayama and M. Yasuda, *Bunko Kenkyu*, **29**, 178 (1980).
[74] S. Z. Mohamed and A. Petrakiev, *Spectrosc. Lett.*, **14**, 47 (1981).
[75] D. F. Bartley, T. Hurst, J. D. Norris and T. S. West, *Rev. Anal. Chem.*, **4**, 19 (1978).
[76] J. D. Norris and T. S. West, *Spectrosc. Lett.*, **11**, 707 (1978).
[77] P. E. Walters, F. M. Hamm and K. J. Smith, *J. Phys.*, **E12**, 820 (1979).
[78] S. Chilukuri and W. L. Lichten, *Rev. Sci. Instrum.*, **50**, 256 (1979).
[79] W. B. Barnett, *Atom. Absorp. Newslett.*, **12**, 142 (1973).
[80] J. M. Hall and C. Woodward, *Spectrosc. Lett.*, **2**, 113 (1969).
[81] H. C. Hoare, R. A. Mostyn and B. T. N. Newland, *Anal. Chim. Acta*, **40**, 181 (1968).
[82] D. D. Siemer, *Appl. Spectrosc.*, **34**, 487 (1980).
[83] D. S. Gough and J. V. Sullivan, *Anal. Chim. Acta*, **108**, 347 (1979).
[84] D. S. Gough and J. V. Sullivan, *Anal. Chim. Acta*, **124**, 259 (1981).
[85] A. I. Bezlepkin, A. S. Khomyak and V. V. Aleksandrov, *Zavodsk. Lab.*, **47**, 29 (1981).
[86] J. B. Willis, *Appl. Opt.*, **7**, 1295 (1968).
[87] J. E. Chester, R. M. Dagnall and M. R. G. Taylor, *Analyst*, **95**, 705 (1970).
[88] L. de Galan and G. F. Samaey, *Spectrochim. Acta*, **25B**, 245 (1970).
[89] R. F. Suddendorf and M. B. Denton, *Appl. Spectrosc.*, **28**, 814 (1974).
[90] R. J. Reynolds and D. S. Lagden, *Analyst*, **96**, 319 (1971).
[91] B. M. Gatehouse and J. B. Willis, *Spectrochim. Acta*, **17**, 710 (1961).
[92] J. E. Allen, *Spectrochim. Acta*, **18**, 259 (1962).
[93] J. Aggett and G. O'Brien, *Analyst*, **106**, 497 (1981).
[94] J. Aggett and G. O'Brien, *Analyst*, **106**, 506 (1981).
[95] K. Kitagawa, M. Yanagisawa and T. Takeuchi, *Anal. Chim. Acta*, **115**, 121 (1980).
[96] K. C. Thompson and K. Wagstaff, *Analyst*, **105**, 641 (1980).

[97] J. A. Burrows, J. C. Heerdt and J. B. Willis, *Anal. Chem.*, 37, 579 (1965).

[98] J. A. Dean, *Analyst*, 85, 621 (1960).

[99] A. P. D'Silva, R. N. Kniseley and V. A. Fassel, *Anal. Chem.*, 36, 1287 (1964).

[100] V. A. Fassel, R. B. Myers and R. N. Kniseley, *Spectrochim. Acta*, 19, 1187 (1963).

[101] R. N. Kniseley, A. P. D'Silva and V. A. Fassel, *Anal. Chem.*, 35, 911 (1963).

[102] J. H. Medlin, N. H. Suhr and J. B. Bodkin, *Atom. Absorp. Newslett.*, 8, 25 (1969).

[103] J. E. Allan, *Spectrochim. Acta*, 17, 459 (1961).

[104] W. Slavin and D. C. Manning, *Anal. Chem.*, 35, 253 (1963).

[105] M. D. Amos and P. E. Thomas, *Anal. Chim. Acta*, 32, 139 (1965).

[106] D. C. Manning, *Atom. Absorp. Newslett.*, 3, 84 (1964).

[107] J. W. Robinson, *Anal. Chem.*, 33, 1067 (1961).

[108] D. J. David, *Spectrochim. Acta*, 20, 1185 (1964).

[109] J. B. Willis, *Nature*, 207, 715 (1965).

[110] M. D. Amos and J. B. Willis, *Spectrochim. Acta*, 22, 1325 (1966).

[111] J. A. Bowman and J. B. Willis, *Anal. Chem.*, 39, 1210 (1967).

[112] S. R. Koirtyohann and E. E. Pickett, *Spectrochim. Acta*, 23B, 673 (1968).

[113] D. C. Manning, *Atom. Absorp. Newslett.*, 5, 127 (1966); 6, 35 (1967); 6, 75 (1967); 9, 97 (1970).

[114] W. Slavin, A. Venghiattis and D. C. Manning, *Atom. Absorp. Newslett.*, 5, 84 (1966).

[115] J. W. Husler, *Atom. Absorp. Newslett.*, 10, 60 (1971).

[116] V. A. Fassel, J. O. Rasmuson, R. N. Kniseley and T. G. Cowley, *Spectrochim. Acta*, 25B, 559 (1970).

[117] R. J. Jaworowski, R. P. Weberling and D. J. Bracco, *Anal. Chim. Acta*, 37, 285 (1967).

[118] G. F. Wallace, B. K. Lumas, F. J. Fernandez and W. B. Barnett, *Atom. Spectrosc.*, 2, 130 (1981).

[119] G. F. Kirkbright, T. S. West and P. J. Wilson, *Atom. Absorp. Newslett.*, 11, 53 (1972).

[120] D. C. Manning, *Atom. Absorp. Newslett.*, 5, 127 (1966).

[121] R. M. Dagnall, K. C. Thompson and T. S. West, *Analyst*, 93, 153 (1968).

[122] J. B. Willis, V. A. Fassel and J. A. Fiorino, *Spectrochim. Acta*, 24B, 157 (1969).

[123] S. K. Hughes and R. C. Fry, *Appl. Spectrosc.*, 35, 26 (1981).

[124] R. J. Lukasiewicz, P. H. Berens and B. E. Buell, *Anal. Chem.*, 47, 1045 (1975).

[125] R. J. Lukasiewicz and B. E. Buell, *Appl. Spectrosc.*, 31, 541 (1977).

[126] J. C. M. Pau, E. E. Pickett and S. R. Koirtyohann, *Analyst*, 97, 860 (1972).

[127] R. J. Lukasiewicz, *Anal. Chem.*, 51, 1621 (1979).

[128] G. P. Gracheva, N. N. Norozov, L. K. Popyalkovskaya and Yu. D. Skudaev, *Zh. Prikl. Spektrosk.*, 31 211 (1979).

[129] L. Capacho-Delgado and D. C. Manning, *Atom. Absorp. Newslett.*, 4, 317 (1965).

[130] L. Capacho-Delgado and D. C. Manning, *Spectrochim. Acta*, 22, 1505 (1966).

[131] J. E. Schallis and H. L. Kahn, *Atom. Absorp. Newslett.*, 7, 75 (1968).

[132] J. H. Taylor, *Atom. Absorp. Newslett.*, 8, 95 (1969).

[133] A. E. Boling, *Spectrochim. Acta*, 22, 425 (1966).

[134] H. L. Kahn and J. E. Schallis, *Atom. Absorp. Newslett.*, 7, 5 (1968).

[135] T. Nakahara, M. Muremori and S. Musha, *Anal. Chim. Acta*, 62, 267 (1972).

[136] I. Rubeska and M. Miksovsky, *Atom. Absorp. Newslett.*, 11, 57 (1972).

[137] K. A. Saturday and G. M. Hieftje, *Anal. Chem.*, 49, 2013 (1977).

[138] V. N. Morozov, G. P. Gracheva, Ya. D. Skudaev and L. K. Popyalkovskaya, *Zh. Prikl. Spectrosk.*, 27, 400 (1977).

[139] V. A. Razumov, *J. Appl. Spectrosc. (USSR)*, 29, 1143 (1978).

[140] R. Tsujino, A. Ogawa and S. Muska, *Bull. Chem. Soc. Japan*, 52, 1219 (1979).

[141] J. E. Allan, *Analyst*, 83, 466 (1958).

[142] D. J. David, *Analyst*, 83, 655 (1958).

[143] J. Monvoisin and R. Mavrodineanu, *Spectrochim. Acta*, 4, 152 (1950).

[144] W. Slavin, *Atom. Absorp. Newslett.*, **2**, 1 (1963).
[145] F. J. Fernandez, B. Lumas and M. M. Beaty, *Atom. Spectrosc.*, **1**, 55 (1980).
[146] V. C. O. Schüller and G. S. Janes, *J. S. Afr. Inst. Min. Metall.*, **62**, 786 (1962).
[147] L. R. P. Butler, *J. S. Afr. Inst. Min. Metall.*, **62**, 796 (1962).
[148] P. B. Zeeman and L. R. P. Butler, *Appl. Spectrosc.*, **16**, 120 (1962).
[149] V. C. O. Schüller and A. V. Jansen, *J. S. Afr. Inst. Min. Metall.*, **62**, 790 (1962).
[150] W. Lang and R. Herrmann, *Optik.*, **19**, 422 (1962).
[151] O. E. Clinton, *Spectrochim. Acta*, **16**, 985 (1960).
[152] J. B. Willis, *Anal. Chem.*, **33**, 556 (1961).
[153] R. A. G. Rawson, *Analyst*, **91**, 630 (1966).
[154] R. Friedman and W. C. Johnson, *J. Appl. Phys.*, **21**, 791 (1950).
[155] G. F. Kirkbright, M. Sargent and T. S. West, *Talanta*, **16**, 1467 (1969).
[156] G. F. Kirkbright, M. Sargent and T. S. West, *Atom. Absorp. Newslett.*, **8**, 34 (1969).
[157] G. F. Kirkbright and L. Ranson, *Anal. Chem.*, **43**, 1238 (1971).
[158] G. F. Kirkbright and M. Marshall, *Anal. Chem.*, **44**, 1288 (1972).
[159] G. F. Kirkbright and M. Marshall, *Anal. Chem.*, **45**, 1610 (1973).
[160] R. E. Russo and G. M. Hieftje, *Spectrochim. Acta*, **36B**, 231 (1981).
[161] J. M. Ramsey, *Anal. Chem.*, **52**, 2141 (1980).
[162] S. R. Koirtyohann, *Atom. Absorp. Newslett.*, **6**, 77 (1967).
[163] W. Slavin, *Atom. Absorp. Newslett.*, **6**, 9 (1967).
[164] F. A. Boling, *Spectrochim. Acta*, **23B**, 495 (1968).
[165] J. A. Dean and W. J. Carnes, *Anal. Chem.*, **34**, 192 (1962)
[166] R. C. Fry and M. B. Denton, *Anal. Chem.*, **49**, 1413 (1977).
[167] R. S. Babington, *U.S. Patents* 3421692, 3425058, 3425059 and 3504859.
[168] W. R. Wood and A. L. Loomis, *Phil. Mag. Ser. VII*, **4**, 22 (1927).
[169] H. C. Hoare, R. A. Mostyn and B. T. Newland, *Anal. Chim. Acta*, **40**, 181 (1968).
[170] J. Stupar and J. B. Dawson, *Appl. Opt.*, **7**, 1351 (1968).
[171] C. D. West and D. N. Hume, *Anal. Chem.*, **36**, 412 (1964).
[172] H. Dunkin, G. Pforr, W. Mikkeleit and K. Geller, *Z. Chem.*, **3**, 196 (1963).
[173] R. H. Wendt and V. A. Fassel, *Anal. Chem.*, **37**, 920 (1965).
[174] W. J. Kirsten and G. O. Bertilsson, *Anal. Chem.*, **38**, 648 (1966).
[175] R. H. Wendt and V. A. Fassel, *Anal. Chem.*, **38**, 338 (1966).
[176] C. D. West, *Anal. Chem.*, **40**, 247 (1968).
[177] H. Dunkin, G. Pforr, W. Mikkeleit and K. Geller, *Spectrochim. Acta*, **20**, 1531 (1964).
[178] K. W. Olson, W. J. Haas and V. A. Fassel, *Anal. Chem.*, **49**, 632 (1977).
[179] L. R. Layman and F. E. Lichte, *Anal. Chem.*, **54**, 638 (1982).
[180] J. B. Willis, *Spectrochim. Acta*, **25B**, 487 (1970).
[181] J. B. Willis, *Spectrochim. Acta*, **23A**, 811 (1967).
[182] G. M. Hieftje and H. V. Malmstadt, *Anal. Chem.*, **40**, 1860 (1968); **41**, 1735 (1969).
[183] N. C. Clampitt and G. M. Hieftje, *Anal. Chem.*, **44**, 1211 (1972); **46**, 382 (1974).
[184] G. J. Bastiaans and G. M. Hieftje, *Anal. Chem.*, **46**, 901 (1974).
[185] C. B. Boss and G. M. Hieftje, *Anal. Chem.*, **49**, 2112 (1977); **51**, 1897 (1979).
[186] C. B. Boss and G. M. Hieftje, *Appl. Spectrosc.*, **32**, 377 (1978).
[187] P. J. T. Zeegers, R. Smith and J. D. Winefordner, *Anal. Chem.*, **40**, No. 13, 26A (1968).
[188] K. P. Li, *Anal. Chem.*, **48**, 2050 (1976); **50**, 628 (1978).
[189] C. D. Allemand and R. M. Barnes, *Appl. Spectrosc.*, **31**, 434 (1977).
[190] J. A. Holcombe, R. H. Eklund and K. E. Grice, *Anal. Chem.*, **50**, 2097 (1978).
[191] K. W. Olson and R. K. Skogerboe, *Appl. Spectrosc.*, **32**, 181 (1978).
[192] J. W. Novak and R. F. Browner, *Anal. Chem.*, **52**, 287 (1980); **52**, 792 (1980).
[193] M. S. Cresser and R. F. Browner, *Spectrochim. Acta*, **35B**, 73 (1980).
[194] M. S. Cresser and R. F. Browner, *Anal. Chim. Acta*, **113**, 33 (1980).

[195] M. S. Cresser and R. F. Browner, *Appl. Spectrosc.*, **34**, 364 (1980).
[196] R. Kelly and P. J. Padley, *Nature*, **216**, 258 (1967).
[197] S. Nukiyama and Y. Tamasawa, *Trans. Soc. Mech. Engrs. Japan*, **5**, 63 (1938).
[198] N. Mohamed, R. C. Fry and D. L. Wetzel, *Anal. Chem.*, **53**, 639 (1981).
[199] L. de Galan and J. D. Winefordner, *J. Quant. Spectr. Radiat. Transfer*, **7**, 251 (1967).
[200] R. Belcher, R. M. Dagnall and T. S. West, *Talanta*, **11**, 1257 (1964).
[201] E. Pungor, B. Weszprémy and M. Pályi, *Mikrochim. Acta*, 436 (1961).
[202] H.-C. Liu, *Fen Hsi Hua Hsueh*, **8**, 379 (1980).
[203] K. Szivos, *Period. Polytech. Chem. Eng.*, **25**, 121 (1981).
[204] A. W. Boorn, M. S. Cresser and R. F. Browner, *Spectrochim. Acta*, **35B**, 823 (1980).
[205] R. N. Savage and G. M. Hieftje, *Rev. Sci. Instrum.*, **49**, 1418 (1978).
[206] J. W. Berry, D. G. Chappell and R. B. Barnes, *Ind. Eng. Chem., Anal. Ed.*, **18**, 19 (1946).
[207] S. Eckhard and A. Püschel, *Z. Anal. Chem.*, **182**, 334 (1960).
[208] R. Püschel, L. Simon and R. Herrmann, *Optik*, **21**, 441 (1964).
[209] G. Uny and J. Spitz, *Spectrochim. Acta*, **25B**, 391 (1970).
[210] R. Herrmann, *Optik*, **18**, 422 (1961).
[211] J. D. Winefordner and T. J. Vickers, *Anal. Chem.*, **36**, 1939 (1964).
[212] S. Greenfield, H. McD. McGeachin and P. B. Smith, *Anal. Chim. Acta*, **84**, 67 (1976).
[213] H. Howarth, T. N. Mckenzie and M. W. Routh, *Appl. Spectrosc.*, **35**, 164 (1981).
[214] M. W. Routh, *Appl. Spectrosc.*, **35**, 170 (1981).
[215] L. R. Layman, J. G. Crock and F. E. Lichte, *Anal. Chem.*, **53**, 747 (1981).
[216] H. Berndt and W. Slavin, *Atom. Absorp. Newslett.*, **17**, 109 (1978).
[217] W. R. Wolf and K. K. Stewart, *Anal. Chem.*, **51**, 1201 (1979).
[218] C. G. Fisher, W. B. Barnett and D. L. Wilson, *Atom. Absorp. Newslett.*, **17**, 33 (1978).
[219] K. Fukamachi and N. Ishibashi, *Anal. Chim. Acta*, **119**, 383 (1980).
[220] J. A. Koropchak and G. N. Coleman, *Anal. Chem.*, **52**, 1252 (1980).
[221] R. C. Fry, S. J. Northway and M. B. Denton, *Anal. Chem.*, **50**, 1719 (1978).
[222] T. Uchida, I. Kojima and C. Iida, *Bunseki Kagaku*, **27**, T 44 (1978).
[223] R. Avni and C. T. Alkemade, *Mikrochim. Acta*, 460 (1960).
[224] R. E. Bernstein, *South African J. Med. Sci.*, **20**, 57 (1955).
[225] R. D. Caton and R. W. Bremner, *Anal. Chem.*, **26**, 805 (1954).
[226] R. Herrmann and W. Lang *Optik*, **19**, 208 (1962).
[227] P. Porter and G. Wyld, *Anal. Chem.*, **27**, 733 (1955).
[228] F. A. Williams, *Eighth Symposium on Combustion, Pasadena* (1960).
[229] L. Simon, *Optik*, **19**, 621 (1962).
[230] J. H. Gibson, W. Grossman and W. D. Cooke, *Anal. Chem.*, **35**, 266 (1963).
[231] M. P. Parsons and J. D. Winefordner, *Anal. Chem.*, **38**, 1593 (1966).
[232] K. Nakano and T. Takada, *Nippon Kagaku Zasshi*, **88**, 109 (1967).
[233] R. A. G. Rawson, *Analyst*, **19**, 630 (1966).
[234] J. E. Allan, *Spectrochim. Acta*, **17**, 467 (1961).
[235] F. J. Feldman and W. C. Purdy, *Anal. Chim. Acta*, **37**, 273 (1965).
[236] J. B. Willis, *Anal. Chem.*, **34**, 614 (1962).
[237] F. J. Feldman, R. E. Bosshart and G. D. Christian, *Anal. Chem.*, **34**, 1175 (1967).
[238] J. W. Robinson and R. J. Harris, *Anal. Chim. Acta*, **26**, 439 (1962).
[239] A. J. Lemonds and B. E. McClellan, *Anal. Chem.*, **45**, 1455 (1973).
[240] F. J. Feldman and G. D. Christian, *Can. Spectrosc.*, **13**, 2 (1968).
[241] K. Nakano and T. Takada, *Nippon Kagaku Zasshi*, **88**, 109 (1967).
[242] D. J. Halls and A. Townshend, *Anal. Chim. Acta*, **36**, 278 (1966).
[243] M. Margoshes and B. L. Vallee, *Anal. Chem.*, **28**, 1066 (1956).
[244] J. Debras-Guedon and I. A. Voinovitch, *Chem. Anal. (Warsaw)*, **5**, 193 (1960).
[245] F. J. Wallace, *Analyst*, **88**, 259 (1963).

[246] C. T. J. Alkemade and M. H. Voorhuis, *Z. Anal. Chem.*, **163**, 91 (1958).
[247] T. R. Andrew and P. N. R. Nichols, *Analyst*, **87**, 25 (1962).
[248] M. Margoshes and B. L. Vallee, *Anal. Chem.*, **28**, 180 (1956).
[249] I. Rubeska and B. Moldan, *Anal. Chim. Acta*, **37**, 421 (1967).
[250] R. Herrmann, *Z. Anal. Chem.*, **212**, 1 (1965).
[251] B. V. L'vov, *Atomic Absorption Spectral Analysis*, Soviet Academy of Sciences, Moscow, 1966.
[252] J. O. Rasmuson and A. N. Hambly, *Anal. Chem.*, **37**, 879 (1965).
[253] J. O. Rasmuson and A. N. Hambly, *Spectrochim. Acta*, **31B**, 229 (1976).
[254] D. S. Smyly, W. P. Townshend, P. J. T. Zeegers and J. D. Winefordner, *Spectrochim. Acta*, **26B**, 531 (1971).
[255] J. D. Winefordner, C. T. Mansfield and T. J. Vickers, *Anal. Chem.*, **35**, 1607 (1963).
[256] E. Hinnov and H. Kohn, *J. Opt. Soc. Am.*, **47**, 156 (1957).
[257] E. L. Grove, C. W. Scott and F. Jones, *Talanta*, **12**, 327 (1965).
[258] F. W. Hofmann and H. Kohn, *J. Opt. Soc. Am.*, **51**, 512 (1961).
[259] R. W. Reid and T. W. Sugden, *Discussions Faraday Soc.*, **33**, 213 (1962).
[260] R. Mavrodineanu and H. Boiteux, *Flame Spectroscopy*, Wiley, New York, 1965.
[261] J. Rose, *Dynamic Physical Chemistry*, Pitman, London, 1961.
[262] C. L. Chakrabarti, G. R. Lyles and F. B. Dowling, *Anal. Chim. Acta*, **29**, 489 (1963).
[263] J. A. Dean, J. C. Burger, T. C. Rains and H. E. Zittel, *Anal. Chem.*, **33**, 1722 (1961).
[264] S. Eckhard and A. Püschel, *Z. Anal. Chem.*, **172**, 334 (1960).
[265] V. A. Fassel, R. H. Curry and R. N. Kniseley, *Spectrochim. Acta*, **18**, 1127 (1962).
[266] R. Mavrodineanu, *Spectrochim. Acta*, **17**, 1016 (1961).
[267] V. G. Mossotti and V. A. Fassel, *Spectrochim. Acta*, **20**, 1117 (1964).
[268] W. Slavin, S. Sprague and D. C. Manning, *Atom. Absorp. Newslett.*, **15**, 1 (1963).
[269] J. B. Willis, *Spectrochim. Acta*, **16**, 259 (1960).
[270] R. Mavrodineanu (ed.), *Analytical Flame Spectroscopy, Selected Topics*, Philips Technical Library, MacMillan, London, 1970.
[271] D. T. Coker, J. M. Ottaway and N. K. Pradhan, *Nature*, **233**, 69 (1971).
[272] N. V. Mossholder, V. A. Fassel and R. N. Kniseley, *Anal. Chem.*, **45**, 1614 (1973).
[273] D. T. Coker and J. M. Ottaway, *Nature*, **230**, 156 (1971).
[274] D. J. Halls, *Spectrochim. Acta*, **32B**, 221 (1977).
[275] G. F. Kirkbright, A. Semb and T. S. West, *Spectrosc. Lett.*, **1**, 7 (1967).
[276] A. N. Hayhurst and N. R. Telford, *J. Chem. Soc. Faraday Trans. I*, 237 (1972).
[277] T. Hollander, P. J. Kalff and C. T. J. Alkemade, *J. Chem. Phys.*, **39**, 2558 (1963).
[278] D. E. Jensen and P. J. Padley, *Trans. Faraday Soc.*, **62**, 2140 (1966).
[279] R. Kelly and P. J. Padley, *Trans. Faraday Soc.*, **65**, 355 (1969).
[280] R. Kelly and P. J. Padley, *Proc. Roy. Soc.*, **A327**, 345 (1972).
[281] A. N. Hayhurst and N. R. Telford, *Trans. Faraday Soc.*, **66**, 2784 (1970).
[282] D. E. Jensen, *J. Chem. Phys.*, **51**, 4674 (1966).
[283] E. M. Bulewicz and P. J. Padley, *Combust. Flame*, **5**, 331 (1961).
[284] A. Fontijn, *Pure Appl. Chem.*, **39**, 287 (1974).
[285] R. Kelly and P. J. Padley, *Trans. Faraday Soc.*, **67**, 1384 (1971).
[286] A. G. Gaydon and H. G. Wolfhard, *Flames, Their Structure, Radiation and Temperature*, 3rd Ed., Chapman and Hall, London, 1970.
[287] J. Lawton and F. J. Weinberg, *Electrical Aspects of Combustion*, Clarendon Press, Oxford, 1969.
[288] M. N. Saha, *Phil. Mag.*, **40**, 472 (1920).
[289] G. R. Kornblum and L. de Galan, *Spectrochim. Acta*, **28B**, 139 (1973).
[290] C. Woodward, *Spectrosc. Lett.*, **4**, 191 (1971).
[291] D. E. Jensen, *Combust. Flame*, **12**, 261 (1968).
[292] B. V. L'vov, *Spectrochim. Acta*, **33B**, 153 (1978).

[293] B. V. L'vov and L. K. Polzik, *J. Anal. Chem. (USSR)*, **33**, 1143 (1978).

[294] R. E. Sturgeon and C. L. Chakrabarti, *Prog. Anal. Atom. Spectrosc.*, **1**, 5 (1978).

[294a] M. W. Routh, P. S. Doidge, J. Chidzey and B. Frary, *Am. Lab.*, **14**, No. 6, 80 (1982).

[295] A. S. King, *Astrophys. J.*, **28**, 300 (1908).

[296] B. V. L'vov, *J. Eng. Phys.*, **2**, 44 (1959).

[297] B. V. L'vov, *J. Eng. Phys.*, **2**, 56 (1959).

[298] B. V. L'vov, *Spectrochim. Acta*, **17**, 761 (1961).

[299] B. V. L'vov, *Zavodsk. Lab.*, **28**, 931 (1962).

[300] B. V. L'vov, *Opt. Spektrosk.*, **19**, 507 (1965).

[301] B. V. L'vov and G. G. Lebedev, *Zh. Prikl. Spectrosk.*, **7**, 264 (1967).

[302] B. V. L'vov, *Spectrochim. Acta*, **24B**, 53 (1969).

[303] R. Woodriff and G. Ramelow, *Spectrochim. Acta*, **23B**, 665 (1968).

[304] R. Woodriff, R. W. Stone and A. M. Held, *Appl. Spectrosc.*, **22**, 408 (1968).

[305] R. W. Morrow and R. J. McElhaney, *Appl. Spectrosc.*, **27**, 287 (1973).

[306] B. Welz, *CZ-Chem.-Tech.*, **1**, 455 (1972); *Chem. Abstr.*, **78**, 37450h (1973).

[307] J. W. Robinson and D. K. Wolcott, *Anal. Chim. Acta*, **74**, 43 (1975).

[308] H. Massmann, *Spectrochim. Acta*, **23B**, 215 (1968).

[309] H. Massmann, *Methodes Physiques d'Analyse*, **4**, 193 (1968).

[310] C. J. Pickford and G. Rossi, *Analyst*, **97**, 647 (1972).

[311] D. A. Katskov, I. G. Burtseva, I. L. Grinshtein and L. P. Kruglikova, *Zh. Analit. Khim.*, **35**, 2289 (1980).

[312] M. Hoenig, R. Vanderstappen and P. Van Hoeyweghen, *Analusis*, **6**, 433 (1978).

[313] W. Ishibashi and R. Kikuchi, *Bunseki Kagaku*, **29**, 165 (1980).

[314] W. Slavin, S. A. Myers and D. C. Manning, *Anal. Chim. Acta*, **117**, 267 (1980).

[315] K. R. Sperling and B. Bahr, *Z. Anal. Chem.*, **299**, 206 (1979).

[316] W. Slavin, D. C. Manning and G. R. Carnrick, *Anal. Chem.*, **53**, 1504 (1981).

[317] S. A. Clyburn, T. Kantor and C. Veillon, *Anal. Chem.*, **46**, 2213 (1974).

[318] K. C. Thompson, R. G. Godden and D. R. Thomerson, *Anal. Chim. Acta*, **74**, 289 (1975).

[319] R. W. Morrow and R. J. McElhaney, *Atom. Absorp. Newslett.*, **13**, 45 (1974).

[320] D. D. Siemer, R. Woodriff and B. Watne, *Appl. Spectrosc.*, **28**, 582 (1974).

[321] D. C. Manning, F. J. Fernandez and G. E. Peterson, *Ind. Res.*, **19**, 82 (1977).

[322] D. C. Manning and R. D. Ediger, *Atom. Absorp. Newslett.*, **15**, 42 (1976).

[323] F. J. Szydlowski, E. Peck and B. Bax, *Appl. Spectrosc.*, **32**, 1978 (1978).

[324] C. D. Wall, *Atom. Absorp. Newslett.*, **17**, 61 (1978).

[325] J. R. Montgomery and G. N. Peterson, *Anal. Chim. Acta*, **117**, 397 (1980).

[326] K. Julshamn, *Atom. Absorp. Newslett.*, **16**, 149 (1977).

[327] S. R. Koirtyohann, E. D. Glass and F. E. Lichte, *Appl. Spectrosc.*, **35**, 22 (1981).

[328] G. Buzzelli and A. W. Mosen, *Talanta*, **24**, 383 (1977).

[329] J. F. Alder and D. A. Hickman, *Atom. Absorp. Newslett.*, **16**, 110 (1977).

[330] R. H. Eklund and J. A. Holcombe, *Anal. Chim. Acta*, **109**, 97 (1979).

[331] C. W. Fuller, *Anal. Chim. Acta*, **62**, 261 (1972).

[332] Z. Slovák and B. Dočekal, *Anal. Chim. Acta*, **130**, 203 (1981).

[333] I. A. Kuzovlev, Y. N. Kuznetsov and O. A. Sverdlina, *Zavodsk. Lab.*, **39**, 428 (1973).

[334] H. M. Ortner and E. Kantuscher, *Talanta*, **22**, 581 (1975).

[335] J. H. Runnels, R. Merryfield and H. B. Fisher, *Anal. Chem.*, **47**, 1258 (1975).

[337] H. Fritzsche, W. Wegscheider, G. Knapp and H. M. Ortner, *Talanta*, **26**, 219 (1979).

[338] T. M. Vickrey, G. V. Harrison and G. J. Ramelow, *Atom. Spectrosc.*, **1**, 116 (1980).

[339] T. M. Vickrey, G. V. Harrison, G. J. Ramelow and J. C. Carver, *Anal. Lett.*, **13**, 781 (1980).

[340] T. M. Vickrey, M. E. Howell, G. V. Harrison and G. J. Ramelow, *Anal. Chem.*, **52**, 1743 (1980).

[341] T. M. Vickrey, G. V. Harrison and G. J. Ramelow, *Anal. Chem.*, **53**, 1573 (1981).
[342] V. J. Zatka, *Anal. Chem.*, **50**, 538 (1978).
[343] E. Norval, M. G. C. Human and L. R. P. Butler, *Anal. Chem.*, **51**, 2045 (1979).
[344] J. Takahashi, T. Kitahara, N. Hirabayashi and K. Fuwa, *Bunskei Kagaku*, **30**, 90 (1981).
[345] E. J. M. DeHaas and F. A. Wolff, *Clin. Chem.*, **27**, 205 (1981).
[346] B. V. L'vov, L. A. Pelieva and A. I. Shamopolskii, *Zh. Prikl. Spectrosk.*, **27**, 395 (1977).
[347] B. V. L'vov, *Spectrochim. Acta*, **33B**, 153 (1978).
[348] R. E. Sturgeon, *Anal. Chem.*, **49**, 1255A (1977).
[349] D. C. Gregoire and C. L. Chakrabarti, *Anal. Chem.*, **49**, 2018 (1977).
[350] E. J. Hinderberger, M. L. Kaiser and S. R. Koirtyohann, *Atom. Spectrosc.*, **2**, 1 (1981).
[351] W. J. Price, T. C. Dymott and P. J. Whiteside, *Spectrochim. Acta*, **35B**, 3 (1980).
[352] F. J. Fernandez, M. M. Beaty and W. B. Barnett, *Atom. Spectrosc.*, **2**, 16 (1981).
[353] W. Slavin, D. C. Manning and G. R. Carnrick, *Atom. Spectrosc.*, **2**, 137 (1981).
[354] M. L. Kaiser, S. R. Koirtyohann and E. J. Hinderberger, *Spectrochim. Acta*, **36B**, 773 (1981).
[355] J. A. Holcombe, M. T. Sheehan and T. Rettburg, *Pittsburgh Conference on Analytical Chemistry and Applied Spectroscopy*, Atlantic City, 1981.
[356] C. D. Wall, *Talanta*, **24**, 755 (1977).
[357] D. A. Katskov and I. L. Grinshtein, *Zh. Prikl. Spektrosk.*, **28**, 968 (1978).
[358] B. V. L'vov and L. A. Pelieva, *Can. J. Spectrosc.*, **23**, 1 (1978).
[358a] V. P. Garnys and L. E. Smythe, *Anal. Chem.*, **51**, 62 (1979).
[359] D. C. Manning, W. Slavin and S. Myers, *Anal. Chem.*, **51**, 2375 (1979).
[360] M. H. West, J. F. Molina, C. L. Yuan, D. G. Davis and J. V. Chauvin, *Anal. Chem.*, **51**, 2370 (1979).
[361] V. N. Nuzgin, E. P. Pilpenko, Yu. P. Lyashenko and Yu. B. Atnagie, *Zh. Prikl. Spektrosk.*, **29**, 364 (1978).
[362] E. W. Wolff, M. P. Landy and D. A. Peel, *Anal. Chem.*, **53**, 1566 (1981).
[363] Z. Grobenski, R. Lehmann and B. Welz, *Pittsburgh Conference on Analytical Chemistry and Applied Spectroscopy*, Atlantic City, 1981.
[364] H. Berndt and J. Messerschmidt, *Spectrochim. Acta*, **36B**, 809, 845 (1981).
[365] S. Guecer and H. Berndt, *Talanta*, **28**, 334 (1981).
[366] D. C. Manning and W. Slavin, *Anal. Chim. Acta*, **118**, 301 (1980).
[367] A. Cedergren, W. Frech, E. Lundberg and J. A. Persson, *Anal. Chim. Acta*, **128**, 1 (1981).
[368] E. J. Czobik and J. P. Matousek, *Spectrochim. Acta*, **35B**, 741 (1980).
[369] G. E. Batley and J. P. Matousek, *Anal. Chem.*, **52**, 1570 (1980).
[370] B. Holen, R. Bye and W. Lund, *Anal. Chim. Acta*, **130**, 257 (1981); **131**, 37 (1981).
[371] G. Torsi, E. Desimoni, F. Palmisano and L. Sabbatini, *Anal. Chim. Acta*, **124**, 143 (1981).
[372] T. S. West and X. K. Williams, *Anal. Chim. Acta*, **45**, 27 (1969).
[373] M. D. Amos, *Am. Lab.*, No. 8, 33 (1970).
[374] M. D. Amos, P. A. Bennett, K. G. Brodie, P. W. Y. Lung and J. P. Matousek, *Anal. Chem.*, **43**, 211 (1971).
[375] M. Glenn, J. Savory, L. Hart, T. Glenn and J. D. Winefordner, *Anal. Chim. Acta*, **57**, 263 (1971).
[376] E. D. Prudnikov, *Zh. Prikl. Spektrosk.*, **33**, 622 (1980).
[377] E. D. Prudnikov, L. P. Kolosova, V. K. Kalachev and Yu. A. Bychkov, *Vestn. Leningr. Univ.*, 1980, *Geol. Geog.*, No. 1, 120.
[378] Yu. I. Belyaev, V. I. Shcherbakov and A. V. Karyakin, *Zh. Analit. Khim.*, **35**, 2074 (1980).
[379] B. N. Ivanov, V. F. Grigor'ev and A. V. Karyakin, *Zh. Prikl. Spektrosk.*, **33**, 12 (1980).

[380] V. Sychra, D. Kolihová, O. Vyskocilová, R. Hlavac and P. Püschel, *Anal. Chim. Acta*, 105, 263 (1979).

[381] O. Vyskocilová, V. Sychra, D. Kolihová and P. Püschel, *Anal. Chim. Acta*, 105, 271 (1979).

[382] K. Ohta and M. Suzuki, *Anal. Chim. Acta*, 96, 77 (1978); 104, 293 (1979); 107, 245 (1979); 108, 69 (1979); 110, 49 (1979).

[383] K. Ohta and M. Suzuki, *Talanta*, 26, 207 (1979).

[384] P. Püschel, Z. Formánek, R. Hlaváč, D. Kolihová and V. Sychra, *Anal. Chim. Acta*, 127, 109 (1981).

[385] R. D. Hudson, *Phys. Rev.*, 135A, 1212 (1964).

[386] M. Suzuki, K. Ohta and T. Yamakita, *Anal. Chim. Acta*, 133, *Comput. Tech. Optim.*, 5, 203 (1981).

[387] J. W. Robinson and S. Weiss, *Spectrosc. Lett.*, 13, 685 (1980).

[388] Y. Hoshino, T. Utsunomiya and F. Fukui, *Rept. Res. Eng. Mater. Tokyo Inst. Technol.*, 109 (1980).

[389] Yu. B. Atnashev, V. N. Muzgin, Yu. P. Lyashenko, V. E. Korepanov and E. P. Pilipenko, *Zh. Analit. Khim.*, 35, 2156 (1980).

[390] N. T. Faithfull, *Lab. Pract.*, 27, 105 (1978).

[391] A. Cedergren, W. Frech, E. Lundberg and J. A. Persson, *Anal. Chim. Acta*, 128, 1 (1981).

[392] H. M. Donega and T. E. Burgess, *Anal. Chem.*, 42, 1521 (1970).

[393] W. G. Jones and A. Walsh, *Spectrochim. Acta*, 16, 249 (1960).

[394] B. M. Gatehouse and A. Walsh, *Spectrochim. Acta*, 16, 602 (1960).

[395] J. A. Goleb and J. K. Brody, *Anal. Chim. Acta*, 28, 457 (1963).

[396] B. W. Gandrud and R. K. Skogerboe, *Appl. Spectrosc.*, 25, 243 (1971).

[397] C. G. Bruhn and W. W. Harrison, *Anal. Chem.*, 50, 16 (1978).

[398] K. Tsujii and K. Kuga, *Bunseki Kagaku*, 27, 415 (1978).

[399] D. A. McCamey and T. M. Niemczyk, *Appl. Spectrosc.*, 34, 692 (1980).

[400] M. Cooksey and W. B. Barnett, *Atom. Absorp. Newslett.*, 18, 101 (1979).

[401] G. Lundgren, L. Lundmark and G. Johansson, *Anal. Chem.*, 46, 1028 (1974).

[402] D. C. Manning, J. D. Kerber and M. D. Amos, *Pittsburgh Conference on Analytical Chemistry and Applied Spectroscopy*, Cleveland, 1977.

[403] R. E. Sturgeon and C. L. Chakrabarti, *Prog. Anal. Atom. Spectrosc.*, 1, 5 (1978).

[404] J. P. Erspamer and T. M. Niemczyk, *Appl. Spectrosc.*, 35, 512 (1981).

[405] H. L. Kahn, M. K. Conley and J. J. Sotera, *Am. Lab.*, 12, No. 8, 72 (1980).

[406] C. D. Wall and T. Catterick, *Talanta*, 25, 705 (1978).

[407] A. Kawase, N. Fudagawa and S. Nakamura, *Bunseki Kagaku*, 27, T 30 (1978).

[408] B. V. L'vov, *Zh. Prikl. Spektrosk.*, 8, 517 (1968).

[409] B. V. L'vov, D. A. Katskov and G. G. Lebedev, *Zh. Prikl. Spektrosk.*, 9, 558 (1968).

[410] D. A. Katskov and B. V. L'vov, *Zh. Prikl. Spektrosk.*, 15, 783 (1971).

[411] D. Clark, R. M. Dagnall and T. S. West, *Anal. Chim. Acta*, 63, 11 (1973).

[412] J. P. Matoušek and K. G. Brodie, *Anal. Chem.*, 45, 1606 (1973).

[413] J. W. McLaren and R. C. Wheeler, *Analyst*, 102, 542 (1977).

[414] E. Lundberg, *Appl. Spectrosc.*, 32, 276 (1978).

[415] J. G. T. Regan and J. Warren, *Atom. Absorp. Newslett.*, 17, 89 (1978).

[416] W. Slavin and D. C. Manning, *Anal. Chem.*, 51, 261 (1979).

[417] S. Bäckman and R. W. Karlsson, *Analyst*, 104, 1017 (1979).

[417a] K. Takada and K. Hirokawa, *Talanta*, 29, 849 (1982).

[418] S. G. Salmon, R. H. Davis, Jr. and J. A. Holcombe, *Anal. Chem.*, 53, 324 (1981).

[419] L. Ebdon, G. F. Kirkbright and T. S. West, *Anal. Chim. Acta*, 58, 39 (1972).

[420] R. E. Sturgeon, C. L. Chakrabarti, I. S. Maines and P. C. Bertels, *Anal. Chem.*, 47, 1240 (1975).

[421] E. J. Czobik and J. P. Matoušek, *Anal. Chem.*, **50**, 3 (1978).
[422] W. B. Barnett and M. M. Cooksey, *Atom. Absorp. Newslett.*, **18**, 61 (1979).
[423] G. Torsi and G. Tessari, *Anal. Chem.*, **45**, 1812 (1973).
[424] G. Tessari and G. Torsi, *Anal. Chem.*, **47**, 842 (1975).
[425] G. Tessari and G. Torsi, *Ann. Chim. (Rome)*, **68**, 967 (1978).
[426] C. W. Fuller, *Analyst*, **99**, 739 (1974); **100**, 229 (1975); **101**, 798 (1976).
[427] S. L. Paveri-Fontana, G. Torsi and G. Tessari, *Anal. Chem.*, **46**, 1032 (1974).
[428] S. L. Paveri-Fontana, G. Torsi and G. Tessari, *Ann. Chim. (Rome)*, **66**, 691 (1976).
[429] J. Aggett and A. Sprott, *Anal. Chim. Acta*, **72**, 49 (1974).
[430] W. C. Campbell and J. M. Ottaway, *Talanta*, **21**, 837 (1974).
[431] D. J. Johnson, B. L. Sharp, T. S. West and R. M. Dagnall, *Anal. Chem.*, **47**, 1234 (1975).
[432] R. E. Sturgeon, C. L. Chakrabarti and C. H. Langford, *Anal. Chem.*, **48**, 1792 (1976).
[433] B. V. L'vov, D. A. Katskov, L. P. Kruglikova and L. K. Polzik, *Spectrochim. Acta*, **31B**, 49 (1976).
[434] D. A. Katskov, *Zh. Prikl. Spektrosk.*, **26**, 598 (1976).
[435] J. Zsakó, *Anal. Chem.*, **50**, 1105 (1978).
[436] B. Smets, *Spectrochim. Acta*, **35B**, 33 (1980).
[437] J. M. Ottaway and F. Shaw, *Analyst*, **101**, 582 (1976).
[438] D. Littlejohn and J. M. Ottaway, *Analyst*, **104**, 208 (1979).
[439] M. S. Epstein, T. C. Rains and T. C. O'Haver, *Appl. Spectrosc.*, **30**, 324 (1976).
[440] R. E. Sturgeon, S. S. Berman and S. Kashyap, *Anal. Chem.*, **52**, 1049 (1980).
[441] R. E. Sturgeon and S. S. Berman, *Anal. Chem.*, **53**, 632 (1981).
[442] J. V. Sullivan and A. Walsh, *Spectrochim. Acta*, **21**, 727 (1965); **22**, 1843 (1966).
[443] J. A. Bowman, *Anal. Chim. Acta*, **37**, 465 (1967).
[444] B. S. Rawling and J. V. Sullivan, *Trans. Inst. Min. Met.*, **76**, C 238 (1967).
[445] P. L. Boar and J. V. Sullivan, *Fuel*, **46**, 230 (1967).
[446] J. V. Sullivan and A. Walsh, *Appl. Optics*, **7**, 1271 (1968).
[447] R. Stephens, *Can. J. Chem.*, **58**, 1621 (1980).
[448] J. A. Bowman, J. V. Sullivan and A. Walsh, *Spectrochim. Acta*, **22**, 205 (1966).
[449] R. M. Lowe, *Spectrochim. Acta*, **24B**, 191 (1969).
[450] N. A. Sebestyen, *Spectrochim. Acta*, **25B**, 261 (1970).
[451] R. L. Cochran and G. M. Hieftje, *Anal. Chem.*, **50**, 791 (1978).
[451] T. W. Barnard, *Anal. Chem.*, **51**, 1172A (1979).
[453] F. W. Willmott and I. Mackenzie, *Anal. Chim. Acta*, **103**, *Comput. Tech. Optimiz.*, **2**, 401 (1978).
[454] C. G. Fisher, W. B. Barnett and D. L. Wilson, *Atom. Absorp. Newslett.*, **17**, 33 (1979).
[455] M. W. Gaumer, S. Sprague and W. Slavin, *Atom. Absorp. Newslett.*, **5**, 58 (1966).
[456] G. H. Keats, *Atom. Absorp. Newslett.*, **4**, 319 (1965).
[457] S. Sprague and W. Slavin, *Atom. Absorp. Newslett.*, **4**, 367 (1965).
[458] S. Slavin and W. Slavin, *Atom. Absorp. Newslett.*, **5**, 106 (1966).
[459] H. L. Kahn, *Am. Lab.*, No. 8, 52 (1969).
[460] W. B. Barnett and R. D. Ediger, *Atom. Absorp. Newslett.*, **17**, 125 (1979).
[461] F. X. Deloye, I. Voinovitch, M. Chatelier and J. M. Bergue, *Actual. Chim.*, No. 9, 39 (1979).
[462] T. J. Johnson, *Intern. Lab.*, **10**, September, 59 (1980).
[463] L. de Galan, *Chem. N. Z.*, **44**, 94 (1980).
[464] M. R. Harris and N. W. Lepp, *Analyst*, **106**, 283 (1981).
[465] J. M. Ottaway, L. Bezur and J. Marshall, *Analyst*, **105**, 1130 (1980).
[466] L. E. Holboke, *Chem. Biomed. Environ. Instrum.*, **11**, 27 (1981).
[467] W. Brunner, *Spez. Ber. Kernforschungsanlage Jülich*, Jül-spez-46 (1979).
[468] M. Cooksey and W. B. Barnett, *Atom. Absorp. Newslett.*, **18**, 1 (1979).

3

General methods and techniques

3.1 MEASUREMENT METHODS

Historically, the height of the absorption signal displayed on a meter or chart recorder has been measured and related to concentration. With electrothermal atomization peak area and peak height both provide reliable performance in certain situations and most modern instruments allow direct measurements of both. It has been reported that measurement of the area gives a significantly more linear calibration curve (Fig. 3.1) and eliminates interferences due to alteration of the atomization rate by the matrix [1].

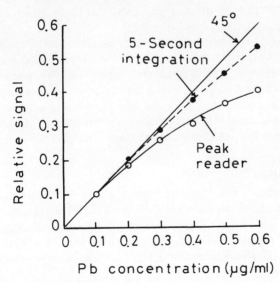

Fig. 3.1. Lead calibration curves obtained by reading the peak height and integrating the area under the peak.
[Reprinted from *American Laboratory*, **7**, No. 8, 43 (1975). Copyright 1975 by International Scientific Communications, Inc.]

3.1.1 Calibration Graph Method

In this method, the concentration of the analyte is determined by comparison with a standard solution of known concentration. A series of at least 6 standards of different known metal content, giving absorbance values between 0.2 and 0.8, is prepared by diluting a stock solution, and atomized under the experimental conditions to be used. The calibration graph is established by plotting absorbance against concentration. It should be linear with zero intercept and reproducible slope. A non-zero intercept is a function of non-analyte signals (or of analyte present as impurity in the reagents), whereas the slope is related to both fundamental and instrumental factors, such as the spectral profile of the lamp line and the absorption line as well as their displacements relative to each other. A least-squares fit to the calibration data is most easily done by a computer, which can also be used to estimate the noise [2] and thus improve the results. Instrumental modification [3] and the use of a digital read-out system [4] have been suggested to extend and improve the analytical calibration graph.

The relationship between concentration and expected standard deviation has been discussed [5]. It should be noted that many calibration graphs consist of an almost linear portion near the origin, followed by a part which bends to a greater or lesser extent towards the concentration axis. Such deviation from linearity is often more pronounced at high analyte concentrations and high lamp current and arises when the spectral width of the lamp emission line is not small compared with that of the absorption line. Theoretical and experimental aspects of the convex curvature of the analytical curve due to these factors have been discussed [6–14].

Limbek *et al.* [15] described a calibration algorithm for the synthesis of a mathematical curve that matches the curvature of all types of calibration curves exceptionally well, even at high absorbance values and with severe curvature. The success of this method demonstrates that the power of an intelligently used microprocessor in atomic absorption can lead to improved accuracy, precision and usable dynamic range. The method also provides ease of operation, extends the analytical capability and adds more flexibility.

3.1.2 Factor Method

When the calibration graph is linear, a much simpler procedure may be used. This involves the use of two identical aliquots (A and B) of the sample solution, of volume V_x. A small volume (V_s) of the standard (concentration C_s) is added to aliquot (A) and the same volume of solvent is added to aliquot (B), followed by measurement of the absorbance of both solutions. The unknown sample concentration (C_x) is then given by

$$C_x = S_B V_s C_s / (S_A - S_B) V_x \qquad (3.1)$$

where S_A and S_B are the absorbances due to solutions A and B, respectively.

This technique involves a single-point calibration and may be used with flame and electrothermal atomization [16], but as the latter is often characterized by a relatively narrow linear range, serious errors may arise in its use.

3.1.3 Bracketing Standard Method
This method partially compensates for non-linearity in the analytical curve and needs only a preliminary calibration curve, which need not be regularly repeated. The analyte solution is measured and its approximate concentration estimated from the calibration curve. Next, the absorbances of two standards close to the estimated analyte concentration are measured, and the analyte concentration is calculated on the assumption that the calibration graph is linear between the two standards. The method is useful for improving precision, provided that the standards are made up to be very close to the estimated analyte concentration and scale expansion is used.

3.1.4 Multiple Standard Addition Method
In this method, the analyte solution is divided into four or more aliquots, all but one of which are spiked with identical volumes of standards of increasing concentration (C_1, C_2, C_3...) and the last aliquot is diluted with the same spike volume of pure solvent. Under these conditions, all the solutions will differ in analyte concentration but have the same matrix composition, so the influence of the matrix will be the same for all. The absorbances are then measured and plotted *vs.* concentration of added standard. The concentration of analyte in the sample is determined by extrapolating this plot either graphically (Fig. 3.2) or by a least-squares fit. The following relation may also be used for determining the concentration.

$$C = [(C + C_3)A_C]/[A_{C+C_s} - A_C]$$ (3.2)

Interference due to matrix effects can be detected by comparing the slopes of the curves for the spiked analyte solutions and for pure standards. In the absence of interference both slopes are the same. In effect, the method is equivalent to preparing a standard calibration curve with exact matrix matching.

Some instruments have automatic zero setting. In this case the instrument can be set so that the output signal for the sample solution reads zero. The signal from a sample spiked with a known concentration of analyte gives a read-out corresponding to the concentration of added standard. If a blank solution is then tested, the signal displayed will be negative and correspond to the negative of the concentration of the analyte [17]. This approach is limited by the short negative read-out range available, and an alternative has been designed that gives a positive read-out [18].

Applications and limitations of the standard addition method have been discussed [19].

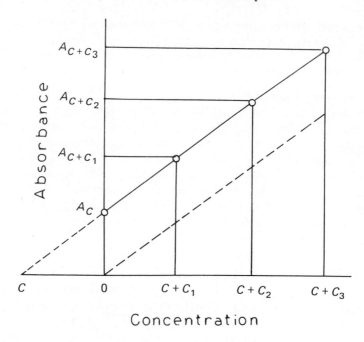

Fig. 3.2. Method of standard additions.

3.1.5 Analyte Dilution Method

This method is used to overcome interference effects by the matrix or the excessive ionization of the analyte. In principle, the analyte and standard solutions are diluted equally with a solution containing a fixed excess of a substance which either causes interference or is more easily ionized than the analyte. Under these conditions, the relatively small but variable amounts of interfering substance(s) already present in the analyte sample, but not in the standard, will have a proportionately smaller effect. This method is useful provided that the interfering substance(s) will cause enhancement of the absorbance rather than suppression, and that the analyte concentration after dilution will still be within the measurement range

3.1.6 Internal Standard Method

This method is based on the addition of a standard reference element that absorbs at a different wavelength from the analyte. The reference standard is added in the same concentration to all analyte and blank samples. The ratios of the analyte signals A_a to that of the reference standard A_r are then measured and the calibration curve is prepared by plotting A_a/A_r vs. the concentration of analyte in the standards [20]. Two requirements must be fulfilled for successful application of this method: (a) absence of the reference element in the test sample; (b) identical behaviour of both the analyte and the reference element in the atom cell. These requirements are difficult to achieve, but the method is useful in double-channel atomic-absorption spectrometry.

3.1.7 Absorption Inhibition-Titration Method

'Titration' methods based on the inhibition effect of some anions on atomic absorption or emission by metal ions have been described [21–25]. The method involves addition of a standard metal solution as a titrant to a stirred solution of the anion and measurement of the absorption (or emission) signal during the titration. The type of titration curve obtained is dependent upon the atomization conditions and the nature of the analyte in solution. The unusual shape of these titration curves (Fig. 3.3) results from complicated processes occurring as a result of the evaporation of the analyte droplets. This method is procedurally a

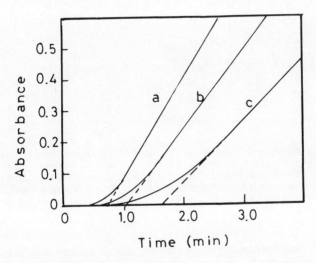

Fig. 3.3. Titration curves for titrations of 2.0 (a), 4.0 (b), and 6.0 (c) μg/ml sulphate with 50 μg/ml magnesium solution. Titrant flow-rate 1.49 ml/min; chart speed 2.0 in./min; hydrogen, air: 30, 10 ft³/hr. [According to R. Looyenga and C. O. Huber, *Anal. Chim. Acta,* **55,** 179 (1971). By permission of the copyright holders, Elsevier Scientific Publishing Co.]

titration but does not necessarily involve a stoichiometric reaction between the titrant and analyte. In practice, a volume of the solution containing the anion is placed in a beaker and stirred magnetically. The titrant delivery and aspiration tubes are inserted into the anion solution through a cover (which serves to keep the two tubes separated) (Fig. 3.4). After aspiration has started, titrant flow and signal-recording are initiated by a common switch. Titrant delivery is continued until the appropriate end-point signals have been recorded. The method has been used for determination of many species as well as for investigation of flame atomization processes [26–30].

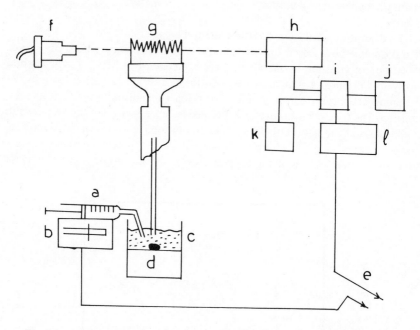

Fig. 3.4. Apparatus for atomic-absorption inhibition titration: a, titrant; b, constant-rate infusion pump; c, analyte solution; d, magnetic stirrer; e, common switch; f, hollow-cathode lamp; g, flame; h, monochromator; i, amplifier; j. meter; k, printer; l, recorder. [Reprinted with permission from J. R. Sand and C. O. Huber, *Anal. Chem.,* **48**, 1331 (1976). Copyright 1976 American Chemical Society.]

The use of microprocessors again provides ease of operation, greater versatility and improved performance. Capabilities such as automatic calibration over a very wide absorbance range, the facility to use bracketing standards, and the ability to select variable integration times provide superior analytical data in a shorter time [31].

3.2. CONTROL OF ANALYTICAL INTERFERENCES

Although AAS is regarded as a specific technique, different types of interferences, falling into the following categories, are experienced: physical, chemical, ionization, spectral and background. Statistical methods for the practical evaluation of AAS interferences have been discussed [32]. Interferences in both flame and electrothermal atomization devices and their elimination and control have been reviewed [33-38]. Generally speaking, physical, chemical and atomization interferences can be circumvented by careful sample preparation and/or use of the standard addition technique. Interferences due to background absorption can be automatically compensated for by instrumental accessories utilizing a continuum source or the Zeeman effect.

3.2.1 Physical Interferences

This type of interference arises from any change in viscosity, surface tension and specific gravity of either the analyte or the standard reference solution, which in turn can affect the nebulization efficiency and influence the number of free atoms in the flame. Organic solvents have been used to increase both the nebulization efficiency and the flame temperature [39]. However, the presence of organic substances is by far the most complex and manifold source of interference, as they raise the background emission and cause fluctuation in the flame temperature [40].

On the other hand, physical interferences do not commonly occur in the electrothermal atomization methods, because the analyte is directly placed in the atomizer without nebulization. However, in contrast to their behaviour in flame systems, certain organic solvents (e.g. xylene, methyl isobutyl ketone and hexane) can cause problems when solutions containing them are injected into graphite tube atomizers. These solvents produce background absorption spectra at 200-250 nm and 2500°C, even after drying at 100°C and ashing at 800°C [41]. Moreover, the analyte is occluded in the interfering matrix [42,43]. However, physical interferences can be eliminated by (a) matching the physical properties and matrix composition of both the standard and test samples; (b) use of the standard addition technique for measurement; (c) dilution of the analyte sample; (d) the use of a sample pump with a fixed feed rate, instead of the nebulizer [44,45].

3.2.2 Chemical Interferences

The number of free atoms available for excitation is influenced by any chemical reaction that takes place in the atomization cell. A generalized theory demonstrating the effects of diffusion and reaction kinetics on the distribution of atoms in the flame has been proposed to account for chemical interferences[46]. Formation of thermally stable oxides, hydroxides, carbides and nitrides, interaction with anions or cations [47] from the matrix, and the effect of some mineral acids [48] are the most common causes of interference. However, most of these sources of interference can be avoided by (a) raising the atomization temperature to dissociate the less volatile species; (b) addition of the interfering species to both the standard reference and the analyte solution; (c) use of protective agents to form volatile species with the analyte; (d) the use of releasing agents to react preferentially with the interferent [49-56]. It should be noted that with the electrothermal atomization technique, where an inert or reducing atmosphere is used, various chemical interferences caused by flames are unlikely to occur.

3.2.3 Ionization Interferences

Some metals are easily ionized in various flames, to different degrees [57]. Ionization is greater at lower metal concentrations than at high. Such ionization decreases the number of free atoms available for excitation. The effect of ionization can be eliminated, however, by (a) lowering the atomization temperature; (b) addition of a large excess (0.2–10 g/l.) of an ionization suppressor (i.e. an easily ionized element) to the standard reference and the analyte solutions [58]. Analyte ionization in electrothermal atomization cells is less than that in the flame [59]. Addition of easily ionized elements to analyte solutions atomized by electrothermal devices may cause loss of sensitivity [60].

3.2.4 Spectral Interferences

The presence of a resonance line of a matrix element in the vicinity of the resonance line of the analyte can cause overlapping of the two signals. The overlap occurs when the two wavelengths differ by less than 0.05 nm. Spectral interferences have been reported in some cases and are more likely to occur when multielement light-sources are used [61–69]. If a modulated light-source and an amplifier turned to the modulation frequency are used, actual overlapping becomes extremely rare.

3.2.5 Background Interference

This is a non-specific absorption originating from light-scattering by solid particles or liquid droplets in the atom cell, and light-absorption by molecules or radicals in the sample matrix. To compensate for it, the background absorption is usually measured and subtracted from the total absorption. This background correction can be performed by using either a continuum source or the Zeeman effect.

3.2.5.1 Background correction with a continuum source. Deuterium, hydrogen or tungsten–halogen lamps, switched into the optical system, are used to measure the background contribution to the absorption signals [70]. This correction can be made automatically by passing the light from the AAS light-source and light from the continuum source through the atom cell alternately and in rapid sequence by means of a rotating-mirror chopper, and then to the monochromator (Fig. 3.5). The electronic circuit then shows the difference between the radiation intensities. All manufacturers of atomic-absorption instruments offer simultaneous background-correction systems, and dual-channel [71,72] and double-beam [73] instruments based on this principle are now available. Several monitoring devices for background correction have also been suggested [74–77].

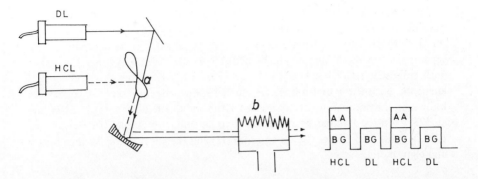

Fig. 3.5. Diagrammatic representation of automatic background correction. DL, deuterium lamp; HCL, hollow-cathode lamp; a, rotating chopper; b, flame; AA, atomic-absorption signal; BG, background absorption signal.

Although background correction by use of a continuum source is adequate with the flame technique it has the following limitations: (a) several continuum sources are required to cover both the ultraviolet and visible regions; (b) the signal-to-noise ratio is subject to some degradation; (c) the intensities of the AAS source and continuum source cannot be matched exactly; (d) the correction is inaccurate if the background is structured; (e) very good alignment between the optical beams of the AAS source and the continuum lamp is required; (f) under or over compensation for the background signal occurs with samples having high background level; (g) the reliability is limited in application to electrothermal atomization systems. Thus, the need for improved background-correction performance for electrothermal atomization systems has generated considerable interest in application of the Zeeman effect.

3.2.5.2. Background correction by the Zeeman effect. A magnetic field of several kilogauss splits the atomic spectral lines into several polarized components with different energy states. The transitions between the new states are given by the selection rule ($\Delta M_J = 0 \pm 1$, where M_J is the magnetic quantum number) which holds for both emission and absorption. This phenomenon is known as the Zeeman effect The spectral line of an atom exhibiting a normal Zeeman effect is replaced by the $\Delta M_J = 0$ component (π line) and $\Delta M_J = \pm 1$ components ($\sigma_\pm$ lines). The π line appears at the original wavelength of the spectral line, but with half the original intensity, whereas the $\sigma_\pm$ lines appear at equally displaced longer and shorter wavelengths than the original line, with intensity a quarter of the original. The π component is linearly polarized with the electric vector parallel to the magnetic field, whereas the $\sigma_\pm$ components are circularly polarized about the lines of force, with the electric vector linearly polarized perpendicular to the magnetic field. The σ_+ and σ_- component vectors rotate in opposite directions.

An example of such a type of effect can be demonstrated with the 285.2 nm line of magnesium or the 553.6 nm line of barium. When the former spectral line is subjected to a magnetic field strength of 10 kG, the $\sigma_{\pm}$ components are separated from the π component by about 3.8×10^{-3} nm. A symmetrical triplet with intensity ratio 1:2:1 is displayed (Fig. 3.6). However, for some other elements both the π and σ lines are further split to give several lines, and this is known as the anomalous Zeeman effect. The sodium D lines and the silver line at 328.1 nm are familiar examples. Sodium D_1 splits into 4 lines (two π and two $\sigma_{\pm}$), D_2 splits into 6 lines (two π and four $\sigma_{\pm}$) and the silver line splits into four π and eight $\sigma_{\pm}$ lines.

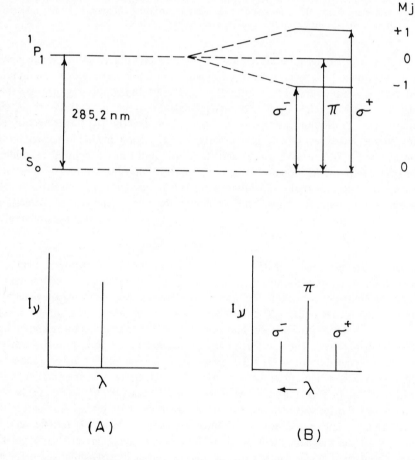

Fig. 3.6. Normal Zeeman effect for magnesium. A, in the absence of a magnetic field; B, in a magnetic field of 10 kG.

The use of the Zeeman effect for background correction is based on the fact that the background absorption is largely due to molecular scattering and absorption [70,78-80] which, contrary to the atomic absorption, is unaffected by the presence of a magnetic field. The first report on the application of the Zeeman effect for background correction was published in 1971 by Hadeishi and McLaughlin [81] in connection with determination of mercury. Applications and development of this technique have been reviewed [82-90]. In principle, a magnetic field is applied either to the radiation source (source-shift Zeeman background correction) or to the atom cell (analyte-shift Zeeman background correction). The magnetic field may be fixed or modulated, and aligned in a direction transverse or longitudinal to the optical path. A fixed or rotating polarizer placed before or after the atom cell may be used. Though significantly influenced by the position or type of these components, that is not the case regarding sensitivity and analytical range.

With source-shift Zeeman background correction [91-99], the spectral-source line is split into its π and $\sigma_\pm$ components. In passing through the atomized sample, the π component is absorbed by both the analyte and background, where-as the $\sigma_\pm$ components are absorbed only by the background. This technique has been adapted for use with both flame and electrothermal atomization systems. The major advantage of this mode of operation is the ease with which it can be applied to any atomization system. On the other hand, difficulties have been experienced in obtaining stable output without high noise from light-sources operated in a magnetic field. Some workers, however, have shown that a hollow-cathode lamp can be operated at an rf modulation frequency of 100 MHz in a 3.8-kG field and provide a stable output signal [100].

In analyte-shift Zeeman background correction, the conventional light-sources are used and the atomization cell is placed in the magnetic field [101-109]. The radiation emitted from the light-source is allowed to pass through a rotating polarizer which splits it into two linearly polarized light-beams, parallel and perpendicular to the magnetic field. These beams pass alternately into the atomic vapour of the analyte in the atom cell, which is in the magnetic field. During the first cycle, the parallel line undergoes atomic absorption as well as attenuation by background absorption and scattering. During the next cycle, the perpendicular line is attenuated by only the background absorption and scattering (Fig. 3.7). Since the parallel and perpendicular lines are attenuated exactly the same by the background, subtraction of the two signals produces the true absorption by the analyte.

The use of a variable-field electromagnet around the atomization cell offers improved performance [100-112]. In this approach, the π components are eliminated by a static quartz polarizer placed after the magnet. The commercially available Zeeman-500 system (Perkin-Elmer) is based on this principle [110] (Fig. 3.8). The source lamp is pulsed at 120 Hz and the current to the magnet is modulated at 60 Hz. When the field is off, both the analyte and background

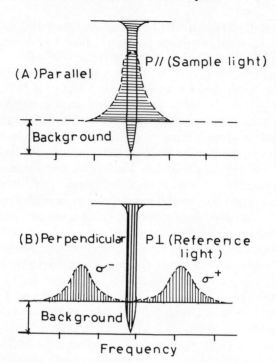

Fig. 3.7. Relationship between parallel and perpendicular polarized light beams and absorption line.

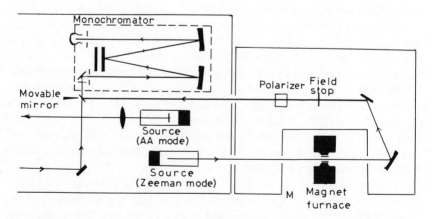

Fig. 3.8. Optical diagram of Zeeman-5000 atomic-absorption spectrometer (according to F. J. Fernandez, W. Bohler, M. M. Beaty and W. B. Barnett, *Atom. Spectrosc.*, **2**, 73 (1981). (by permission of the copyright holders, the Perkin-Elmer Corporation.)

absorptions are measured at the unshifted resonance line. However, when the field is on, only the background is measured, since the σ absorption line profiles are shifted away from the emission line. Thus, background correction is achieved by subtracting the "field on" from the "field off" signal. Many other designs have been devised [113,114].

Regardless of which Zeeman design is used, curvature of the calibration graphs is greater than for conventional atomic absorption. Results of detailed study of the shape of analytical graphs obtained in Zeeman-AAS have been discussed [115-117]. The complexity of the particular Zeeman pattern has a significant effect on the detection limit. A comparison of the detection limit offered by this technique with that of the standard atomic-absorption mode (Table 3.1) indicates a slight increase in the limit. It should be noted, however, that background correction by the most common method (deuterium lamp) is frequently associated with an increase in the limit of detection by a factor of 5-50. Thus the detection limit in the Zeeman mode is considered comparable or even superior to that of the continuum-source correction system. Many examples demonstrating this superiority have been reported [118-120]. A background

Table 3.1 Comparison of the detection limits of some elements without and with background correction by use of the Zeeman effect [89] [reprinted from *American Labotatory*, **11**, No. 11, 35 (1979). Copyright 1979 by International Scientific Communications, Inc.]

Element	Detection limit (g)	
	Zeeman effect	Conventional method
Aluminium	2×10^{-11}	5×10^{-12}
Antimony	3×10^{-10}	2×10^{-11}
Arsenic	1×10^{-11}	1×10^{-11}
Barium	8×10^{-11}	1×10^{-10}
Bismuth	3×10^{-11}	1×10^{-11}
Cadmium	3×10^{-13}	3×10^{-13}
Chromium	9×10^{-12}	1×10^{-11}
Cobalt	2×10^{-11}	4×10^{-11}
Copper	1×10^{-11}	5×10^{-12}
Iron	4×10^{-12}	2×10^{-12}
Lead	4×10^{-12}	5×10^{-12}
Lithium	2×10^{-11}	3×10^{-11}
Manganese	3×10^{-12}	1×10^{-12}
Nickel	3×10^{-11}	1×10^{-10}
Scandium	1×10^{-9}	5×10^{-11}
Silver	9×10^{-13}	5×10^{-13}
Titanium	4×10^{-9}	2×10^{-9}
Vanadium	4×10^{-10}	5×10^{-10}
Zinc	1×10^{-13}	1×10^{-13}

absorbance of 2.0 can be corrected by the Zeeman system to an absorbance below 0.005, which is better by a factor of 4 than the correction obtained with a continuum source [101].

A problem common to all Zeeman-AAS has recently been reported, namely the bending of the calibration graph towards the concentration axis at concentrations $\sim 10^4$ times the lower limit of detection. This is probably due to the fact that the calibration graphs are usually based on the differential absorption between the π and σ components, and that the absorption saturation of the π components occurs at lower concentrations than for the σ components [121]. As a result, double valued calibration curves are obtained (Fig. 3.9). This type of curve seriously affects the results obtained by using flame Zeeman-AAS, but does not cause problems with the electrothermal atomization technique. Koizumi *et al.* [121] studied these transient absorption peaks and showed that their detection warns the analyst that sample dilution is necessary if correct values are to be obtained.

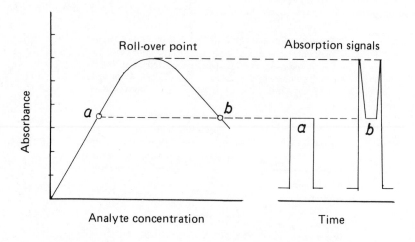

Fig. 3.9. Double-valued calibration curve in flame Zeeman-AAS. [Reprinted with permission from H. Koizumi, H. Sawakabu and M. Koga, *Anal. Chem.*, **54**, 1029 (1982). Copyright 1982 American Chemical Society.]

In general, background correction based on the Zeeman effect has the following advantages: (a) useful correction over a wide range of wavelengths; (b) elimination of the interferences caused by spectral overlap of the absorption lines; (c) baseline stability even with the variable radiation intensity; (d) precise correction of the background at exactly the same wavelength of the resonance line, and retention of the advantages of the double-beam optics since both the reference and sample beams pass through the analyte vapour; (e) efficient correction of high background levels; (f) elimination of spectral interferences.

3.3 MATRIX MODIFICATION

In the electrothermal atomization technique, difficulties are usually encountered in removing the interfering matrix during the charring step, without volatilization of the analyte at the same time. Addition of certain inorganic salts or organic substances to the analyte allows chemical changes to occur during the drying and ashing steps and is known as matrix modification. These substances cause (a) decrease of the volatility of the analyte by formation of high melting point salts, oxides and intermetallic compounds which permit use of a higher charring temperature and enable the removal of most of the matrix, which would otherwise interfere, and (b) increase of the volatility of the matrix, to promote its removal before analyte atomization.

Ediger *et al.* [122] were the first to report the use of ammonium nitrate to reduce the interfering effect of sodium chloride on the determination of copper and cadmium. In this case, sodium chloride is converted into sodium nitrate and ammonium chloride, which are driven off along with the excess of ammonium nitrate, in the ashing step, at temperatures below 500°C. Similarly, ammonium nitrate [123,124], ammonium hydrogen phosphate [125], a mixture of both [126], and ammonium persulphate [127] have been suggested for reducing the matrix interferences in the determination of lead. This technique has also been used with the L'vov platform for the determination of lead in fish tissues [128]. Magnesium nitrate is used as matrix modifier in the determination of aluminium [129]. During the charring step magnesium salts form magnesium oxide, in which the analyte is embedded and is not released at higher temperatures. In other words, the aluminium signal is delayed until the magnesium oxide is vaporized (Fig. 3.10).

Selenium, arsenic and tellurium, which start volatilization and loss at 400°C, 600°C and 800°C, respectively, display no losses in the presence of nickel salts up to a charring temperature of at least 1200°C. Nickel has a similar effect on antimony [130,131], bismuth [132] and arsenic [133-137]. Mercury is not commonly determined by the electrothermal atomization technique, owing to the ease of mercury loss at very low charring temperatures. Addition of sulphide forms mercury sulphide and allows use of charring temperatures of up to 300°C. Lanthanum salts are used as a matrix modifier in the determination of lead and phosphorus. The salt is added to the analyte, or a graphite tube treated with a lanthanum salt can be used [136,138-140].

It has also been reported that some mineral and organic acids effectively act as matrix modifiers. The sensitivity for some elements (e.g. gallium) changes dramatically with matrix composition. The highest sensitivity is obtained by addition of oxidizing acids, probably because of the formation of relatively stable oxides during the charring [136]. Nitric acid has also been suggested for cadmium [141]. Citric acid, histidine, lactic acid, ascorbic acid and EDTA promote low-temperature atomization of cadmium and significantly reduce the interferences from volatilization of other matrix components [142-144]. Table 3.2 shows some matrix modifiers.

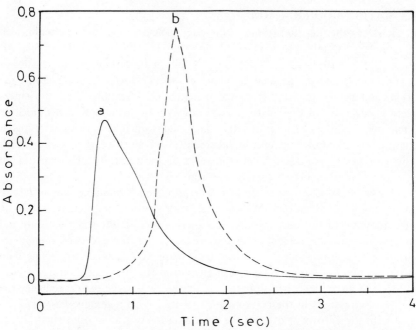

Fig. 3.10. Effect of magnesium on the absorption signal of aluminium. a, in the absence of magnesium; b, in the presence of 50 μg of magnesium nitrate (according to W. Slavin, G. R. Carnick and D. C. Manning, Pittsburgh Conference on Analytical Chemistry and Applied Spectroscopy, Atlantic City, 1982).

Table 3.2 Matrix modifiers used for AAS determination
of some elements

Element	Matrix modifier	Reference
Aluminium	$Mg(NO_3)_2$	129
Antimony	$Ni(NO_3)_2$	130,131
Arsenic	$Ni(NO_3)_2$	133–137
Beryllium	$Mg(NO_3)_2$	
Cadmium	$(NH_4)_2HPO_4$	141–144
	NH_4F, EDTA,	
	ascorbic acid,	
	citric acid, HNO_3	
Chromium	$Mg(NO_3)_2$	
Gallium	$HClO_4$	136
Lead	$(NH_4)_2HPO_4$	125,128
Mercury	$(NH_4)_2S$	
Manganese	$Mg(NO_3)_2$	
Phosphorus	$La(NO_3)_3$	136,138,140
Selenium	$Ni(NO_3)_2$	
Tellurium	$Ni(NO_3)_2$	
Thallium	H_2SO_4	

3.4 USE OF ORGANIC REAGENTS

Preconcentration of the analyte and elimination of the interferences of some elements in AAS have generally been approached by using organic reagents. Enhancement and suppression of the absorbance of some elements in the presence of organic reagents are known to be dependent on the nature of the reagent and the method of atomization. Komárek and Sommer have recently reviewed the application of various organic reagents in AAS [145].

The use of organic reagents in the flame offers the advantages of: (a) formation of metal derivatives which are more volatile and more readily dissociable than the corresponding oxides; (b) decomposition of the reagents to give products which improve the characteristic properties and increase the temperature of the flame [146]. Reduction of the analyte to pure metal or metal carbide, however, affects the sensitivity of the method, as the volatility of such species decreases in the order oxide > metal > carbide [147,148]. Insufficient dissociation and incomplete volatilization of some metal chelates have also been reported. The use of organic reagents in electrothermal atomization is not always useful. Volatilization of some chelates at temperatures lower than that required for atomization causes loss of the analyte and decreases the absorption signals. Decomposition of the organic reagents into carbon in the flameless atomizers, though it enhances the reduction of metal oxides to free atoms, is offset by the formation of metal carbides [149,150]. Organic solvents, if present, interact with the graphite surface of the atomizer and are not released until high temperatures are applied, which may cause interference in the metal atomization step.

3.4.1 Elimination of Interferences

One of the applications of organic reagents in AAS is to eliminate the interference effect of some elements, probably by complex formation with the analyte and/or the interferent so that the analyte is protected from interaction with the interferent and prevented from forming thermally stable species. It should be noted that the stability of the chelates in solution is not a decisive factor, since even weak chelating agents such as glucose, sucrose, ethylene glycol, glycerol and mannitol are also effectively used [151-153]. Another explanation of the effect of organic reagents is based on the fact that formation of metal atoms from their chelates is frequently exothermic, causing disintegration of the droplets, whereas the production of metal atoms from inorganic species is always endothermic.

Among the various types of reagent, 8-hydroxyquinoline has been used for the elimination of the interference due to beryllium [154], bismuth [155], chromium [156], cobalt [157] and iron [158] in the determination of calcium and magnesium [159-166]; EDTA for the removal of the interferences in the determination of magnesium, calcium, copper, manganese and nickel [151, 166-171]; ascorbic, maleic, adipic, glutaric and tartaric acids [172-176] for removing the effect of chloride in the determination of lead, copper and manganese; 5-sulphosalicylic acid for removing the interferences caused in the determination of magnesium, calcium [166,167], chromium, aluminium, iron [178], titanium [179], molybdenum, beryllium, tungsten and tantalum [180,181].

3.4.2 Enhancement and Suppression of Absorption
The use of organic reagents in AAS can produce either enhancement or suppression of atomic absorbance, depending on the nature of the reagent and the atomization technique. Aliphatic amines (e.g. trimethylamine, diethylamine, methylamine) suppress the absorbance of various elements in an air–acetylene flame, owing to insufficient dissociation of the metal complexes [182]. These amines however, enhance the absorbance of zirconium, hafnium and titanium in the nitrous oxide–acetylene flame [183]. Organic acids such as formic, acetic, propionic and butyric enhance the absorbance of titanium whereas reducing acids such as oxalic, citric and ascorbic acids suppress it [179,184,185]. Aromatic sulphonic acids in large excess enhance the absorbance of titanium and molybdenum in the nitrous oxide–acetylene flame in the presence of sodium and potassium [179-181]. The role of the surfactants sodium dodecylsulphate and dodecyltrimethylammonium chloride in AAS has been investigated with emphasis on the droplet size in aerosols produced by pneumatic nebulizers [186,187]. The marked effectiveness of the surfactants in improving the sensitivity and reducing interference is related to production of finer aerosols, because of the lowering of the surface tension. 8-Hydroxyquinoline, dithizone, diethyldithiocarbamate, pyrrolidine dithiocarbamate and cupferron have no effect on the absorbance of many elements when electrothermal atomization is used [188].

3.4.3 Extraction and Concentration of Metals
Samples containing low concentrations of the analyte or high concentrations of the interferent can be determined after a prior extraction step with a suitable organic chelating agent [189-191]. Table 3.3 shows some applications of these reagents.

3.5 METHODOLOGICAL APPROACHES
3.5.1 Direct Methods
Most metals are easily determined directly by measuring their absorbance at the principal resonance lines, provided that the atomization conditions and instrumental parameters are carefully controlled. Samples containing low concentrations of the analyte can be analysed after a prior concentration step. Extraction [192-244], co-precipitation [245], electro-deposition [246-250] and ion-exchange chromatography [251] are usually used for preconcentration of the analyte and to remove it from the sample matrix, thereby extending the concentration range of the atomic-absorption measurements and lowering the

Organic reagent	Solvent	Metal extracted	Reference
Acetylacetone	Chloroform	Pb	192
Amberlite LA-1	Xylene	Cr, Pb	193,194
Amberlite LA-2	Chloroform	Mo, Re, W	195
	Toluene	Cu	196
Ammonium pyrrolidine dithiocarbamate (APDC)	MIBK	Ag, As, Cd, Co, Cr, Cu, Fe, Mn, Ni, Pb, Re, Se, Zn,	197–202
4-Benzoyl-3-methyl-1-phenyl-5-pyrazolone	Nitrobenzene	As	203
Diethyldithiocarbamate (DDC)	MIBK	Pb	204
	MIBK	Cd, Co, Cr, Cu, Mn, Pb	205–207
Diphenylcarbazone	Pyridine–toluene	Cd, Co, Cu, Ni, Pb, Zn	208
Diphenylthiourea	Chloroform	Ag, Cu, Tl	209
Dipivaloylmethane	MIBK	Cd	210
Dithizone	Chloroform	Ag, Cd, In, Ni, Co, Cu, Pb, Zn	211–213
	MIBK	Be, Cu	214,215
	Ethyl nitrate	Ag	216
Hexamethylenimine hexamethylene dithiocarbamate	Amyl acetate	Pb	217
	Butyl acetate	Bi, ln, Pb	218
1-Nitroso-2-naphthol	MIBK	Co	219
n-Octylamine	Toluene	Ir, Pd, Pt, Rh, Ru	220,221
Potassium ethyl xanthate	MIBK	Cd, Cu, Mn	222
Pyridine-2-aldehyde 2-pyridylhydrazone	Amyl alcohol	Fe, Zn	223
1-(2-Pyridylazo)-2-naphthol	Benzene or MIBK	Zn	224
Tetraoctylammonium	Dichloroethane	Ir, Pd, Pt, Ru	225
Tricaprylmethylammonium	Xylene	Cd	226
Tri-n-octylamine	Benzene	Cu, Fe, In, Sn	227
	Benzene	Ag, Au, Bi, Cd, Hg, Pb	228
Tri-n-octylphosphine	MIBK	Bi, Pb, Sb, Sn	229
Trioctylamine or trihexylamine	Xylene, MIBK	Si	230
Trioctylphosphine oxide	MIBK	Sb	231
Triphenylphosphine	Benzene	Au	232
Zephiramines	MIBK	Cr	233
	Ethyl acetate	Cd	234
Zinc dibenzyldithiocarbamate	MIBK	Cu	235,236

MIBK = methyl isobutyl ketone

determination limit. A monograph on preconcentration has recently appeared [251a]. However, the principle atomic resonance lines of most non-metals occur in the vacuum ultraviolet region of the spectrum, where air, flame gases and the quartz optics of the spectrometer absorb significantly. This renders direct application of the common AAS procedures impossible for determination of these elements.

3.5.2 Methods Based on Indirect Reactions

Many varied indirect methods have been proposed.

(a) Precipitation of the analyte with a metal ion, which is then measured either in the filtrate or the precipitate. Sulphate [252] and oxalate [253] are determined by precipitation with barium and calcium, respectively.

(b) Reaction of the analyte with a metal chelate to form an ion-association complex, followed by extraction and measurement of its metal content. Phthalic acid [254] and nitrate [255] are determined by reaction with the Cu(I)–neocuproine complex to form $[Cu(I)(neocuproine)_2C_6H_4(COOH)\text{-}COO^-]$ and $[Cu(I)(neocuproine)_2NO_3]$, followed by extraction of these species and measurement of copper at 324.7 nm.

(c) Formation of heteropoly acids followed by extraction and measurement of the metal content. Phosphate [256] and alkaloids [257] are converted into the corresponding phosphomolybdate compounds and the twelve molybdenum atoms then associated with each original phosphorus atom or three alkaloid molecules are determined at 313 nm.

(d) Solubilization or volatilization of metal ions from pure metals. Cyanide ions [258] and nitro compounds [259] are determined by their reaction with silver and cadmium metals, respectively, to release equivalent amounts of silver and cadmium ions. Halide solutions are allowed to flow through a burner with a chimney packed or lined with indium [260], silver [261] or copper [262] to give the corresponding volatile metal halides.

(e) Reduction to the element, followed by its isolation and measurement. Iodide and aldehydes are determined by reaction with selenium(IV) and silver(I) to give elemental selenium and silver, respectively. The elements are isolated, washed, dissolved in acids and atomized [263].

(f) Selective extraction or precipitation of one oxidation state of an element in the presence of another. Iodide and iodate are determined by reaction with chromium(VI) and iron(II), followed by extraction of the excess of chromium(VI) or the iron(III) produced, respectively [263]. Diols are determined by oxidation with periodate, followed by precipitation with silver of the equivalent amount of iodate released [264].

(g) Displacement of some metal ions from their complexes. Fluoride can be determined by displacement of iron(III) from its thiocyanate complex, followed by extraction and measurement of the excess of metal complex [265].

3.5.3 Methods Based on Enhancement and Suppression of Absorption

Some anions, cations and organic compounds have either enhancing or depressive effects on the absorption of some metals in the flame. These effects are mainly due to the formation of easily or difficultly dissociated species in the atomization step.

3.5.3.1 Cation-anion interaction. The absorption of magnesium in an air-coal gas flame is significantly depressed by fluoride ions [266]. The magnitude of the depression at the 285.2 nm line of magnesium is linearly related to the fluoride concentration. On the other hand, fluoride ions enhance the atomic absorption of zirconium in a luminous nitrous oxide-acetylene flame [267]. Thus the absorption of zirconium at 360 nm is linearly increased by increase in the fluoride ion concentration. A similar effect has been reported for magnesium with phosphate and silicate ions.

3.5.3.2 Cation-cation interaction. The absorption of iron in a fuel-rich air-acetylene flame is enhanced by the presence of aluminium or titanium. The absorption of strontium and chromium is suppressed by titanium and iron, respectively.

3.5.3.3 Cation-molecule interaction. Ammonia and organic amino-compounds enhance the absorption of zirconium and can be measured by procedures based on this phenomenon [268]. Glucose at concentration levels less than $10^{-6} M$ causes a marked depression in the absorbance of calcium. Ribonuclease and glucose oxidase enzymes can also be determined by measuring their depressive effect on the absorption of calcium [263].

3.5.4 Methods Based on Metal Volatilization

The volatilization technique provides a number of significant benefits in atomic-absorption spectrometry. It involves separation of the analyte by volatilization at room temperature or high temperature, in the form of the free element or its derivatives. By this technique, it is possible to measure very low analyte concentrations because the volatilized species can be collected from a large solution volume, leaving behind the matrix and interfering substances.

3.5.4.1 Generation of free metal. Several authors [269-273] have documented the advantages of the volatilization technique for the determination of volatile elements. Although the principle upon which this technique is based was originally introduced a long time ago by Bunsen, it attracted awakened interest in the 1950s. Nanogram to microgram amounts of volatile elements such as arsenic, bismuth, cadmium, selenium and thallium, in relatively non-volatile matrices, have been determined by volatilization at temperatures up to a practical limit of 1400°C. The volatilized element is separated in a stream of carrier gas, condensed and measured [274,275].

Elemental mercury, however, can be produced from mercury compounds by reduction with tin(II) chloride or sodium borohydride at room temperature. The metal has an appreciable vapour pressure and gives a monatomic gas. The mercury is swept out of solution with a carrier gas and directed into a long-path absorption tube for absorption measurement at 263.7 nm, without the need for either a flame or an electrothermal atomizer [276-297]. This technique is known as cold-vapour atomic-absorption spectrometry and the instrumentation has been optimized to provide a detection limit of 0.002 ng/ml for mercury. Practically, the mercury from a given volume of a solution is either swept once through the absorption cell or recirculated through the cell to produce a continuous signal. Oda and Ingle [298] described three different designs for continuous flow reduction vessels for the ultratrace determination of mercury. Samples and reductant solutions are continuously fed to the reduction vessels, where the mercury is reduced, then stripped out from a thin stream of solution with a countercurrent flow of gas over the solution stream. Automated systems for cold-vapour atomic-absorption spectrometry have also been described [299-301]. Torsi *et al.* [302] described an electrostatic accumulation furnace for electrothermal atomic-absorption spectrometric determination of mercury.

3.5.4.2 Generation of metal hydride. Several elements (e.g. As, Bi, Ge, Pb, Se, Sb, Sn and Te) are known to form volatile covalent hydrides suitable for subsequent AAS measurement. In principle, the solution containing such elements is reduced with zinc, magnesium or aluminium metal, or sodium borohydride or borocyanide to form the corresponding hydrides, which are allowed to pass, either directly or after collection in a special device, to the atomizer [303-309]. The hydrides of these elements are dissociated into free metal atoms, the absorption of which is measured at the resonance lines. The sensitivity of these procedures is 50-200 times greater than that of the flame methods, and 5-10 times that of the electrothermal atomization methods (Table 3.4). The technique is

Table 3.4 Detection limits for some hydride-forming elements by the various AAS techniques

Element	Detection limit (μg/l.)		
	Direct flame	Direct graphite furnace	Hydride generation
Antimony	30	0.15	0.10
Arsenic	140	0.20	0.02
Bismuth	20	0.10	0.02
Selenium	70	0.50	0.02
Tellurium	19	0.10	0.02
Tin	110	0.20	0.50

relatively free from interferences, as it involves separation of the metals as gases from the associated materials. The technique has received considerable attention during the last five years, and has been the subject of detailed studies.

3.5.4.3 Generation of metal carbonyls and alkyls. Because the hydride generation techniques cannot be applied to all metals, other volatile derivatives have been suggested. Nanogram and subnanogram quantities of nickel have been determined by carbonyl generation. The metal is reduced with sodium borohydride to its elemental form, which is then made to combine with carbon monoxide to form nickel carbonyl [310]. The nickel carbonyl is stripped from the solution with a helium–carbon monoxide gas stream, collected in a liquid-nitrogen trap and atomized in a quartz-tube burner in an atomic-absorption spectrophotometer. The limit of detection of the method is 0.05 ng of nickel and the precision for 3 ng is 4.5%.

Although lead can be converted into the hydride PbH_4, a method has been described for its AAS determination after its methylation. Lead is extracted into chloroform as the dithiocarbamate complex, the solvent is evaporated and the residue is methylated with methyl-lithium to form tetramethyl-lead. The analyte vapour is trapped on a short column of Porapak Q, from which it is eluted into a quartz-furnace AAS instrument [311]. A relative standard deviation of 6.8% is obtainable at the 50-ng/ml level. It is important to notice that because of disproportionation, the maximum conversion of lead into tetramethyl-lead is only 50%:

$$2Pb^{2+} + 4LiCH_3 \rightarrow Pb(CH_3)_4 + Pb + 4Li^+ \qquad (3.3)$$

3.6 SAMPLE HANDLING

3.6.1 Atomization of Micro Samples in the Flame

Aspiration of as little as 100 μl of analyte solution has recently been described [312-319]. Venturi sampling for aspiration of both aqueous and organic solvents into a flame without lift-off can also be used [320]. Introduction of sample solutions into the flame by means of electrically heated metal loops minimizes the volume needed for atomization and greatly improves the detection limit. This technique of sample introduction avoids the use of nebulizers, with its inherently low efficiency, so the detection limits for volatile elements are substantially lower than those attainable by conventional flame AAS. Sample volumes less than 40 μl can be analysed by the metal loop methods [321,322].

On the other hand, several devices have been used in conjunction with the conventional flame to improve the residence time of the free atoms, and hence permit the use of micro samples. The flame (total consumption or turbulent) is directed to one end of a 40–100 cm long tube, made of quartz, alumina or a ceramic, the axis of which coincides with the optical axis of the spectrometer

[323-331]. This technique allows the use of very small sample volumes, and decreases the detection limit for many elements by a factor of 10-100 from that of the direct flame method, but is applicable only to metals which do not readily form oxides in the flame. Since light transmission in the tube depends on multiple reflections at the inner surface, deposition of solid materials from the flame impairs the sensitivity. To overcome this difficulty, the tube is isolated, or heated from the outside with a second burner or an electrically heated jacket [332-335].

3.6.2 Automated Atomization of Analyte Solutions
A multichannel peristaltic pump with pneumatic nebulizer has been used for automated atomization [336]. The samples taken from an auto-sampler table are passed by the pump to the nebulizer at a constant selected rate. Another channel of the pump can be used to deliver an ionization buffer and/or diluent simultaneously to the nebulizer, thus reducing sample preparation time. The use of such pumps ensures a constant delivery rate of sample irrespective of changes in nebulizer gas flow or sample viscosity.

3.6.3 Flow Injection of Analyte Solutions
Flow-injection analysis (FIA) is a well defined analytical technique used mainly for increasing sample output. This technique, first described by Stewart *et al.* [337] and Růžička and Hansen [338], has found wide applications in conjunction with many electrochemical and spectroscopic detection systems. FIA involves rapid injection of the test sample, either manually through a septum (Fig. 3.11), or automatically (Fig. 3.12), into a continuously moving non-segmented carrier stream of pure water, reagent solution or organic solvent.

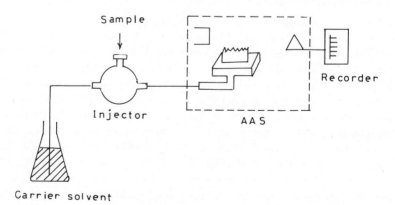

Fig. 3.11. Flow diagram of the flow-injection–AAS analysis apparatus. [According to F. Fukamachi and N. Ishibashi, *Anal. Chim. Acta,* **119,** 383 (1908). By permission of the copyright holders, Elsevier Scientific Publishing Co.]

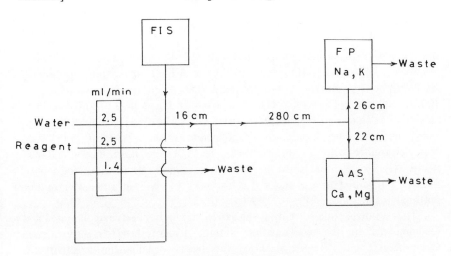

Fig. 3.12. Flow system for the simultaneous determination of Na, K, Ca and Mg. Sampling rate 124/hr. FIS, flow injection samples; FP, flame photometer. [According to W. D. Basson and J. F. Van Staden, *Z. Anal. Chem.*, **302**, 370 (1980). By permission of the copyright holders, Springer Verlag.]

The technique has been adapted for sample introduction in flame AAS by injecting the analyte into a carrier stream continuously pumped into the nebulizer by peristaltic pump and recording the absorption as a function of time. The term flow-injection sample manipulator (FISM) may be used to describe the function of this technique as an interface.

A simple model has been proposed to account for the absorbance–time curves obtained by FIA–AAS [339]. Three basic modes of analyte dispersion (low, medium and high dispersion) can be used [340]. The low-dispersion system provides the analogue of discrete nebulization with a peak base-width of 5 sec and a precision better than 1%. The medium-dispersion system provides the analogue of the standard addition method and gives a peak base-width of about 8 sec. The high-dispersion system provides a known concentration–time profile that may be useful for calibration purposes. The use of a rapidly moving carrier stream and a short sampling period (< 2 sec) allows the analysis of as many as 500 samples per hour, but the practical working rate is lower, being dictated by sample handling. Sample preparation, dilution, and mixing with suppressive or enhancing reagents can thus easily be automatically performed [341].

It should be noted that FIA is a dynamic method, yielding the analytical read-out while two kinetic processes, the physical mixing of the analyte and reagent and the simultaneous chemical reactions, are still in progress. This results in a transient signal. In spite of this non-equilibrium condition, the technique yields easily reproducible and useful signals provided that the carrier flow is constant and a consistent injection technique is used. The degree of sample dispersion can be controlled by varying the flow-rate, the sample size and the dimensions of the capillary tube leading to the nebulizer.

Wolf and Stewart [342] presented the use of FI sample manipulations with AAS for the determination of copper and zinc. Aliquots of the samples (25–300 μl, equivalent to 3–4 ng/ml) are injected at a rate of 120–180 per hour. Similarly, calcium and magnesium in urine and serum [343], lithium in serum [344], and calcium, magnesium and potassium in plant materials [345] have been determined with a typical sample rate of 300 per hour. Basson and Van Staden [346] have described a method for simultaneous determination of sodium, potassium, magnesium and calcium, by FI–AAS, at a sampling rate of 500 analyses per hour.

The use of organic solvents as the carrier stream has been demonstrated and recommended for the determination of trace elements [347]. n-Butyl acetate and methyl isobutyl ketone are effective solvents for enhancing sensitivity for many metals in the flame. An integrated AAS instrument, with a flow-injection sample manipulator as the interface between a high-pressure liquid chromatograph and the spectrometer, has been devised to allow both the chromatographic and atomic-absorption detectors to operate under optimum conditions for both [348]. The system is used for separation and determination of protein-bound calcium and magnesium in clinical samples.

3.6.4 Atomization of Solids

Because of the disadvantages of sample decomposition by dry ashing and wet digestion techniques, there has recently been a trend towards direct analysis of solid, freeze-dried or slurried samples [349–356]. This offers the advantages of (a) elimination of the time-consuming decomposition step; (b) omission of any prior separation or concentration procedure; (c) minimization of losses and reagent contamination.

3.6.4.1 Atomization of powders. Solid samples are powdered and placed in a small platinum, tantalum or graphite boat and inserted into the flame or graphite tube furnace [357–362]. As the sample is volatilized, atomic vapour passes upwards and a transient absorption signal is observed. Instead of these boats, a small nickel cup, the 'Delves cup', can be used [363]. An absorption tube made of nickel or ceramic (10–12 cm long, 1 cm diameter) with a 5 mm hole midway along the wall of the tube and aligned in the optical path over the flame, can also be used. When the cup is pushed into the flame, atomic vapour enters the absorption tube through the hole and the atoms are retained in the absorption path [364–374]. Automation of the micro-sampling Delves-cup technique has also been described [375–378]. Results obtained for the determination of metals in some solid samples are shown in Table 3.5.

Powdered samples may be converted into an aerosol of fine particles and

Table 3.5 Analysis of some solid samples by
graphite AAS [378a]

Sample	Element	μg/g		
		Certified		Found
Bovine liver	Cd	0.27	± 0.04	0.29
	Mn	10.30	± 1.00	9.80
Orchard leaves	Cd	0.11	± 0.01	0.11
	Cu	12.00	± 1.00	11.9
	Ni	1.30	± 0.200	1.2
Oyster tissue	Cr	0.69	± 0.27	0.63
	V	2.8		2.8
Spinach	Pb	1.2	± 0.2	1.1
	Rb	12.1	± 0.2	11.8
Wheat flour	Cd	0.032	± 0.007	0.033
	Cu	2.0	± 0.3	2.1

measured with an air-argon flame by sampling with a d.c. arc or a high-voltage spark [379-382]. Laser irradiation of solid samples to form aerosols with concentrations proportional to the content of analyte in the solid is also used for AAS determination of some elements [383-394]. The temperatures produced in the aerosol by these systems, however, are considerably higher than those attainable in flames or with electrothermal atomizers. Consequently, the ratio of neutral atoms to excited or ionized atoms is less favourable for many metals.

3.6.4.2 Atomization of suspensions and slurries. Solid samples can be atomized after dilution by mixing with calcium carbonate [395], sodium chloride [396–400] or solid fuel [401-404] and burning just below the radiation beam. Suspensions of powders in organic solvents [405], water [406,407], or aqueous solutions of dispersing agents [408,409] have also been used. Large-bore total-consumption burners based on the Babington principle of aerosol generation [410] have been used for atomization of ground and sieved powder suspensions, but these burners have excessive flicker noise and give extremely poor sensitivity [411-413]. Fry and Denton [414,415] introduced the first clog-free nebulizer capable of handling high-solid food slurries. Similar methods have been suggested for the AAS of slurries and insoluble materials [416,417]. In a recent publication, Mohamed and Fry [418] described a method for atomization of homogenized animal tissues, reduced to a slurry by rapid homogenization by sonic cavitation, followed by direct AAS measurement (Fig. 3.13).

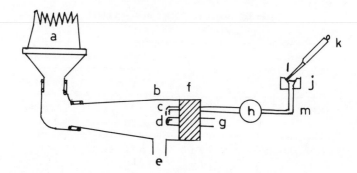

Fig. 3.13. High-solids clog-free nebulizer design and spray chamber. a, flame; b, fuel delivery at side of chamber; c, sample tube; d, gas orifice; e, drain; f, Teflon plug; g, oxidant gas; h, peristaltic pump; j, Teflon sample cone; k, 500-μl Eppendorf pipette; l, tissue homogenate; m, Tygon tubing, 19 mm i.d. [Reprinted with permission from N. Mohamed and R. C. Fry, *Anal. Chem.*, **53**, 450 (1981). Copyright 1981 American Chemical Society.]

REFERENCES

[1] P. Schramel, *Anal. Chim. Acta*, **72**, 414 (1974).

[2] J. D. Kerber and W. B. Barnett, *Atom. Absorp. Newslett.*, **8**, 113 (1969).

[3] J. M. Harnly and T. C. O'Haver, *Anal. Chem.*, **53**, 1291 (1981).

[4] B. Meddings and H. Kaiser, *Atom. Absorp. Newslett.*, **6**, 28 (1967).

[5] E. D. Prudnikov, H. Bradaczek and H. Labischinski, *Z. Anal. Chem.*, **308**, 342 (1981).

[6] J. D. Winefordner, *Appl. Spectrosc.*, **17**, 109 (1963).

[7] L. de Galan, W. W. McGee and J. D. Winefordner, *Anal. Chim. Acta*, **37**, 436 (1967).

[8] Z. van Gelder, *Spectrochim. Acta*, **25B**, 669 (1970).

[9] F. Kaufman and D. A. Parkes, *Trans. Faraday Soc.*, **66**, 1579 (1970).

[10] H. Prugger, *Optik*, **21**, 320 (1964).

[11] C. S. Rann, *Spectrochim. Acta*, **23B**, 245 (1968).

[12] I. Rubeška and V. Svoboda, *Anal. Chim. Acta*, **32**, 352 (1965).

[13] H. C. Wagenaar and L. de Galan, *Spectrochim. Acta*, **30B**, 361 (1975).

[14] H. C. Wagenaar, I. Novotný and L. de Galan, *Spectrochim. Acta*, **29B**, 301 (1974).

[15] B. E. Limbek, C. J. Rowe, J. Wilkinson and M. W. Routh, *Lab. Instrum. Spectrosc., Series III*, Vol. 11, 17 (1981).

[16] C. L. Chakrabarti, C. C. Wan, H. A. Hamed and P. C. Bertels, *Can. Res.*, **13**, 31 (1980).

[17] C. W. Fuller, *Atom. Absorp. Newslett.*, **11**, 65 (1972).

[18] J. R. Hall, R. G. Godden and D. R. Thomerson, *Analyst*, **105**, 820 (1980).

[19] J. P. Franke, *Z. Anal. Chem.*, **308**, 351 (1981).

[20] F. J. Feldman, *Anal. Chem.*, **42**, 719 (1970).

[21] K. C. Singhal, R. P. C. Sinha and B. K. Banerjee, *Technology (Sindri-India)*, **6**, 219 (1969).

[22] R. W. Looyenga and C. O. Huber, *Anal. Chem.*, **43**, 498 (1971).

[23] W. E. Crawford, C. I. Lin and C. O. Huber, *Anal. Chim. Acta*, **64**, 387 (1973).

[24] C. I. Lin and C. O. Huber, *Anal. Chem.*, **44**, 2200 (1972).

[25] J. R. Sand and C. O. Huber, *Anal. Chem.*, **48**, 1331 (1976).

[26] J. Posta and J. Lakatos, *Spectrochim. Acta*, **35B**, 601 (1980).

[27] J. Posta and J. Lakatos, *Magy. Kem. Foly.*, **86**, 284 (1980).
[28] D. D. Stojanovic and J. D. Winefordner, *Anal. Chim. Acta*, **124**, 295 (1981).
[29] M. Taddia, *Anal. Chim. Acta*, **129**, 259 (1981).
[30] R. Looyenga and C. O. Huber, *Anal. Chim. Acta*, **55**, 179 (1971).
[31] F. J. Fernandez and J. D. Kerber, *Am. Lab.*, **8**, No. 3, 49 (1976).
[32] M. Feinberg and C. Ducanze, *Analusis*, **8**, 185 (1980).
[33] I. Rubeška and J. Musil, *Prog. Anal. Atom. Spectrosc.*, **2**, 309 (1979).
[34] E. J. Czobik and J. P. Matousek, *Anal. Chem.*, **50**, 2 (1978).
[35] E. Roesler, *Chem. Tech. (Heidelberg)*, **9**, 549 (1980).
[36] J. P. Maney and V. J. Luciano, *Anal. Chim. Acta*, **125**, 183 (1981).
[37] C. L. Chakrabarti, C. C. Wan, H. A. Ahmed and P. C. Bertels, *Anal. Chem.*, **53**, 444 (1981).
[38] J. M. Harnly and T. C. O'Haver, *Anal. Chem.*, **49**, 2187 (1977).
[39] J. E. Allan, *Spectrochim. Acta*, **17**, 459 (1961).
[40] R. Avni and C. Th. J. Alkemade, *Mikrochim. Acta*, 460 (1960).
[41] M. Betz, S. Guecer and F. Fuchs, *Z. Anal. Chem.*, **303**, 4 (1980).
[42] R. B. Cruz and J. C. VanLoon, *Anal. Chim. Acta*, **72**, 231 (1974).
[43] J. Smeyers-Verbeke, Y. Michotte, P. Vanden Winkel and D. L. Massart, *Anal. Chem.*, **48**, 125 (1976).
[44] A. H. Jones, *Atom. Absorp. Newslett.*, **9**, 1 (1970).
[45] H. A. Rohleder, F. Dietl and B. Sansoni, *Spectrochim. Acta*, **29B**, 19 (1974).
[46] K-P. Li, *Anal. Chem.*, **53**, 317 (1981).
[47] A. Hulanicki, R. Karwowska and J. Sowinski, *Talanta*, **28**, 455 (1981).
[48] A. M. Pashadzhanov, L. V. Morieskaya and A. A. Nemodruk, *Zh. Analit. Khim.*, **36**, 1343 (1981).
[49] J. B. Willis, *Nature*, **186**, 249 (1960).
[50] J. B. Willis, *Spectrochim. Acta*, **16**, 259 (1960).
[51] A. C. West and W. D. Cooke, *Anal. Chem.*, **32**, 1471 (1960).
[52] I. Rubeška and B. Molden, *Anal. Chim. Acta*, **37**, 421 (1967).
[53] F. J. Wallace, *Analyst*, **88**, 259 (1963).
[54] M. D. Amos and J. B. Willis, *Spectrochim. Acta*, **22**, 1325 (1966).
[55] R. H. Wendt and V. A. Fassel, *Anal. Chem.*, **38**, 337 (1966).
[56] J. A. Bowman and J. B. Willis, *Anal. Chem.*, **39**, 1210 (1967).
[57] C. W. Woodward, *Spectrosc. Lett.*, **4**, 191 (1971).
[58] G. R. Kornblum and L. de Galan, *Spectrochim. Acta*, **28B**, 139 (1973).
[59] R. E. Sturgeon and S. S. Berman, *Anal. Chem.*, **53**, 632 (1981).
[60] W. Luecke, F. E. Schermann, U. Lennartz and A. J. Papastamutaki, *Jb. Miner. Abh.*, **120**, 178 (1974).
[61] W. R. Kelly and C. B. Moore, *Anal. Chem.*, **45**, 1274 (1973).
[62] V. A. Fassel, J. A. Rasmuson and T. G. Cowley, *Spectrochim. Acta*, **23B**, 579 (1968).
[63] S. Slavin and T. W. Sattur, *Atom. Absorp. Newslett.*, **7**, 99 (1968).
[64] J. D. Norris and T. S. West, *Anal. Chem.*, **46**, 1423 (1974).
[65] J. E. Allan, *Spectrochim. Acta*, **24B**, 13 (1969).
[66] K. C. Thompson, *Analyst*, **95**, 1043 (1970).
[67] C. W. Frank, W. G. Schrenk and C. E. Meloan, *Anal. Chem.*, **38**, 1005 (1966).
[68] D. C. Manning and F. Fernandez, *Atom. Absorp. Newslett.*, **7**, 24 (1968).
[69] J. D. Norris and T. S. West, *Anal. Chem.*, **45**, 2148 (1973).
[70] S. R. Koirtyohann and E. E. Pickett, *Anal. Chem.*, **38**, 585 (1966).
[71] S. R. Koirtyohann, *Anal. Chem.*, **37**, 601 (1965).
[72] B. V. L'vov, *Spectrochim. Acta*, **24B**, 53 (1969).
[73] H. L. Kahn, *Atom. Absorp. Newslett.*, **7**, 40 (1968).
[74] R. F. M. Herber and J. L. M. De Boer, *Anal. Chim. Acta*, **109**, 177 (1979).

[75] D. D. Siemer, *Anal. Chim. Acta,* **119**, 379 (1980).
[76] F. J. M. Maessen and F. D. Posma, *Anal. Chem.,* **46**, 1439 (1974).
[77] P. del Castilho and R. F. M. Herber, *Anal. Chim. Acta,* **94**, 269 (1977).
[78] S. R. Koirtyohann and E. E. Pickett, *Anal. Chem.,* **38**, 1087 (1966).
[79] J. W. Robinson, G. D. Hindman and P. J. Slevin, *Anal. Chim. Acta,* **66**, 165 (1970).
[80] H. L. Kahn and D. C. Manning, *Am. Lab.,* **4**, No. 8, 51 (1972).
[81] T. Hadeishi and R. D. McLaughlin, *Science,* **174**, 404 (1971).
[82] S. D. Brown, *Anal. Chem.,* **49**, 1269 (1977).
[83] R. Stephens, *CRC Crit. Rev. Anal. Chem.,* **4**, 1 (1978).
[84] M. T. C. de Loos-Vollebregt and L. de Galan, *Spectrochim. Acta,* **33B**, 495 (1978).
[85] R. Stephens, *CRC Crit. Rev. Anal. Chem.,* **9**, 167 (1980).
[86] K. Yasuda, H. Koizumi, K. Ohishi and T. Noda, *Prog. Anal. Atom. Spectrosc.,* **3**, 299 (1980).
[87] M. T. C. de Loos-Vollebregt, *INIS Atomindex,* **11**, (1980), Abstract Wo 553016.
[88] C. Van Nieuwenhuyzen, *Chem. Weekbl.,* No. 2, 106 (1976).
[89] J. D. Miller and H. Koizumi, *Am. Lab.,* **11**, No. 11, 35 (1979).
[90] A. Petrakiev and Ts. Iotov, *Wiss.Z. Karl-Max. Univ. Leipzig, Math.-Naturwiss. Reihe,* **28**, 347 (1979).
[91] T. Hadeishi, *Appl. Phys. Lett.,* **21**, 438 (1972).
[92] T. Hadeishi and R. D. McLaughlin, *Anal. Chem.,* **48**, 1009 (1976).
[93] T. Hadeishi and R. D. McLaughlin, *Am. Lab.,* **7**, No. 8, 57 (1975).
[94] H. Koizumi and K. Yasuda, *Spectrochim. Acta,* **31B**, 237 (1976).
[95] H. Koizumi and K. Yasuda, *Anal. Chem.,* **47**, 1679 (1975).
[96] H. Koizumi and K. Yasuda, *Anal. Chem.,* **48**, 1178 (1976).
[97] R. Stephens and D. E. Ryan, *Talanta,* **22**, 655 (1975).
[98] D. E. Veinot and R. Stephens, *Talanta,* **23**, 849 (1976).
[99] R. Stephens, *Talanta,* **24**, 233 (1977); **25**, 435 (1978); **26**, 57 (1979).
[100] K. Oishi, Y. Arai, S. Kayama, S. Murayama and K. Fukuda, *Spectrochim. Acta,* **35B**, 155 (1980).
[101] E. Grassman, J. B. Dawson and D. J. Ellis, *Analyst,* **102**, 804 (1977).
[102] H. Koizumi and K. Yasuda, *Spectrochim. Acta,* **31B**, 523 (1976).
[103] C. Parker and A. Pearl, *British Patent* 1,385,791 (5 January 1972).
[104] Y. Uchida and S. Hattori, *Oyo Butsuri,* **44**, 852 (1975).
[105] J. B. Dawson, E. Grassam, D. J. Ellis and M. J. Keir, *Analyst,* **101**, 315 (1976).
[106] H. Koizumi, K. Yasuda and H. Katayama, *Anal. Chem.,* **49**, 1106 (1977).
[107] H. Koizumi, *Anal. Chem.,* **50**, 1101 (1978).
[108] H. Koizumi, H. Yamada, K. Yasuda, K. Uchmo and K. Oishi, *Spectrochim. Acta,* **36B**, 603 (1981).
[109] K. G. Brodie and P. R. Liddell, *Anal. Chem.,* **52**, 1059 (1980).
[110] W. Bohler, F. J. Fernandez, M. M. Beaty and W. B. Barnett, *Atom. Spectrosc.,* **2**, 73 (1981).
[111] P. R. Liddell and K. G. Brodie, *Anal. Chem.,* **52**, 1256 (1980).
[112] M. T. C. de Loos-Vollebregt and L. de Galan, *Spectrochim. Acta,* **35B**, 495 (1980).
[113] H. Koizumi and K. Yasuda, *Bunko Kenkyu,* **23**, 290 (1974).
[114] V. Otruba, J. Jambor, J. Komárek, J. Horák and L. Sommer, *Anal. Chim. Acta,* **101**, 367 (1978).
[115] M. T. C. de Loos-Vollebregt and L. de Galan, *Appl. Spectrosc.,* **33**, 616 (1979).
[116] M. T. C. de Loos-Vollebregt and L. de Galan, *Appl. Spectrosc.,* **34**, 464 (1980).
[117] R. Stephens, *Talanta,* **25**, 723 (1978).
[118] F. J. Fernandez, S. A. Myers and W. Slavin, *Anal. Chem.,* **52**, 741 (1980).
[119] K. G. Brodie and P. R. Liddell, *Anal. Chem.,* **52**, 1059 (1980).
[120] J. D. Miller and H. Koizumi, *Lab. Instrum. Spectrosc., Series III,* Vol. III, 33 (1981).

[121] H. Koizumi, H. Sawakabu and M. Koga, *Anal. Chem.*, **54**, 1029 (1982).
[122] R. D. Ediger, G. E. Peterson and J. D. Kerber, *Atom. Absorp. Newslett.*, **13**, 61 (1974).
[123] D. C. Manning and W. Slavin, *Anal. Chem.*, **50**, 1234 (1978).
[124] R. E. Sturgeon, S. S. Berman and D. S. Russell, *Anal. Chem.*, **51**, 2364 (1979).
[125] M. L. Kaiser, S. R. Koirtyohann, E. J. Hinderberger and H. E. Taylor, *Spectrochim. Acta*, **36B**, 773 (1981).
[126] C. Bruhn and G. Navarrete, *Anal. Chim. Acta*, **130**, 209 (1981).
[127] K. R. Sperling, *Z. Anal. Chem.*, **287**, 23 (1977).
[128] T. W. May and W. G. Brumbough, *Anal. Chem.*, **54**, 1032 (1982).
[129] W. Slavin, G. R. Cornick and D. C. Manning, *Pittsburgh Conference on Analytical Chemistry and Applied Spectroscopy*, Atlantic City, N.J., March 1982.
[130] K. C. Thompson and D. R. Thomerson, *Atom. Absorp. Newslett.*, **15**, 122 (1976).
[131] K. C. Thompson and D. R. Thomerson, *Analyst*, **99**, 595 (1974).
[132] E. S. Gladney, *Atom. Absorp. Newslett.*, **16**, 114 (1977).
[133] D. Chakraborti, W. De Jonghe and F. Adams, *Anal. Chim. Acta*, **119**, 331 (1980).
[134] R. B. Denyszyn, P. M. Groshe and D. E. Wagoner, *Anal. Chem.*, **50**, 1094 (1978).
[135] Y. Odanaka, O. Matano and S. Goto, *Bunseki Kagaku*, **28**, 517 (1979).
[136] R. E. Ediger, *Atom. Absorp. Newslett.*, **14**, 127 (1975).
[137] H. Freeman, J. F. Uthe and B. Flemming, *Atom. Absorp. Newslett.*, **15**, 49 (1076).
[138] M. S. Vigler, A. Strecker and A. Varnes, *Appl. Spectrosc.*, **32**, 60 (1978).
[139] D. J. Driscoll, D. A. Clay, C. H. Rogers, R. H. Jungers and F. E. Butler, *Anal. Chem.*, **50**, 767 (1978).
[140] F. J. Slikkerveer, A. A. Braad and P. W. Hendrikse, *Atom. Spectrosc.*, **1**, 30 (1980).
[141] A. Le Bihan and J. Courtot-Coupez, *Analusis*, **3**, 59 (1975).
[142] R. Guevremont, R. E. Sturgeon and S. S. Berman, *Anal. Chim. Acta*, **115**, 163 (1980).
[143] R. Guevremont, *Anal. Chem.*, **52**, 1574 (1980).
[144] M. Hoenig, R. Vanderstappen and P. Van Hoeyweghen, *Analusis*, **7**, 17 (1979).
[145] J. Komárek and L. Sommer, *Talanta*, **29**, 159 (1982).
[146] M. W. Routh and M. B. Denton, *Appl. Spectrosc.*, **30**, 344 (1976).
[147] I. Rubeška, *Chem. Anal. (Warsaw)*, **22**, 403 (1977).
[148] I. Rubeška, *Anal. Chem.*, **48**, 1640 (1976).
[149] J. Komárek, D. Kolčava and L. Sommer, *Collection Czech. Chem. Commun.*, **45**, 3313 (1980).
[150] D. C. Hilderbrand and E. E. Pickett, *Anal. Chem.*, **47**, 424 (1975).
[151] P. B. Adams and W. O. Passmore, *Anal. Chem.*, **38**, 630 (1966).
[152] T. C. Rains, H. E. Zittel and M. Ferguson, *Talanta*, **10**, 367 (1963).
[153] T. C. Rains, H. E. Zittel and M. Ferguson, *Anal. Chem.*, **34**, 778 (1962).
[154] T. Maruta, T. Takeuchi and M. Suzuki, *Anal. Chim. Acta*, **58**, 452 (1972).
[155] B. Fleet, K. V. Liberty and T. S. West, *Talanta*, **17**, 203 (1970).
[156] J. M. Ottaway and N. K. Pradhan, *Talanta*, **20**, 927 (1973).
[157] M. Suzuki, K. Hayashi and W. E. C. Wacker, *Anal. Chim. Acta*, **104**, 389 (1979).
[158] J. M. Ottaway, D. T. Coker, W. B. Rowston and D. R. Bhattarai, *Analyst*, **95**, 567 (1970).
[159] F. J. Wallace, *Analyst*, **88**, 259 (1963).
[160] M. Yanagisawa, M. Suzuki and T. Takenshi, *Talanta*, **14**, 933 (1967).
[161] S. Tardon and M. Balcárková, *Chem. Listy*, **60**, 334 (1966).
[162] J. Komárek, J. Jambor and L. Sommer, *Z. Anal. Chem.*, **262**, 91 (1972).
[163] I. Rubeška and B. Moldan, *Acta Chim. Acad. Sci. Hung.*, **44**, 367 (1965).
[164] I. Rubeška and B. Moldan, *Anal. Chim. Acta*, **37**, 421 (1967).
[165] G. B. Marshall and T. S. West, *Talanta*, **14**, 823 (1967).

[166] J. Komárek, J. Jambor and L. Sommer, *Scr. Fac. Sci. Nat. UJEP Brunensis Chemia*, 2, 11 (1972).
[167] A. C. West and W. D. Cooke, *Anal. Chem.*, 32, 1471 (1960).
[168] T. V. Ramakrishna, J. W. Robinson and P. W. West, *Anal. Chim. Acta*, 36, 57 (1966).
[169] I. Janoušek and M. Malát, *Anal. Chim. Acta*, 58, 448 (1972).
[170] T. V. Ramakrishna, P. W. West and J. W. Robinson, *Anal. Chim. Acta*, 40, 347 (1968).
[171] D. Halířová and J. Musil, *Hutnicke Listy*, 27, 888 (1972).
[172] I. Rubeška and J. Korečková, *Chem. Listy*, 73, 1009 (1979).
[173] C. W. Fuller, *Atom. Absorp. Newslett.*, 16, 106 (1977).
[174] J. G. T. Regan and J. Warren, *Atom. Absorp. Newslett.*, 17, 89 (1978).
[175] A. Šefflová and J. Komárek, *Chem. Listy*, 74, 971 (1980).
[176] D. J. Hydes, *Anal. Chem.*, 52, 959 (1980).
[177] C. Rocchiccioli and A. Townshend, *Anal. Chim. Acta*, 41, 93 (1968).
[178] A. Šmrhová, J. Janáček, M. Tomášová and J. Komárek, *Hutnicke Listy*, 31, 668 (1976).
[179] J. Komárek, M. Vrchlabský and L. Sommer, *Z. Anal. Chem.*, 278, 121 (1976).
[180] J. Komárek, V. Mahr and L. Sommer, *Collection Czech. Chem. Commun.*, 46, 708 (1981).
[181] J. Komárek, V. Mahr and L. Sommer, *Chem. Listy*, 75, 877 (1981).
[182] F. R. Hartlage, Jr., *Anal. Chim. Acta*, 39, 273 (1967).
[183] A. M. Bond and J. B. Willis, *Anal. Chem.*, 40, 2087 (1968).
[184] N. Nakahara, M. Munemori and S. Musha, *Bull. Chem. Soc. Japan*, 46, 1172 (1973).
[185] V. K. Panday, *Anal. Chim. Acta*, 57, 31 (1971).
[186] M. Kodama, S. Shimizu, M. Sato and T. Tominaga, *Anal. Lett.*, 10, 591 (1977).
[187] M. Kodama and S. Miyagawa, *Anal. Chem.*, 52, 2358 (1980).
[188] J. Aggett and T. S. West, *Anal. Chim. Acta*, 57, 15 (1971).
[189] J. E. Allan, *Spectrochim. Acta*, 17, 467 (1961).
[190] A. J. Lemonds and B. E. McClellan, *Anal. Chem.*, 45, 1455 (1973).
[191] J. L. Aznarez-Alduan and J. R. Gastillo-Suarez, *An. Quim.*, 73, 699 (1977).
[192] B. D. Balraadjsing, *Commun. Soil. Sci. Plant Anal.*, 5, 25 (1974).
[193] T. Goto, *Bunseki Kagaku*, 23, 1165 (1974).
[194] T. Goto and S. Ginba, *Bunseki Kagaku*, 23, 517 (1974).
[195] C. H. Kim, P. W. Alexander and L. E. Smythe, *Talanta*, 22, 739 (1975).
[196] D. A. Tinsley and R. Iddon, *Talanta*, 21, 633 (1974).
[197] J. D. Kinrade and J. C. VanLoon, *Anal. Chem.*, 46, 1894 (1974).
[198] K. Kremling and H. Petersen, *Anal. Chim. Acta*, 70, 35 (1974).
[199] F. D. Pierce, M. J. Gortatowski, H. D. Mecham and R. S. Fraser, *Anal. Chem.*, 47, 1132 (1975).
[200] H. Sakurai, *Bunseki Kagaku*, 24, 52 (1975).
[201] S. A. Popova, L. Bezur and E. Pungor, *Z. Anal. Chem.*, 271, 269 (1974).
[202] S. S. Leitner, *Anal. Chim. Acta*, 74, 133 (1975).
[203] Y. Yamamoto and T. Kamada, *Bunseki Kagaku*, 25, 567 (1976).
[204] Y. Akama, T. Nakai and F. Kawamura, *Bunseki Kagaku*, 25, 496 (1976).
[205] S. Nishigaki, Y. Tamura, T. Miki, H. Yamada, K. Toba, Y. Shimamura and Y. Kimura, *Tokyo Toritsu Eisei Kenkyusho Kenkyo Nempo*, 24, 231 (1972).
[206] E. A. Childs and J. N. Gaffke, *J. Assoc. Off. Anal. Chem.*, 57, 360 (1974).
[207] D. A. Shearer, R. C. Cloutier and M. Kidiroglou, *J. Assoc. Off. Anal. Chem.*, 60, 155 (1977).
[208] I. A. Brovko, Sh. N. Nazarov and K. A. Rish, *Zh. Analit. Khim.*, 29, 2387 (1974).
[209] S. Sukiman, *Anal. Chim. Acta*, 84, 419 (1976).
[210] N. Lekehal, M. Hanocq and M. Helson-Cambier, *J. Pharm. Belg.*, 32, 76 (1977).
[211] H. C. Green, *Analyst*, 100, 640 (1975).

[212] H. Armannson, *Anal. Chim. Acta*, 88, 89 (1977).

[213] A. M. Ure and M. C. Mitchell, *Anal. Chim. Acta*, 87, 283 (1976).

[214] K. Hiro, T. Tanaka and A. Kawahara, *Osaka Kogyo Gijutsu Shkiensho Kiho*, 25, 187 (1974).

[215] K. Matsusaki, *Bunseki Kagaku*, 24, 442 (1974).

[216] N. S. Kim and B. T. Li, *Punsok Hwahak*, 14, 79 (1976).

[217] V. M. Byr'ko, L. F. Prishchepov and I. A. Shikheeva, *Zavodsk. Lab.*, 41, 525 (1975).

[218] N. I. Tarasevich, G. V. Kozyreva and Z. P. Portugal'skaya, *Vestn. Mosk. Univ., Khim.*, 16, 241 (1975).

[219] A. Mizuike, M. Hiraido and T. Suzuki, *Bunseki Kagaku*, 26, 72 (1977).

[220] T. V. Lanbina, I. G. Yudelevich, A. A. Vasil'eva, L. M. Gindin and T. Z. Almanova, *Izv. Sib. Otd. Akad. Nauk SSSR, Ser. Khim. Nauk, 1974*, 83.

[221] A. A. Vasil'eva, I. G. Yudelevich, L. M. Gindin, T. V. Lanbina, R. S. Shulman, I. L. Kotlarevskii and V. N. Andrievskii, *Talanta*, 22, 745 (1975).

[222] M. Aihara and M. Kiboku, *Bunseki Kagaku*, 23, 505 (1974); 24, 447 (1975).

[223] K. Lee and E. Jacob, *Mikrochim. Acta*, 65 (1974).

[224] J. Komárek, J. Horák and L. Sommer, *Collection Czech. Chem. Commun.*, 39, 92 (1974).

[225] T. V. Lanbina, I. G. Yudelevich, A. A. Vasil'eva and L. M. Gindin, *Izv. Sib. Otd. Akad. Nauk, SSSR, Ser. Khim. Nauk, 1974*, 915.

[226] G. J. Worrell, T. J. Vickers and F. D. Williams, *Anal. Chim. Acta*, 75, 453 (1975).

[227] S. De Moraes, M. Cipriani and A. Abrao, *Publ. IEA*, No. 406 (1976); *Chem. Abstr.*, 85, 136836 (1976).

[228] S. De Moraes and A. Abrao, *Anal. Chem.*, 46, 1812 (1974).

[229] K. Thornton and K. E. Burke, *Analyst*, 99, 469 (1974).

[230] K. Mizuno, T. Suzuki and Y. Miyagawa, *Nagoya-Shi Kogyo Kenkyusho Kentyu Hokoku, 1974*, 34.

[231] E. P. Welsch and T. T. Chao, *Anal. Chim. Acta*, 76, 65 (1975).

[232] B. L. Serebryani, N. L. Fishkova, O. M. Petrukhin and E. E. Rakovskii, *J. Anal. Chem. (USSR)*, 28, 2070 (1974).

[233] K. Fukamachi, N. Furuta, M. Yanagawa and M. Morimoto, *Bunseki Kagaku*, 23, 187 (1974).

[234] D. Voyce and H. Zeitlin, *Anal. Chim. Acta*, 69, 27 (1974).

[235] N. Ichinose, *Anal. Chim. Acta*, 70, 222 (1974).

[236] N. Ichinose, *Bunseki Kagaku*, 23, 348 (1974).

[237] O. M. Talapova and I. S. Levin, *Zh. Khim. Ref.*, 19GD, Abstract No. 15G 148 (1981).

[238] N. T. Voskrenenskaya, N. F. Pchelintseva and T. I. Tsekhonya, *Zh. Analit. Khim.*, 36, 667 (1981).

[239] M. Aihara and M. Kiboku, *Bunseki Kagaku*, 30, 394 (1981).

[240] I. G. Yudelevich and E. A. Startseva, *Zavodsk. Lab.*, 47, 24 (1981).

[241] Y. Akama, T. Ishii, T. Nakai and F. Kawawura, *Bunseki Kagaku*, 28, 196 (1979).

[242] Y. Sasaki and M. Kawae, *Bunseki Kagaku*, 30, 577 (1981).

[243] M. El-Shaarawy, *Clin. Lab. (Rome)*, 4, 206 (1980).

[244] K. Yasuda, S. Toda, C. Igarashi and S. Tamura, *Anal. Chem.*, 51, 161 (1979).

[245] J. A. Nichols and R. Woodriff, *J. Assoc. Off. Anal. Chem.*, 63, 500 (1980).

[246] C. Fairless and A. J. Bard, *Anal. Lett.*, 5, 433 (1972).

[247] Y. Thomassen, B. V. Larson, F. J. Langmyhr and W. Lund, *Anal. Chim. Acta*, 83, 103 (1976).

[248] G. Torsi, *Ann. Chim. (Rome)*, 67, 557 (1977).

[249] G. E. Batley and J. P. Matousek, *Anal. Chem.*, 49, 203 (1977); 52, 1570 (1980).

[250] D. A. Frick and D. E. Tallman, *Anal. Chem.*, 54, 1217 (1982).

[251] Zs. Horváth, K. Falb and M. Varju, *Atom. Absorp. Newslett.*, 16, 152 (1977).

[251a] J. Minczewski, J. Chwastowska and R. Dybczyński, *Separation and Preconcentration Methods in Inorganic Trace Analysis*, Horwood, Chichester, 1982.

[252] R. Dunk, R. A. Mostyn and H. C. Hoare, *Atom. Absorp. Newslett.*, 8, 79 (1969).

[253] R. Menache, *Clin. Chem.*, 20, 1444 (1974).

[254] T. Kumamaru, *Anal. Chim. Acta*, 43, 19 (1968)

[255] T. Kumamaru, E. Tao, N. Okamoto and Y. Yamamoto, *Bull. Chem. Soc. Japan*, 38, 2204 (1965).

[256] W. S. Zaugg and R. J. Knox, *Anal. Chem.*, 38, 1759 (1966).

[257] S. J. Simon and D. F. Boltz, *Microchem. J.*, 20, 468 (1975).

[258] E. Jungreis and F. Ain, *Anal. Chim. Acta*, 88, 191 (1977).

[259] S. S. M. Hassan and F. Tadros, *Microchem. J.*, 28, 20 (1983).

[260] P. T. Gilbert, *Anal. Chem.*, 38, 1920 (1966).

[261] M. Maruyama, S. Seno and K. Hasegawa, *Z. Anal. Chem.*, 307, 21 (1981).

[262] D. F. Tomkins and C. W. Frank, *Anal. Chem.*, 46, 1187 (1974).

[263] G. D. Christian and F. J. Feldman, *Anal. Chim. Acta*, 40, 173 (1968).

[264] P. J. Oles and S. Siggia, *Anal. Chem.*, 46, 2197 (1974).

[265] Y. Kidani and E. Ito, *Bunseki Kagaku*, 25, 57 (1976).

[266] A. M. Bond and T. A. O'Donnell, *Anal. Chem.*, 40, 560 (1968).

[267] M. D. Amos and J. B. Willis, *Spectrochim. Acta*, 22, 1325 (1966).

[268] A. M. Bond and J. B. Willis, *Anal. Chem.*, 40, 2087 (1968).

[269] W. Geilmann, *Z. Anal. Chem.*, 160, 410 (1958).

[270] W. Geilmann and K. H. Neeb, *Z. Anal. Chem.*, 165, 251 (1959).

[271] H. Heinrichs, *Z. Anal. Chem.*, 294, 345 (1979).

[272] J. Erzinger and H. Puchelt, *Geostandards Newslett.*, 4, 13 (1980).

[273] A. Meyer, Ch. Hofer and G. Tölg, *Z. Anal. Chem.*, 290, 292 (1978).

[274] G. Tölg, *Talanta*, 21, 327 (1974).

[275] H. Heinrichs and H. Keltsch, *Anal. Chem.*, 54, 1211 (1982).

[276] H. Brandenberger and H. Bader, *Helv. Chim. Acta*, 50, 1409 (1967).

[277] H. Brandenberger and H. Bader, *Atom. Absorp. Newslett.*, 7, 53 (1968).

[278] W. R. Hatch and W. L. Ott, *Anal. Chem.*, 40, 2085 (1968).

[279] D. C. Stuart, *Anal. Chim. Acta*, 101, 429 (1978); 106, 411 (1979).

[280] W. R. Simpson and G. Nickless, *Analyst*, 102, 86 (1977).

[281] N. S. Poluektov, R. A. Vitkun and Y. V. Zelyukova, *Zh. Analit. Khim.*, 19, 937 (1964).

[282] J. V. O'Gorman, N. H. Suhr and P. L. Walker, *Appl. Spectrosc.*, 26, 44 (1972).

[283] M. L. Kokot, *Minerals Sci. Eng.*, 6, 236 (1974).

[284] J. W. Wimberley, *Anal. Chim. Acta*, 76, 337 (1975).

[285] A. Bouchard, *Atom. Absorp. Newslett.*, 12, 115 (1973).

[286] G. Tuncel and O. Y. Ataman, *Atom. Spectrosc.*, 1, 126 (1980).

[287] S.-L. Tong, C. K. Chu and S. H. Goh, *Mikrochim. Acta*, 99 (1981 I).

[288] A. M. Ure, *Anal. Chim. Acta*, 76, 1 (1975).

[289] J. E. Hawley and J. D. Ingle, Jr., *Anal. Chem.*, 47, 719 (1975).

[290] S. Chilov, *Talanta*, 22, 205 (1975).

[291] C. E. Oda and J. D. Ingle, Jr., *Anal. Chem.*, 53, 2305 (1981).

[292] J. Murphy, *Atom. Absorp. Newslett.*, 14, 151 (1975).

[293] D. Littlejohn, G. S. Fell and J. M. Ottaway, *Clin. Chem.*, 22, 1719 (1976).

[294] I. Share, *Analyst*, 97, 184 (1972).

[295] J. H. Hwang, P. A. Ullucci and A. L. Malenfant, *Can. Spectrosc.*, 16, 100 (1971).

[296] N. P. Kubasik, H. E. Sine and M. T. Volosin, *Clin. Chem.*, 18, 1326 (1972).

[297] U. Ebbestad, N. Gundersen and T. A. Torgrimsen, *Atom. Absorp. Newslett.*, 14, 142 (1975).

[298] C. E. Oda and J. D. Ingle, Jr., *Anal. Chem.*, 53, 2030 (1981).

[299] P. Coyle and T. Hartley, *Anal. Chem.*, 53, 354 (1981).

[300] R. A. Richardson, *Clin. Chem.*, **22**, 1604 (1976).
[301] A. A. El-Awady, R. B. Miller and M. J. Carter, *Anal. Chem.*, **48**, 110 (1976).
[302] G. Torsi, E. Desimoni, F. Palmisano and L. Sabbatini, *Analyst*, **107**, 96 (1982).
[303] D. R. Roden and D. E. Tallman, *Anal. Chem.*, **54**, 307 (1982).
[304] T. Nakahara, S. Kobayashi and S. Musha, *Anal. Chim. Acta*, **101**, 375 (1978).
[305] R. Smith, *Atom. Spectrosc.*, **2**, 155 (1981).
[306] T. Kubota and T. Ueda, *Bunseki Kagaku*, **27**, 692 (1978).
[307] M. O. Andreae and P. N. Froelich, Jr., *Anal. Chem.*, **53**, 287 (1981).
[308] L. Ebdon, J. R. Wilkinson and K. W. Jackson, *Anal. Chim. Acta*, **136**, 191 (1982).
[309] R. G. Godden and D. R. Thomerson, *Analyst*, **105**, 1137 (1980).
[310] D. S. Lee, *Anal. Chem.*, **54**, 1182 (1982).
[311] T. W. Brueggemeyer and J. A. Caruso, *Anal. Chem.*, **54**, 872 (1982).
[312] D. C. Manning, *Atom. Absorp. Newslett.*, **14**, 99 (1975).
[313] T. Uchida, I. Kojima and I. Chuzo, *Bunseki Kagaku*, **27**, T 44 (1978).
[314] C. Fry, S. J. Northway and M. B. Denton, *Anal. Chem.*, **50**, 1719 (1978).
[315] E. Sebastiani, K. Ohls and G. Riemer, *Z. Anal. Chem.*, **264**, 105 (1973).
[316] M. S. Cresser, *Prog. Anal. Atom. Spectrosc.*, **4**, 219 (1981).
[317] J. M. Malloy, P. N. Keliher and M. S. Cresser, *Spectrochim. Acta*, **35B**, 833 (1980).
[318] T. Makino and K. Takahara, *Clin. Chem.*, **27**, 1445 (1981).
[319] T. Uchida, I. Kojima and C. Iida, *Anal. Chim. Acta*, **116**, 205 (1980).
[320] P. E. Wilson, *Atom. Absorp. Newslett.*, **18**, 115 (1979).
[321] H. Berndt and J. Messerschmidt, *Spectrochim. Acta*, **36B**, 809 (1981).
[322] H. Berndt and J. Messerschmidt, *Anal. Chim. Acta*, **136**, 407 (1982).
[323] K. Fuwa and B. L. Vallee, *Anal. Chem.*, **35**, 942 (1963).
[324] K. Fuwa, P. Pulido, R. McKay and B. L. Vallee, *Anal. Chem.*, **36**, 2407 (1964).
[325] E. J. Agazzi, *Anal. Chem.*, **37**, 365 (1965).
[326] C. L. Chakrabarti, J. W. Robinson and T. S. West, *Anal. Chim. Acta*, **34**, 269 (1966).
[327] T. V. Ramakrishna, J. W. Robinson and P. W. West, *Anal. Chim. Acta*, **37**, 20 (1967).
[328] I. Rubeŝká, *Anal. Chim. Acta*, **40**, 187 (1968).
[329] J. W. Robinson, *Anal. Chim. Acta*, **27**, 456 (1962).
[330] Y. V. Zelyukova and N. S. Poluektov, *Zh. Analit. Khim.*, **18**, 435 (1963).
[331] S. R. Koirtyohann and E. E. Pickett, *Anal. Chem.*, **37**, 601 (1965).
[332] I. Rubeŝká and J. Ŝtupar, *Atom. Absorp. Newslett.*, **5**, 69 (1966).
[333] D. N. Hingle, G. Kirkbright and T. S. West, *Talanta*, **15**, 199 (1968).
[334] I. Rubeŝká and B. Moldan, *Appl. Op.*, **7**, 1341 (1968).
[335] I. Rubeŝká and B. Moldan, *Analyst*, **93**, 148 (1968).
[336] L. R. Lavman, J. G. Crock and F. E. Lichte, *Anal. Chem.*, **53**, 747 (1981).
[337] K. K. Stewart, G. R. Beecher and P. E. Hare, *Fed. Proc. Fed. Am. Soc. Exp. Biol.*, **33**, 1429 (1974).
[338] J. Rŭžička and E. H. Hansen, *Anal. Chim. Acta*, **78**, 145 (1975).
[339] J. F. Tyson and A. B. Idris, *Analyst*, **106**, 1125 (1981).
[340] J. F. Tyson, *Anal. Proc.*, **18**, 542 (1981).
[341] B. D. Mindel and B. Karlberg, *Lab. Pract.*, **30**, 719 (1981).
[342] W. R. Wolf and K. K. Stewart, *Anal. Chem.*, **51**, 1201 (1979).
[343] B. W. Renoe and A. O. O'Brien, *Clin. Chem.*, **26**, 1021 (1980).
[344] B. F. Rocks, R. A. Sherwood and C. Riley, *Clin. Chem.*, **28**, 440 (1982).
[345] E. A. G. Zagatto, F. J. Krug, H. Bergamin, S. S. Jorgensen and B. F. Reis, *Anal. Chim. Acta*, **104**, 279 (1979).
[346] W. D. Basson and J. F. Van Staden, *Z. Anal. Chem.*, **302**, 370 (1980).
[347] K. Fukamachi and N. Ishibashi, *Anal. Chim. Acta*, **119**, 383 (1980).
[348] B. W. Renoe, C. E. Shideler and J. Savory, *Clin. Chem.*, **27**, 1546 (1981).
[349] F. J. Langmyhr, *Analyst*, **104**, 993 (1979).

[350] J. C. VanLoon, *Anal. Chem.*, **52**, 955A (1980).

[351] J. B. Headridge, *Spectrochim. Acta*, **35B**, 785 (1980).

[352] L. Favretto, G. P. Marletta and F. G. Favretto, *Mikrochim. Acta*, 387 (1981 I).

[353] K. W. Jackson, L. Ebdon, D. C. Webb and A. G. Cox, *Anal. Chim. Acta*, **128**, 67 (1981).

[354] P.-O. Berggren, *Anal. Chim. Acta*, **119**, 161 (1980).

[355] F. J. Langmyhr and S. Orre, *Anal. Chim. Acta*, **118**, 307 (1980).

[356] C. L. Chakrabarti, C. C. Wan and W. C. Li, *Spectrochim. Acta*, **35B**, 547 (1980); **35B**, 93 (1980).

[357] U. Kurfuerst and K. H. Grobecker, *Labor Praxis*, **5**, 28 (1981).

[358] E. D. Prudnikov, *Zh. Prikl. Spektrosk.*, **17**, 352 (1972).

[359] U. Ulfvarson, *Acta Chem. Scand.*, **21**, 641 (1967).

[360] H. Brandenburger and H. Bader, *Helv. Chim. Acta*, **50**, 1409 (1967).

[361] H. Brandenburger, *Chimia*, **22**, 449 (1968).

[362] H. Brandenburger and H. Bader, *Atom. Absorp. Newslett.*, **6**, 101 (1967); **7**, 53 (1968).

[363] H. T. Delves, *Analyst*, **95**, 431 (1970).

[364] M. Kahl, D. G. Mitchell, G. L. Kaufman and K. M. Aldous, *Anal. Chim. Acta*, **87**, 215 (1976).

[365] J. D. Kerber and F. J. Fernandez, *Atom. Absorp. Newslett.*, **10**, 78 (1971).

[366] M. M. Joselow and J. D. Bogden, *Atom. Absorp. Newslett.*, **11**, 127 (1972).

[367] D. G. Mitchell, A. F. Ward and M. Kahl, *Anal. Chim. Acta*, **76**, 456 (1975).

[368] H. L. Kahn, F. J. Fernandez and S. Slavin, *Atom. Absorp. Newslett.*, **11**, 42 (1972).

[369] M. M. Joselow and J. D. Bogden, *Am. J. Public Health*, **64**, 238 (1974).

[370] O. W. Law and K. L. Li, *Analyst*, **100**, 430 (1975).

[371] L. Favretto, *Atom. Absorp. Newslett.*, **15**, 98 (1976).

[372] G. L. Favretto, M. G. Pertoldi and L. Favretto, *Atom. Absorp. Newslett.*, **16**, 4 (1977).

[373] D. G. Mitchell, W. N. Mills, A. F. Ward and K. M. Aldous, *Anal. Chim. Acta*, **90**, 275 (1977).

[374] G. F. Carter, *Br. J. Ind. Med.*, **35**, 235 (1978).

[375] K. M. Aldous, D. G. Mitchell and F. J. Ryan, *Anal. Chem.*, **45**, 1990 (1973).

[376] D. G. Pachuta and L. J. C. Love, *Anal. Chem.*, **52**, 438 (1980).

[377] R. Bye, *Z. Anal. Chem.*, **306**, 30 (1981).

[378] D. G. Pachuta and L. J. C. Love, *Anal. Chem.*, **52**, 438 (1980); **52**, 444 (1980).

[378a] Z. Grobenski, R. Lehmann and B. Welz, *Pittsburgh Conference on Anal. Chem. and Appl. Spectrosc.*, Atlantic City, N.J., 1981.

[379] H. G. C. Human, R. H. Scott, A. R. Oakes and C. D. West, *Analyst*, **101**, 265 (1976).

[380] Y. I. Belyaev, L. M. Ivantsov, A. V. Karyakin, Pham Hung Phi and V. V. Shemet, *Zh. Analit. Khim.*, **23**, 508 (1968).

[381] Y. I. Belyaev and V. V. Gordeev, *Oceanology*, **12**, 756 (1972).

[382] V. D. Malykh, V. I. Men'shikov, V. N. Morozov and S. A. Shipitsyn, *Zh. Prikl. Spektrosk.*, **16**, 12 (1972).

[383] T. Kántor, L. Pólos, P. Fodor and E. Pungor, *Talanta*, **23**, 585 (1976).

[384] V. G. Mossotti, K. Laqua and W.-D. Hagenah, *Spectrochim. Acta*, **23B**, 197 (1967).

[385] A. V. Karyakin and V. A. Kaigorodov, *Zh. Analit. Khim.*, **23**, 930 (1968).

[386] E. P. Krivchikova and V. S. Demin, *Zh. Prikl. Spektrosk.*, **14**, 592 (1971).

[387] G. I. Nikolaev and V. I. Podgornaya, *Zh. Prikl. Spektrosk.*, **16**, 911 (1972).

[388] A. V. Karyakin, A. M. Pchelintsev, A. I. Shidlovskii, E. K. Vul'fson and M. N. Tsingarelli, *Zh. Prikl. Spektrosk.*, **18**, 610 (1973).

[389] E. K. Vul'fson, A. V. Karyakin and A. I. Shidlovskii, *Zh. Analit. Khim.*, **28**, 1253 (1973).

[390] D. E. Osten and E. H. Piepmeier, *Appl. Spectrosc.*, **27**, 165 (1973).

[391] E. K. Vul'fson, A. V. Karyakin and A. I. Shidlovskii, *Zavodsk. Lab.*, **40**, 945 (1974).

[392] E. K. Vul'fson, I. F. Gribovskaya, A. V. Karyakin and A. F. Yanushkevich, *Zh. Analit. Khim.*, **30**, 1913 (1975).

[393] J. P. Matoušek and B. J. Orr, *Spectrochim. Acta*, **31B**, 475 (1976).

[394] T. Ishizuka, Y. Uwamino and H. Sunabara, *Anal. Chem.*, **49**, 1339 (1977).

[395] M. A. Coudert and J. M. Vergnaud, *Anal. Chem.*, **42**, 1303 (1970).

[396] K. Govindaraju, G. Mevelle and C. Chouard, *Chem. Geol.*, **8**, 131 (1971).

[397] K. Govindaraju, G. Mevelle and C. Chouard, *Bull. Soc. Fr. Céram.*, **96**, 47 (1972).

[398] K. Govindaraju, R. Herman, G. Mevelle and C. Chouard, *Atom. Absorp. Newslett.*, **12**, 73 (1973).

[399] K. Govindaraju, G. Mevelle and C. Chouard, *Anal. Chem.*, **46**, 1672 (1974).

[400] K. Govindaraju, *Analusis*, **3**, 164 (1975).

[401] A. A. Venghiattis, *Atom. Absorp. Newslett.*, **6**, 19 (1967).

[402] A. A. Venghiattis and L. Whitlock, *Atom. Absorp. Newslett.*, **6**, 135 (1967).

[403] A. A. Venghiattis, *Spectrochim. Acta*, **23B**, 67 (1967).

[404] P. Miaud and J. Robin, *Bull. Soc. Chim. France*, 854 (1968).

[405] M. Kadhiki and S. Oshima, *Anal. Chim. Acta*, **51**, 387 (1970).

[406] M. Ghiglione, E. Eljuri and C. Cuevas, *Appl. Spectrosc.*, **30**, 320 (1976).

[407] W. W. Harrison and P. O. Juliano, *Anal. Chem.*, **43**, 248 (1971).

[408] C. W. Fuller, *Analyst*, **101**, 961 (1976).

[409] C. W. Fuller and I. Thompson, *Analyst*, **102**, 141 (1977).

[410] R. S. Babington, *Popular Science*, No. 5, 102 (1973).

[411] W. W. Harrison and P. O. Juliano, *Anal. Chem.*, **43**, 248 (1971).

[412] P. T. Gilbert, *Anal. Chem.*, **34**, 1025 (1962).

[413] V. I. Lebedev, *Zh. Analit. Khim.*, **24**, 337 (1969).

[414] R. C. Fry and M. B. Denton, *Anal. Chem.*, **49**, 1413 (1977).

[415] R. C. Fry and M. B. Denton, *Appl. Spectrosc.*, **33**, 393 (1979).

[416] J. E. O.Reilly and D. G. Hicks, *Anal. Chem.*, **51**, 1905 (1979).

[417] J. Ramírez-Muñoz, M. E. Roth and W. F. Ulrich, *Pittsburgh Conference on Analytical Chemistry and Applied Spectroscopy*, Cleveland, 1969.

[418] N. Mohamed and R. C. Fry, *Anal. Chem.*, **53**, 450 (1981).

4

Nitrogen compounds

4.1 TOTAL NITROGEN

Determination of nitrogen in organic compounds is one of the most widely used analyses in many laboratories. Flame spectrometry has been suggested for the determination of many organonitrogen compounds. The earlier methods are based on the measurement of either the emission by the CN band at 388.3 nm, with that of the CH band at 389 nm as internal standard, or the emission by the NH band at 336 nm in a hydrogen diffusion flame [1,2]. The intensity of the CN signal is proportional to the concentration of nitrogen present in the atomized solution in the inner cone of the flame. Nanogram levels of amines have been determined by these methods.

Methods utilizing atomic or molecular absorption spectrometry have been suggested for determining organonitrogen compounds after their conversion into ammonia. The Kjeldahl method is the method of choice for converting nitrogen into ammonia. This involves digestion with sulphuric acid at about 370°C for 1 hr in the presence of mercury (or less effectively selenium) as catalyst [3-7]. In the general procedure a 2-5 mg sample of the nitrogen compound is transferred to the bottom of the digestion flask and mixed with 10 mg of selenium powder, 40 mg of copper sulphate-potassium sulphate (1:1) mixture, and 1 ml of concentrated sulphuric acid. The mixture is then boiled over a flame until digestion is complete. The nitrogen is converted into ammonium bisulphate, from which ammonia is quantitatively released.

However, determination of ammonia by AAS cannot be accomplished directly, because the primary resonance line of nitrogen lies in the vacuum ultraviolet region. Ammonia is, however, indirectly determined by measuring its enhancement effect on the atomic absorption of some metal ions, or directly by monitoring its molecular absorption in the flame. It has been reported that ammonia and other nitrogenous compounds enhance the absorption of zirconium atoms in the nitrous oxide-acetylene flame [8]. The degree of enhancement seems to be dependent on the availability of the unshared electron pair of the nitrogen atom. This effect may be due to the fact that zirconium ions, even in the presence of $2M$ hydrochloric acid, exist in an oxyzirconium polymeric form [9] such as $[Zr_4(OH)_8]^{8+}$. In the nitrous oxide-acetylene flame, these species give ZrO_2 (m.p. $\sim 2700°C$), which is difficult to dissociate. In the presence of ammonia or an amine, however, monomeric zirconium adducts are formed [10] which are more efficiently atomized than ZrO_2 in the flame. The atomic absorption of zirconium at 360.1 nm linearly increases with increase in the concentration of nitrogenous compounds originally present, then flattens out to a plateau when the N:Zr molar ratio is greater than about 4.

For determining trace levels of ammonia, a reagent consisting of $5 \times 10^{-3}M$ solution in 0.1-$3M$ hydrochloric acid is used. Measurement of 10^{-4}-$10^{-2}M$ ammonia can also be performed with $0.2M$ zirconium solution. The results obtained with 10^{-4}-$10^{-2}M$ ammonia solutions show an average recovery in the range 90–117%. Methyl-, dimethyl-, trimethyl- and benzylamine, as well as piperidine, quinoline, guanidine, hydrazine and phenylhydrazine, enhance the absorption of zirconium equally as well as ammonia and are similarly determined. Aromatic amines and other nitrogenous compounds have a lower enhancement effect, but can be satisfactorily determined after conversion into ammonia by the Kjeldahl method. It should be noted that digestion of nitrogenous compounds that also contain fluorine releases fluoride ions along with ammonia, and both enhance the absorption of zirconium in the same way. To circumvent the effect of the fluoride ions, $0.01M$ fluoride solution is added to both the test sample and the standard solutions used for constructing the calibration graph. Sulphate and nitrate ions, if present in concentrations as high as 100 times that of the ammonia, cause depression of the absorption of zirconium. Such circumstances are rarely encountered with real organic compounds.

Although ammonia has a similar effect on the AAS of titanium and hafnium, few data are available on the determination of ammonia by use of these metals. A study of the atomic absorption of silver in an air-acetylene flame reveals remarkable enhancement of the silver absorption by the presence of ammonia [11]. This effect is pH-dependent and maximum enhancement takes place at pH 8.5–10.3. In this pH range, the enhancement is linearly related to the ammonia concentration up to 20 $\mu g/l$. Ammonium salt solutions containing 10–20 μg of ammonia per ml are analysed by aspiration of a 10-ml portion in the presence of 0.84 mg of sodium hydroxide and 12.8 μg of silver nitrate. The recovery of ammonia at this concentration level is 98–100% [11].

The co-precipitation of ammonium molybdophosphate with a known amount of thallium molybdophosphate, subsequent dissolution of the precipitate and measurement of the molybdenum provides the basis for another indirect AAS method for the determination of ammonia [12]. The molybdenum content of the precipitate, which is equivalent to both thallium and ammonia, is measured at 313.2 nm in an air-acetylene flame. The method suffers, however, from lack of selectivity, as many metal ions are precipitated by phosphomolybdate.

The well known characteristic ultraviolet absorption band of gaseous ammonia at 201 nm has been used for measurement. The solution is directly nebulized into the flame, and a hydrogen-continuum hollow-cathode lamp is used for excitation. Although a linear relationship between the concentration and absorption of ammonia can be obtained, this technique is not sensitive enough to have any practical analytical value. Displacement of ammonia from solution by alkali, passing the released gas through a 500-mm long cell and subsequent measurement of the molecular absorption of ammonia allows its detection at a level as low as 20 μg/ml, but with poor precision, probably because of adsorption and condensation of some droplets of the aqueous test solution at the end of the absorption cell. These difficulties have been avoided by using a condenser with steam passing through the outer jacket in place of cooling water [13,14]. The apparatus shown in Fig. 4.1 has been used for the measurement of 10–50 μg/ml ammonia-nitrogen levels. The ammonia solution ($\sim$ 2 ml) is treated with 40% sodium hydroxide solution ($\sim$ 20 ml) in the reaction vessel, and the ammonia gas released is swept with a stream of air ($\sim$ 100–200 ml/min) through the heated condenser. Light from a hollow-cathode lamp operated in the d.c. mode is focused with a 100-mm focal length silica lens so that a narrow beam passes through the centre of the condenser onto the entrance slit of the mono-chromator of the AAS instrument.

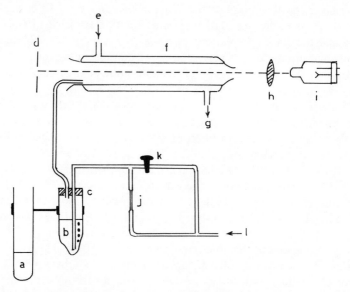

Fig. 4.1. Schematic representation of the apparatus used for the determination of ammonia: a, tube containing acid for flushing period; b, sample; c, centrifuge tube; d, monochromator entrance slit; e, steam inlet; f, condenser; g, steam outlet; h, silica lens; i, hollow-cathode lamp; j, capillary constriction; k, tap; l, air inlet [according to M. S. Cresser, *Anal. Chim. Acta,* **85**, 253 (1976); by permission of the copyright holders, Elsevier Scientific Publishing Co.].

The rate of ammonia evolution from the test solution, and in turn, the magnitude of the absorption maximum increases with increase in the sodium hydroxide concentration. On the other hand, the heat of neutralization when a concentrated alkali solution is added to the reaction mixture causes a noticeable fluctuation in the absorption signal, which necessitates temperature control. This is important in the determination of the ammonia-nitrogen in strongly acidic solutions such as those usually obtained by Kjeldahl digestion. However, application of the method to the determination of organonitrogen in soil and plant materials proved to be suitable, and results in good agreement with those obtained by ammonia distillation and acidimetric titration have been reported [13]. None of the reagents used in the digestion (i.e. perchloric acid, selenium, potassium sulphate, copper sulphate) had any interfering effect.

Muroski and Syty [15] described a similar method, based on the injection of the ammonium salt solution into a strong base, and measurement of the transient absorption signal of the ammonia gas at 194 nm, with an adapted AAS instrument. The burner head from the nebulizer–burner is removed and replaced with a 15-cm quartz-windowed flowthrough glass absorption cell supported by a holder fitted into the neck of the nebulizer–burner. The deuterium arc lamp normally used for background correction is used as the radiation source. A 60-ml reaction vessel similar to that previously devised for hydrogen sulphide measurement [16] has also been utilized (Fig. 4.2). Sodium hydroxide solution is intro-

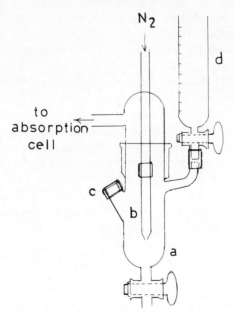

Fig. 4.2. Reaction vessel for evolution of ammonia; a, glass vessel; b, glass tube; c, injection port; d, burette. [Reprinted by permission from A. Syty, *Anal. Chem.*, **51**, 911 (1979). Copyright 1979 American Chemical Society.]

duced into the vessel from a burette, through a side-arm, and the ammonium salt solution is injected into the vessel by means of a 1-ml Hamilton syringe through a rubber septum that covers the injection port. A continuous flow of nitrogen is then introduced into the reaction vessel, through a plain glass tip submerged in the alkali solution, and the spent reagents are removed from the bottom of the vessel by a stopcock.

The sharpness and reproducibility of the ammonia absorption peaks depend on the volume and concentration of the alkali used, as well as on the flow-rate of the nitrogen. For optimum signals, 6 ml of $10N$ sodium hydroxide and a carrier-gas flow-rate of 0.4–2.4 l./min are used. Under these conditions, a linear relationship between absorption and concentration of ammonia up to 1000 μg/ ml is obtainable with a detection limit of about 1 μg/ml, without interference from nitrate, sulphate, sulphite, cyanide, sulphide, thiocyanate, bromide or iodide. The method is applicable to the determination of nitrogen in organic compounds and flour. The sample ($\sim$ 0.69 g) is digested for 2–3 hr with 10 g of potassium sulphate, several crystals of copper sulphate and 25 ml of concentrated sulphuric acid, followed by dilution to 250 ml and measurement of ammonia in a portion of the digest by either the standard addition or calibration method. The results compare well with those obtained by the commonly used distillation procedure. The detection limit for ammonia is 1 μg/ml ($\sim$ 0.03%N). The method permits measurement of 30 samples/hr, as the signal is displayed only 24 sec after sample injection. A similar method has been described, in which the sensitivity is increased by using a 100-cm absorption cell, a higher temperature for ammonia gas generation, and a nitrogen trap for separation of ammonia from water vapour [17]. However, these methods are conducted manually and involve several manipulation steps.

Vijan and Wood [18] described an automated method for determining ammonia and organonitrogen compounds, based on the gas-phase molecular absorption method of Cresser [13,14]. The automation is achieved by using an automatic sampler, a multichannel peristaltic pump, an adjustable heating bath and a gas–liquid separator, connected to an atomic-absorption spectrometer (Fig. 4.3). Instead of the transient signals generated by the manual method, the automated method produces steady-state signals and provides a stable base-line. The automated method is faster and more sensitive than the Cresser method [13,14] and its detection limit for ammonia is 0.26 μg/ml. Muroski and Syty's method [15] is faster, but less sensitive by a factor of 15.

4.2 AMINES

Aliphatic primary amines and diamines react with copper(II) or nickel(II) and salicylaldehyde or its derivatives, in the presence of triethanolamine as a proton acceptor, to form metal complexes [19]. These complexes are quantitatively formed within 10–60 min at room temperature, and measurement of their metal

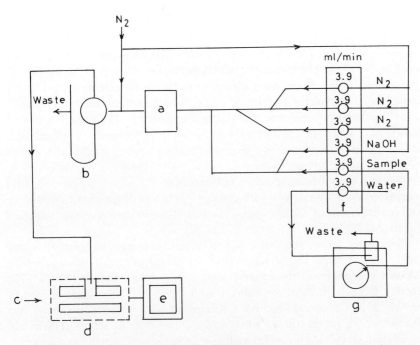

Fig. 4.3. Schematic diagram of AutoAnalyzer–AAS system for ammonia; a, heating bath; b, gas–liquid separator; c, light-beam; d, heated quartz cell in AAS instrument; e, recorder; f, proportioning pump; g, sampler. [Reprinted with permission from P. N. Vijan and G. R. Wood, *Anal. Chem.*, **53**, 1447 (1981). Copyright 1981 American Chemical Society.]

content provides an indirect determination of the amine. Extraction or precipitation of the chelates is necessary to separate the excess of unreacted metal.

$$2RNH_2 + 2 \text{ } \underset{OH}{\overset{CHO}{\bigcirc}} + M^{2+} \rightarrow CH=N \underset{M}{\overset{R \text{ } R}{N}} N=CH \underset{O \text{ } O}{\bigcirc} + 2H^+ + 2H_2O$$

Mitsui and Fujimura [20] described an AAS method of this kind for determining some primary amines by reaction with copper(II) and 5-nitrosalicylaldehyde. The amine solution is allowed to react with the chelate-forming reagent, which is prepared by mixing 3.372 g of triethanolamine, 0.205 g of nitrosalicylaldehyde, 1 ml of 50% acetaldehyde solution and 5 ml of 4% copper sulphate solution, in a total volume of 50 ml. The reaction mixture is diluted with water, the copper chelate is precipitated and filtered off after one hour, then the copper content of the filtrate is measured, in an air–acetylene flame, at 324.7 nm. Alternatively the precipitate is dissolved in 12N nitric acid, the solution diluted and the metal measured. The method was used for determining various amines, generally in amounts ranging from about 0.5 to about 10 mg, with errors of about 2% or less. For some amines smaller amounts (10–70 μg) were tried, but the results tended to be low by 7–20%, depending on the method used. Amines tested included the aminophenols, toluidines, naphthylamines, m- and p-phenylenediamine, o- and m-anisidine, aniline, benzidine, p-chloroaniline, cyclohexylamine, 3,3'-dichlorobenzidine, hexylamine, hexamethylenediamine, isopropylamine, methylamine, procaine and triethylenetetramine. The three nitroaniline isomers and 4-nitro-1-naphthylamine could not be determined by this method. Secondary and tertiary amines in concentrations up to 30 times that of the primary amine did not interfere.

Some aromatic primary *ortho* and *para* diamines, and aminophenols, have been determined by reaction with aqueous potassium dichromate, followed by AAS measurement of the chromium chelates of the oxidation products of these compounds [21].

Secondary amines react with carbon disulphide in aqueous solution to give dialkyldithiocarbamic acid derivatives. In the presence of nickel(II) or copper(II), coloured precipitates of the corresponding metal salts are formed [22-24]. The reaction has been utilized for indirect AAS determination of many secondary amines [25]. The nickel complex is filtered off on a fritted glass funnel, then digested with a 1:1 mixture of nitric acid and hydrochloric acid, and the resulting solution is analysed for nickel by flame AAS.

$$2R_2NH + 2CS_2 + Ni^{2+} \longrightarrow Ni(R_2N\overset{\overset{\displaystyle S}{\|}}{C}\text{-}S)_2 + 2H^+$$

Most secondary amines react quantitatively in this way within 90 min at room temperature. At 30–35°C the reaction is complete within 45–75 min, but higher temperature may cause decomposition of the dialkyldithiocarbamate.

The linear relationship between the atomic absorbance and concentration of nickel permits the determination of 0.68–8.5 μmole of amine (equivalent to 2.25 ppm Ni). The practical detection limit of this reaction, however, is approximately 0.3 μmole of the amine. At above 5 μmole of the amine, the calibration graph becomes flattened. The use of the less sensitive line at 341.5 nm for nickel may be used to extend the upper limit of detection. The average recovery for the amines tested (N-butyl-n-dodecylamine, diethylamine, dicyclohexylamine, di-n-butylamine, di-n-dodecylamine, di-n-hexylamine, di-n-octylamine, N-ethyl-n-butylamine, piperidine) is 99%. Interference by primary amines is circumvented by a prior condensation reaction with 2-ethylhexaldehyde or salicylaldehyde, to give a Schiff's base which does not react with carbon disulphide.

Diethylamine, ephedrine and methamphetamine hydrochlorides are similarly determined by reaction with carbon disulphide and copper(II). The salt solutions are neutralized with aqueous ammonia, and treated with carbon disulphide and excess of copper sulphate solution (at least 4-fold molar excess relative to the amine). The complex is extracted into methyl isobutyl ketone and aspirated into an air–acetylene flame [26]. A linear calibration graph for amine concentrations in the range of 1–10 μg/ml and an average recovery of 93.3–103.2% are obtainable, with no interference from primary aliphatic and aromatic amines.

Aliphatic primary, secondary and tertiary amines have been determined by reaction with cobalt(II) and thiocyanate, followed by measurement of the cobalt content of the complex, by flame AAS [27]. The procedure used involves mixing the sample with 0.5 ml of an acetone solution of potassium thiocyanate (32.6 mg/ml) and 0.5 ml of an acetone solution of cobalt(II) (27.91 mg of $Co(NO_3)_2.6H_2O$ per ml), followed by evaporation of the solvent on a water-bath at 90°C. A benzene solution (5 ml) of the amine (0.01–0.13 mg/ml) is then added. The solution of the reaction product [bis(amine)tetrakis(thiocyanato)-cobaltate] is then dried over sodium sulphate, diluted 5-fold with ethanol and aspirated into the flame. Recoveries of 96–104% are obtainable with cyclohexyl-amine, dibutylamine, diethanolamine, diethylamine, hexylamine, piperidine and triethylamine. Diphenylamine, monoethanolamine, hexamethylenediamine, triethanolamine and ethylenediamine do not react, however.

Pharmaceuticals containing a long-chain alkyl group or a large heterocyclic residue are determined by reaction with a known amount of sodium dioctyl-sulphosuccinate anionic detergent [28]. The excess of the reagent is measured by ion-association complex formation with the copper-1,10-phenanthroline chelate, extraction with methyl isobutyl ketone and flame AAS determination of the copper. Compounds containing ester, amide or hydroxy groups, or heterocyclic sulphur do not give the reaction under these conditions. Another method has been described, for determining benzalkonium chloride and hexa-decyltrimethylammonium toluene-p-sulphonate by passing the solution of these amines through a cation-exchange resin (Dowex 50W) in the calcium form. The equivalent amount of calcium displaced is measured by AAS [29]. The error of the method is ± 2% for sample concentrations in the range 10–30 mg/ml.

Some aminoquinoline antimalarial drugs (amodiaquine, chloroquine and primaquine) have been determined by AAS determination of the cobalt content of an organic extract of the complex formed by the drug with cobalt(II) and thiocyanate. The mean recovery is 100 ± 2% [28a].

4.3 α-AMINO-ACIDS

Beauchene *et al.* [30] have described a flame photometric method for the deter-mination of 16 α-amino-acids, based on reaction with a suspension of insoluble copper phosphate. After separation of the excess of copper reagent by centri-fugation, the copper bound to the amino-acid is measured in the flame. The method can be used to determine not only amino-acid concentrations as low as 0.3 μg/ml but also amino-nitrogen in protein hydrolysates. Hall *et al.* [31] later adapted this method for determination of α-amino-acids in urine. The feasibility of the method for semiautomatic and clinical use has been demonstrated with urine and plasma samples [32,33]. The samples, after deproteination with trichloroacetic acid, are allowed to react with copper phosphate at pH 7.4 (phosphate–borate buffer), and after centrifugation, the solution of copper–amino-acid complex is continuously introduced into an AutoAnalyzer equipped with an AAS instrument as the detector system (Fig. 4.4). The relationship between amino-acid-nitrogen concentration and copper absorbance is linear up to 0.4 $\mu g/ml$, and this covers the normal range for adult urine (amino-N 0.15–0.35 $\mu g/ml$). A commercial automatic sampling device has been used in conjunction with an atomic-absorption spectrometer equipped with a graphite furnace, to detect the copper–amino-acid complexes isolated from human serum [34]. The major copper complexes (i.e. copper–histidine and copper glutamate) are separated on a column of silica gel, and 10–20 μl portions of the effluent are used for copper measurement.

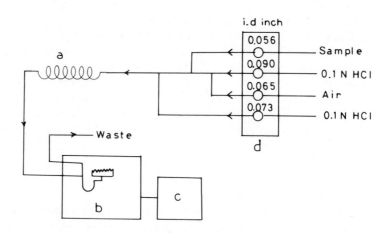

Fig. 4.4. Schematic diagram of AutoAnalyzer–AAS system for amino-acids; a, mixing coil; b, AAS instrument; c, recorder; d, proportioning pump. [Reprinted from B. Bousquet, J. L. Bouvier and C. Dreux, *Clin. Chim. Acta,* **42**, 327 (1972), by permission of the copyright holders, Elsevier Scientific Publishing Co.]

Microdetermination of some α-amino-acids by use of the copper phosphate reaction has also been described [35]. To a solution containing 0.5–3 mg of α-amino-acid, copper phosphate is added and after centrifugation the copper content in the solution is measured at 324.8 nm. The α-amino-acid content is calculated on the assumption that the copper:amino-acid ratio is 1:2. Recoveries for alanine, aspartic acid, glutamic acid, glycine, lysine hydrochloride, proline, serine, threonine, tyrosine and valine were in the range 97.4–102.9%. Lower concentrations of various α-amino-acids have been determined by reaction with salicylaldehyde to give the Schiff's bases, which on reaction with copper(II) at pH 9.5–11.5 give a chelate extractable into methyl isobutyl ketone. AAS measurement of the copper content of the complex, in an air–acetylene flame, gives a linear calibration graph over the α-amino-acid concentration range of 1.5–15 µg/ml, and an average recovery of 99.8% [36]. The copper chelates thus derived from acidic and basic α-amino-acids can be extracted quantitatively at pH 4.3 without interference from other α-amino-acids, which permits analysis of certain mixtures [37]. Binary mixtures (e.g. methionine and histidine, or glycine and glutamic acid) have been satisfactorily analysed in this way. The results show a relative standard deviation of about 0.8% (Table 4.1). Typical organic excipients, such as starch, lactose, stearate and carboxymethylcellulose, do not interfere.

Table 4.1 AAS determination of some α-amino-acids by reaction with salicylaldehyde and copper [36,37] (reprinted by permission of the copyright holders, Japan Society for Analytical Chemistry)

Sample	Recovery (%)		Standard deviation (%)
	pH 10.0	pH 4.3	
Neutral amino-acids:			
Glycine	99.8	–	0.7
L-Valine	97.2	–	
Aromatic amino-acids:			
L-Phenylalanine	98.9	–	
L-Tyrosine	99.7	–	
L-Tryptophan	99.5	–	
Sulphur-containing amino-acids:			
L-Cystine	99.2	–	
L-Methionine	100.1	–	0.7
Acidic amino-acids:			
L-Aspartic acid	14.6	99.3	
L-Glutamic acid	12.3	96.9	
Basic amino-acids:			
L-Arginine	17.8	97.7	
L-Histidine	11.5	98.5	0.7
L-Threonine	–	97.6	
L-Proline	6.5	–	
Amino-acid mixtures:			
Glycine	100.0	–	0.6
L-Glutamic acid	–	98.9	0.8
L-Histidine	–	98.1	0.8
L-Methionine	100.5	–	0.8

4.4 CARBOXYLIC ACID AMIDES

Micromolar quantities of carboxylic acid amides are determined by a flame emission method based on the Hofmann reaction but AAS could also probably be used. This involves reaction of the amide with barium hypobromite (generated by the addition of bromine to an aqueous solution of barium hydroxide) to give the corresponding amine with one less carbon atom, and insoluble barium carbonate. The latter is isolated by filtration and dissolved in nitric acid, then the resulting solution is analysed for barium by either atomic-emission [38] or atomic-absorption spectrometry.

$$RCONH_2 + 2Ba(OH)_2 + Br_2 \longrightarrow RNH_2 + BaCO_3 \downarrow + BaBr_2$$
$$+ 2H_2O$$

The reaction proceeds to completion on heating for 10–20 min at 70°C. Amides substituted with electron-donating groups (e.g. p-methylbenzamide) react more readily than those substituted with electron-attracting groups (e.g. p-nitrobenzamide).

The optimum working range for barium is 2–10 μg/ml which corresponds to approximately 1.5–7.0 μmole of amide when the recommended procedure is used. Addition of potassium chloride solution ($\sim$ 2 g/l.) to both standard and sample suppresses ionization of barium in the nitrous oxide–acetylene flame, thus increasing the sensitivity. However, three main factors have been reported to influence the accuracy of the method. These are the concentration of the hypobromite reagent, the effect of atmospheric carbon dioxide, and the solubility of barium carbonate. High concentrations of bromine cause high recoveries for aromatic amides. This is attributed to bromination of the amides to give insoluble products which precipitate along with the barium carbonate, thus affecting the complete washing of the barium carbonate. The use of a relatively high concentration of barium hydroxide and low concentration of bromine eliminates this problem and gives satisfactory results. Problems associated with the solubility of barium carbonate are circumvented by washing the precipitate with cold water at 0°C. Atmospheric carbon dioxide causes fluctuation of the results within ± 5%, and can be eliminated by working with a CO_2-free atmosphere.

The sensitivity of this method is greater than or equal to that of the most widely used spectrophotometric and titrimetric procedures [39,40]. However, the AAS method offers the advantage of high selectivity. No interferences are caused by secondary and tertiary amides and nitriles, as long as the reaction temperature does not exceed 75°C. High results may be obtained if the reaction is conducted at temperatures > 90°C. An average recovery of 99% is obtainable with many amides under controlled reaction conditions, but 2-chloroacetamide gives < 60% recovery, and salicylamide 150%. All the other amides tested give < 5% error in the determination. The total amide content of mixtures can be assessed by the same procedure. The recovery for 3–6 μmole of some binary mixtures (e.g. acrylamide–methylacrylamide, butyramide–isobutyramide and propionamide–butyramide) is above 95%.

4.5 NITRATE, NITRO AND NITROSO COMPOUNDS

Kumamaru *et al.* [41] described a method for the determination of nitrates, based on the ion-association complex formed by nitrate and the neocuproine (2,9-dimethyl-1,10-phenanthroline) copper(I) chelate and its quantitative extraction into methyl isobutyl ketone (MIBK). The nitrate solution $(10^{-4}M)$ is treated with $10^{-2}M$ copper sulphate, 5% hydroxylammonium chloride solution and $0.25M$ phosphate buffer, followed by dilution with water and extraction with $2 \times 10^{-3}M$ neocuproine solution in MIBK. Aspiration of the organic phase into the flame and measurement of the absorbance of copper at 324.7 nm gives a response that is linearly related to the nitrate concentration in the range 1–8 $\times 10^{-6}M$ in the solution aspirated. Perchlorate, iodide and thiocyanate, if present, affect the stoichiometric extraction of the copper chelate [42]. The method has been modified for analysis of plant material [43].

The method has been used for the determination of both organic nitrates and nitro compounds [44]. These compounds are converted into nitrate ions by oxidation with cerium(IV) sulphate or potassium permanganate solution in 10% sulphuric acid. Though aliphatic nitro and nitrate compounds are quantitatively converted into nitrate ions, less than 10% conversion is obtained for aromatic compounds (e.g. nitrosoaniline and nitrobenzene). The optimal working range is 0.005–5 ppm of nitrogen, the recovery ranges from 95 to 104% and the standard deviation is in the range 0.001–0.012 ppm.

The method described by Cresser [13,14] for the determination of total nitrogen and ammonia-nitrogen has been adapted for determining nitrates [45]. Up to 50 μg of nitrate-nitrogen per ml can be determined by reduction to ammonia with 40% sodium hydroxide solution and 15% titanium(III) sulphate in 23% sulphuric acid. The reaction proceeds to 98% completion virtually immediately, and longer standing merely causes poor precision. The reaction mixture is shaken in a tightly stoppered plastic centrifuge tube for 1 min and the absorbance of the ammonia (displaced by an air flow of 660 ml/min), measured at 201 nm, is a function of nitrate concentration.

Hassan [46] described a method for the determination of organic nitrates, based on reduction with cadmium in acidic medium followed by AAS measurement (228.8 nm, air–acetylene flame) of the resulting cadmium ions. Four equivalents of cadmium ions are reproducibly released per mole of nitrate, by heating with $0.05–0.1M$ hydrochloric acid. The nature of this reaction was verified by sweeping the gaseous reaction products over anhydrous magnesium perchlorate to absorb water and then into an evacuated gas cell fitted to a Beckman infrared spectrometer. The spectrum obtained showed strong absorption bands at 1285 cm^{-1} due to the $N^{+} \rightarrow O^{-}$ stretch, 2224 cm^{-1} due to the $N \equiv N$

stretch and combination bands at 3480 and 2788 cm^{-1} due to bending-stretching vibrations, confirming the presence of nitrous oxide. Further confirmation of the formation of nitrous oxide was obtained by gas-liquid chromatography. These data indicate that the reaction of the nitrates with cadmium metal under slightly acidic conditions proceeds through a 4-electron reduction process. It should be noted that quantitative reduction of nitrate to nitrous oxide has previously only been reported as occurring with some organic reductants [47-49] and reduction of nitrate with cadmium metal in alkaline media proceeds towards the formation of nitrite [50-53].

$$2R-ONO_2 + 4Cd + 8H^+ \rightarrow 2ROH + N_2O + 3H_2O + 4Cd^{2+}$$

The results obtained (Table 4.2) for some organic nitrates in the concentration range $0.5-50\mu M$ show an average recovery of 98.9% (standard deviation 1%). These data compare favourably with those obtained by measuring the cadmium ions by polarography or by titration with EDTA, potentiometrically with a cadmium ion-selective electrode or visually with Eriochrome Black T as indicator. The high sensitivity of the method is attributed to the combination of a favourable conversion factor (1 μg of nitrate-nitrogen $\equiv$ 16 μg of Cd) and the low limit of detection of cadmium by AAS.

Table 4.2 AAS determination of some nitrates by reaction with cadmium metal [46] (reprinted by permission of the copyright holders, Pergamon Press)

Sample	Calculated	Found			
		AAS	Polarography	EDTA (Cd-ISE)	EDTA (visually)
Guanidine nitrate	11.47	11.1	11.4	11.2	11.3
Pentaerythritol tetranitrate	17.72	17.5	17.5	17.6	17.6
Thiamine mononitrate	4.25	3.9	4.0	4.0	4.1
Urea nitrate	11.38	11.1	11.2	11.2	11.2

Hassan and Tadros [54] used the cadmium reaction for the microdetermination of nitro and nitroso compounds. These compounds undergo quantitative reduction to the corresponding amines with the release of six and four equivalents of cadmium ion per nitro and nitroso group respectively. The nature of the substituent groups, whether electron-attracting or electron-donating, has no significant effect on either the degree or rate of reduction. However, reduction of dinitro and polynitro compounds is not complete under the conditions used.

2,4-Dinitrophenol, 3,4-dinitrobenzoic acid, m-dinitrobenzene and hexanitrodiphenylamine show average recoveries in the range of 85–95%.

$$Ar\text{-}NO_2 + 3Cd + 6HCl \longrightarrow Ar\text{-}NH_2 + 3CdCl_2 + 2H_2O$$

$$Ar\text{-}NO + 2Cd + 4HCl \longrightarrow Ar\text{-}NH_2 + 2CdCl_2 + H_2O$$

The results obtained with various nitro and nitroso compounds in the concentration range of 10–100μM (Table 4.3) show an average recovery of 98% (mean standard deviation 1.6%). No interference is caused by many oxygen, sulphur and nitrogen compounds. Groups such as azo, oxime, and quinone, which seriously interfere with the commonly used methods for determination of nitro and nitroso compounds [i.e. reduction with vanadium(II), chromium(II), tin(II), iron(II) and titanium(III)] [55,56] do not interfere. Azo and oxime groups are too weak to oxidize the cadmium metal, whereas the quinone loses its oxidation properties through a 1:4 addition reaction with hydrochloric acid in the reaction medium, to give the chlorohydroquinone derivative [57].

Table 4.3 AAS determination of the nitro group in some organic compounds by reduction with cadmium metal [54] (reprinted by permission of the copyright holders, Academic Press)

Compound	Weight (mg)		Recovery (%)
	Taken	Found	
p-Nitroaniline	5.13	5.05	98.4
	6.24	6.16	98.7
p-Nitroanisidine	4.89	4.85	99.2
	5.97	5.96	99.8
p-Nitroanisole	4.56	4.41	96.7
	6.72	6.48	96.4
m-Nitrobenzaldehyde	7.01	6.60	94.2
	6.25	5.85	93.6
Nitrobenzene	6.73	6.52	96.9
	5.86	5.73	97.8
4-Nitrobiphenyl	4.22	4.16	98.6
	6.75	6.67	98.8
m-Nitrophenol	6.66	6.27	94.1
	4.59	4.38	95.4
Nitronaphthalene	6.00	5.88	98.0
	5.38	5.24	97.4
p-Nitrophenylacetic acid	7.33	7.12	97.1
	6.91	6.76	97.8
3-Nitrophthalic acid	7.82	7.50	95.9
	6.58	6.35	96.5

Mitsui and Kojima [58] determined some nitro compounds by heating with zinc powder and ammonium chloride at $90°C$ for 10 min. The hydroxylamine derivatives formed were treated with Tollen's reagent after separation of the excess of unreacted zinc powder. The silver chloride precipitated was dissolved in aqueous ammonia, and the silver metal (equivalent to the hydroxylamine) was isolated and dissolved in nitric acid, then the solution was diluted and aspirated into an air–acetylene flame. The method, however, suffers from lack of selectivity, as many oxygen, nitrogen and sulphur compounds readily react with silver(I).

$$R-NO_2 + 4H^+ \xrightarrow[NH_4Cl]{Zn} RNHOH + H_2O$$

$$RNHOH + 2Ag(NH_3)_2OH \longrightarrow RNO + 2H_2O + 2Ag + 4NH_3$$

4.6 NITRILES

Hydrolysis of nitriles does not yield cyanide but under controlled conditions gives ammonia or an ammonium salt. However, some nitriles are quantitatively converted into cyanide ions by enzymatic hydrolysis. For example, the β-glucosidase enzyme has been utilized for the hydrolysis of cyanogenic glucosides [59, 60]. Amygdalin (a glycoside from bitter almonds, the kernels of apricots, peaches and plums, cherry and laurel leaves) undergoes hydrolysis with quantitative release of one mole of cyanide per mole of compound [61-63]. Hassan [64] has described a method for determining trace levels of amygdalin by incubation of the compound at room temperature with β-glucosidase (1 mg/ml) at pH 6, followed by AAS determination of the cyanide released. The use of the enzyme permits selective hydrolysis of amygdalin levels as low as 100 ng/ml. At levels below this, the reaction is extremely slow. Increasing the temperature to above $40°C$ results in denaturation of both the enzyme and the amygdalin.

$$\underset{\underset{OC_{12}H_{21}O_{10}}{|}}{C_6H_5-CHCN} + 2H_2O \xrightarrow{Enzyme} 2C_6H_{12}O_6 + C_6H_5CHO + HCN$$

The hydrolysate containing the cyanide is allowed to react at pH 11 with silver wool (to ensure a large surface area and rapid solubilization of an equivalent amount of silver). The silver content of the soluble complex $[Ag(CN)_2]^-$ is measured at 329.1 nm, with electrothermal atomization. The average recovery is 98.2%, the mean standard deviation 1.6%. Similar methods for the determination of cyanide ions in neutral solutions have been proposed. In one method small pieces of silver metal ($5 \times 1.8 \times 1.8$ mm) are soaked in the cyanide solution for 1 hr, then the dissolved silver is determined by flame AAS [65]. In a modified version of this method, silver wool (5 μm diameter) is placed on a membrane filter in a closed system and the cyanide solution is allowed to pass through the filter at a rate of 60–80 drops/min; electrothermal atomization is then used for

AAS determination of the dissolved silver [66]. This permits the determination of cyanide in the 0–100 ng/ml concentration range, without interference from halides, thiocyanate, sulphate and nitrate.

Some other methods have been suggested for cyanide determination. Danchik and Boltz [67] described a micro-method based on the formation of the ion-association complex of cyanide with the tris(1,10-phenanthroline)iron(II) chelate and its extraction into chloroform. After evaporation of the solvent, the chelate is dissolved in ethanol and directly aspirated into an air-acetylene flame for the measurement of the iron at 248.3 nm. Cyanide at the 500 μg/ml level has been determined by its dissolution of copper from basic copper carbonate in alkaline medium, followed by measurement of the copper [68]. Similarly, a cyanide solution is filtered through insoluble copper(I) cyanide on a membrane filter and the copper content of the soluble cyanide complex is measured [69]. The interference of cyanide in the AAS determination of cadmium has also been used for cyanide determination [70], but the method is relatively insensitive, and the detection limit is as high as $10^{-4}M$.

Cyanide in biological samples has been determined by conversion into hydrogen cyanide in the gas phase and passage of this through lead acetate solution to remove any sulphide and volatile thiocyanates and then into $0.1M$ sodium hydroxide and nickel chloride. The excess of nickel chloride is removed by precipitation with sodium sulphide at pH 7–8 and the nickel in the nickel cyanide complex is determined by AAS [71]. Griever and Syty [72] have described a method for cyanide determination based on its conversion into ammonium ions by a two-step procedure. It is first oxidized with an excess of $0.12M$ potassium permanganate at pH > 12.5 to give cyanate, followed by hydrolysis with excess of $14.5N$ sulphuric acid to yield the ammonium ion.

$$CN^- + 2MnO_4^- + 2OH^- \longrightarrow CNO^- + H_2O + 2MnO_4^{2-}$$

$$CNO^- + 2H^+ + H_2O \longrightarrow NH_4^+ + CO_2$$

The reaction is complete in less than 1 min, and ammonia is then released and measured as previously described by Syty [16]. The detection limit is 1.4 μg/ml.

4.7 BIURET

The first official method for the determination of biuret in urea and mixed fertilizers was based on its reaction with copper(II) in alkaline tartrate solution to give the biuret-copper complex which is measured spectrophotometrically at 555 nm [73]. This method has been used extensively for biuret certification in many types of samples. However, ammonia enhances and phosphorus depresses the colour intensity. Determination of biuret in urea and mixed fertilizers by AAS has therefore been advocated [74]. A 1% sample solution is treated with copper hydroxide suspension, and after 5–10 min the solution is centrifuged and the copper in the solution is measured in an air-acetylene flame; standards are

128 Nitrogen Compounds [Ch.

prepared with known amounts of biuret. Although stoichiometric ratios of copper to biuret are not obtainable, the relation between copper absorbance and biuret concentration is linear. It has been reported that freshly prepared copper hydroxide suspension is not suitable, but a suspension aged for 5-20 hr gives good results.

This method has been modified and used for measuring biuret in urea ($<$ 10 mg) and in mixed fertilizers ($<$ 40 mg) [75]. A strong base is added to an alcoholic solution containing copper(II) and biuret. The excess of copper is precipitated as the hydroxide and kept in the precipitate form by addition of starch to deactivate the surface of the precipitate. After filtration and dilution, the biuret complex is measured. The AAS method gives more reliable results than the official A.O.A.C. procedure, which consistently gives higher results. It has been reported that urea enhances the absorbance of the copper-biuret chelate, causing a positive error [76]. A collaborative study [77] of the A.O.A.C. and AAS methods revealed the convenience of the latter method for determining 0.1-3% of biuret in urea and mixed fertilizers.

REFERENCES

[1] M. Honma and C. L. Smith, *Anal. Chem.*, **26**, 458 (1954).
[2] J. M. S. Butcher and G. F. Kirkbright, *Analyst*, **103**, 1104 (1978).
[3] I. K. Phelps, *J. Assoc. Off. Agr. Chem.*, **3**, 306 (1920).
[4] R. L. Shirley and W. W. Becker, *Ind. Eng. Chem., Anal. Ed.*, **17**, 437 (1945).
[5] G. R. Lake, P. McCutchan, R. Van Meter and J. C. Neel, *Anal. Chem.*, **23**, 1634 (1951).
[6] P. R. W. Baker, *Analyst*, **80**, 481 (1955).
[7] R. B. Bradstreet, *The Kjeldahl Methods for Organic Nitrogen*, Academic Press, New York (1965).
[8] A. M. Bond and J. B. Willis, *Anal. Chem.*, **40**, 2087 (1968).
[9] A. Clearfield, *Rev. Pure Appl. Chem.*, **14**, 91 (1964).
[10] D. C. Bradley and I. M. Thomas, *J. Chem. Soc.*, 3857 (1960).
[11] T. Mitsui and Y. Fujimura, *Bunseki Kagaku*, **23**, 449 (1974).
[12] R. S. Danchik, D. F. Boltz and L. G. Hargis, *Anal. Lett.*, **1**, 891 (1968).
[13] M. S. Cresser, *Anal. Chim. Acta*, **85**, 253 (1976).
[14] M. S. Cresser, *Lab. Pract.*, **26**, 19 (1977).
[15] C. C. Muroski and A. Syty, *Anal. Chem.*, **62**, 143 (1980).
[16] A. Syty, *Anal. Chem.*, **51**, 911 (1979).
[17] M. Takahashi, K. Tanabe, A. Saito, K. Matsumoto, H. Haraguchi and K. Fuwa, *Can. J. Spectrosc.*, **25**, 25 (1980).
[18] P. N. Vijan and G. R. Wood, *Anal. Chem.*, **53**, 1447 (1981).
[19] F. E. Critchfield and J. B. Johnson, *Anal. Chem.*, **28**, 436 (1956).
[20] T. Mitsui and Y. Fujimura, *Bunseki Kagaku*, **23**, 1309 (1974).
[21] N. E. Naftchi, M. A. Becker and A. S. Akerkar, *Anal. Biochem.*, **66**, 424 (1975).
[22] F. R. Duke, *Ind. Eng. Chem., Anal. Ed.*, **17**, 196 (1945).
[23] G. R. Umbreit, *Anal. Chem.*, **33**, 1572 (1961).
[24] E. L. Stanley, J. Baum and J. L. Gove, *Anal. Chem.*, **23**, 1779 (1951).
[25] P. J. Oles and S. Siggia, *Anal. Chem.*, **45**, 2150 (1973).
[26] T. Mitsui and Y. Fujimura, *Nippon Kagaku Kaishi*, 1908 (1974).

[27] Y. Minami, T. Mitsui and Y. Fujimura, *Bunseki Kagaku,* **30**, 475 (1981).

[28] J. Alary, J. Rochat, A. Villet and A. Coeur, *Ann. Pharm. Franc.,* **34**, 345 (1976).

[28a] S. M. Hassan, M. E.-S. Metwally and A. A. Abou Ouf, *Analyst,* **107**, 1235 (1982).

[29] G. Bettoni and C. Franchini, *Farmaco, Ed. Prat.,* **31**, 420 (1976).

[30] R. E. Beauchene, A. D. Berneking, W. G. Schrenk, H. L. Mitchell and R. E. Silker, *J. Biol. Chem.,* **214**, 731 (1955).

[31] F. F. Hall, G. A. Peyton and S. D. Wilson, *Tech. Bull. Regist. Med. Technol.,* **39**, No. 4, 89 (1969).

[32] B. Bousquet, J.-L. Bouvier and C. Dreux, *Clin. Chim. Acta,* **42**, 327 (1972).

[33] F. Hall, B. Schneider, T. Culp and C. Ratliff, *Clin. Chem.,* **18**, 34 (1972).

[34] N. Kahn and J. C. VanLoon, *J. Liq. Chromatog.,* **2**, 23 (1979).

[35] Y. A. Gawargious, A. Besada and M. E. M. Hassouna, *Microchem. J.,* **22**, 96 (1977).

[36] Y. Kidani, S. Uno and K. Inagaki, *Bunseki Kagaku,* **25**, 514 (1976).

[37] Y. Kidani, S. Uno and K. Inagaki, *Bunseki Kagaku,* **26**, 158 (1977).

[38] R. P. D'Alonzo and S. Siggia, *Anal. Chem.,* **49**, 262 (1977).

[39] T. Higuchi, C. H. Barnstein, H. Ghassemi and W. E. Perez, *Anal. Chem.,* **34**, 400 (1962).

[40] S. Steuli, *Anal. Chem.,* **31**, 1652 (1959).

[41] T. Kumamaru, E. Tao, N. Okamoto and Y. Yamamoto, *Bull. Chem. Soc. Japan,* **38**, 2204 (1965).

[42] Y. Yamamoto, T. Kumamaru, Y. Hayashi and Y. Otani, *Bunseki Kagaku,* **18**, 359 (1969).

[43] G. Hoshikawa and Y. Fudano, *Kagawa Daigaku Nogakuba Gakujutsu Hokoku,* **27**, 111 (1976).

[44] M. E. Houser and M. I. Fauth, *Microchem. J.,* **15**, 399 (1970).

[45] M. S. Cresser, *Analyst,* **102**, 99 (1977).

[46] S. S. M. Hassan, *Talanta,* **28**, 89 (1981).

[47] W. I. Awad, S. S. M. Hassan and M. T. Zaki, *Talanta,* **18**, 219 (1971).

[48] S. S. M. Hassan, *Mikrochim. Acta,* 1109 (1970).

[49] S. S. M. Hassan, *Anal. Chim. Acta,* **60**, 442 (1972).

[50] R. S. Lambert and R. J. DuBois, *Anal. Chem.,* **43**, 955 (1971).

[51] R. Grau and A. Mirna, *Z. Anal. Chem.,* **158**, 182 (1957).

[52] A. Henriksen and A. Selmer-Olsen, *Analyst,* **95**, 514 (1970).

[53] A. W. Morris and J. P. Riley, *Anal. Chim. Acta,* **29**, 272 (1963).

[54] S. S. M. Hassan and F. S. Tadros, *Microchem. J.,* **28**, 120 (1983).

[55] I. M. Kolthoff and P. J. Elving (eds.), *Treatise on Analytical Chemistry*, Part II, Vol. 16, p. 47, Wiley, New York (1980).

[56] S. Siggia, *Quantitative Organic Analysis via Functional Groups*, 3rd Ed., p. 526, Wiley, New York (1963).

[57] S. S. M. Hassan and M. B. Elsayes, *Mikrochim. Acta,* 333 (1978 II).

[58] T. Mitsui and R. Kojima, *Bunseki Kagaku,* **26**, 317 (1977).

[59] W. Blaedel, D. Easty, L. Anderson and T. Farrell, *Anal. Chem.,* **43**, 890 (1971).

[60] D. Easty, W. Blaedel and L. Anderson, *Anal. Chem.,* **43**, 509 (1971).

[61] G. A. Rechnitz and R. Llenado, *Anal. Chem.,* **43**, 283 (1971).

[62] R. Llenado and G. A. Rechnitz, *Anal. Chem.,* **43**, 1457 (1971).

[63] M. Mascini and A. Liberti, *Anal. Chim. Acta,* **68**, 177 (1974).

[64] S. S. M. Hassan, *Z. Anal. Chem.,* **312**, 354 (1982).

[65] E. Jungreis, *Israel J. Chem.,* **7**, 583 (1969).

[66] E. Jungreis and F. Ain, *Anal. Chim. Acta,* **88**, 191 (1977).

[67] R. Danchik and D. Boltz, *Anal. Chim. Acta,* **49**, 567 (1970).

[68] S. Manahan and R. Kunkel, *Anal. Lett.,* **6**, 547 (1973).

[69] E. Jungreis and S. Kraus, *Mikrochim. Acta,* 413 (1976 I).

[70] M. Newton and D. Davis, *Anal. Lett.,* **6**, 923 (1973).

[71] M. A. Tsougas and A. Kevatsis, *Vet. Hum. Toxicol.*, **21**, 190 (1979).

[72] S. Grieve and A. Syty, *Anal. Chem.*, **53**, 1711 (1981).

[73] *Official Methods of Analysis*, 12th Ed., A.O.A.C., Washington, D.C., Sections 2.072–2.074 (1975).

[74] K. C. Singhal, R. C. P. Sinha and B. K. Banerjee, *Technology (Sindri, India)*, **6**, 95 (1969).

[75] T. C. Woodis, Jr., G. B. Hunter and F. J. Johnson, *J. Assoc. Off. Anal. Chem.*, **59**, 22 (1976).

[76] J. J. Geurts, J. E. Van Stelle and E. G. Brinkman, *Anal. Chim. Acta*, **41**, 113 (1968).

[77] L. F. Corominas, *J. Assoc. Off. Anal. Chem.*, **60**, 1214 (1977).

5

Phosphorus compounds

Phosphorus is one of the most widely and evenly distributed elements on the earth's surface, and also one of the most diversified in its compounds. It is now widely realized that phosphorus-containing organic compounds play a vital role in life processes, and some of them have important pharmacological and pesticidal uses. Determination of these compounds is based, in general, on conversion into phosphate before application of any of the measurement techniques, including AAS. The interaction of phosphorus compounds with flames has been investigated since 1841 [1]. The most significant early work was done by Salet [2-4] who observed that a green colour was imparted to the flame when phosphorus compounds were introduced into a fuel-rich hydrogen flame chilled with falling water or surrounded by a glass tube cooled by a blast of air, or with a sheath of cold flowing air. Procedures in current use for determination of phosphorus by measuring the emission in the flame are based on this phenomenon. Direct and indirect atomic-absorption spectrometry with both flame and electrothermal atomization systems have recently been described.

5.1 DECOMPOSITION OF ORGANOPHOSPHORUS COMPOUNDS

Simons and Robertson [5] decomposed aliphatic phosphates by digestion with concentrated hydriodic acid for 5 min, but a longer time is required for aromatic phosphates. Catalytic oxidation with nitric acid and a mixture of sodium molybdate, concentrated sulphuric acid and perchloric acid was therefore recommended. Ma and McKinley [6] used the Kjeldahl flask for decomposition with a mixture of concentrated sulphuric and nitric acids. The time of digestion is reduced by using 70% perchloric acid with sulphuric acid [7]. Potassium persulphate or hydrogen peroxide, with concentrated sulphuric acid, was proposed for the digestion of organic compounds containing phosphorus and silicon [8]. Fusion with sodium hydroxide and potassium nitrate, in a platinum boat, has been suggested for aliphatic phosphorus compounds [9] and recently for vegetable matter [9a]. The bomb-fusion technique has been applied to the decomposition of a variety of compounds [10]. The samples are fused with sodium peroxide, potassium nitrate and sucrose. Fusion with magnesium may also be used [11]. Fleischer *et al.* [12] recommended decomposition of organic phosphorus compounds by the Schöniger oxygen-flask method. Hassan *et al.* [13–15] used $0.5M$ sodium hydroxide and saturated aqueous bromine solution as the absorbent in the oxygen flask. Sulphuric acid and alkaline hypobromite [16,17] have also been suggested.

5.2 METHODS BASED ON DIRECT ATOMIZATION IN THE FLAME

The principle atomic resonance lines of phosphorus lie at 177.5, 178.3 and 178.8 nm. With most commercial AAS instruments this unusual range of measurement necessitates the use of (a) an optical path and monochromator purged with an inert gas or evacuated; (b) a hot flame and an efficient atomizer to atomize the sample without introducing absorbing gases into the light-path; (c) an intense radiation source. Manning and Slavin [18] described a method for the determination of phosphorus by measuring the absorbance (nitrous oxide–acetylene flame) at the non-resonance doublet (213.55–213.62 nm) and at 214.9 nm with a hollow-cathode lamp as source. An absorbance of 0.3 was obtained for a 20 mg/ml solution of phosphorus. The method has been used for the determination of total phosphorus in detergents [19]. Samples (containing 13-20% phosphorus) are digested with a 1:1 mixture of perchloric and nitric acids followed by hydrofluoric acid to eliminate any silicate. Then the sample solutions are aspirated with a platinum–titanium nebulizer into a nitrous oxide–acetylene flame burning on a premix burner head.

Besides the low sensitivity offered by use of a conventional hollow-cathode lamp, poor signal:noise ratio and stability commonly occur. Kirkbright and Wilson [20] proposed the use of a demountable hollow-cathode lamp for direct determination of phosphorus. If a graphite hollow-cathode source containing 1 ml of red phosphorus is used, weak emission lines are observed at 177.5, 178.3 and 178.8 nm, whereas the non-resonance lines at 213.5, 213.6 and 253.5 nm are emitted with appreciable intensity. This source minimizes the limitations encountered in the use of the hollow-cathode lamp and provides considerable freedom from self-absorption.

Barnett *et al.* [21] compared the analytical performance of the electrodeless discharge lamp for phosphorus with that of the hollow-cathode lamp. In spite of an almost tenfold increase in the intensity offered by the electrodeless discharge lamp, no significant increase in the sensitivity or detection limit was obtained. The nitrogen-separated nitrous oxide-acetylene flame and the vacuum polychromator have been used in conjunction with an electrodeless discharge lamp for the determination of phosphorus in milk powder, beef extract, and yeast extract [22]. The absorbance at 178.3 nm is linearly related to phosphorus concentration in the range 40–400 μg/ml.

5.3 METHODS BASED ON DIRECT ELECTROTHERMAL ATOMIZATION

A logical approach to increase the sensitivity is to use the graphite furnace for atomization of phosphorus. Combination of this device with the electrodeless discharge lamp overcomes many of the problems associated with the use of conventional flame methods and substantially increases the sensitivity to levels as low as 15 ng of phosphorus [23-26]. L'vov and Khartsyzov [23] suggested the use of an electrically heated graphite cuvette for the determination of phosphorus at 177.49 nm. Azuma and Aramaki [27] investigated the sensitivity of the electrothermal atomization technique as a function of the type of phosphorus compound analysed, and the temperatures of ashing and atomization, using an electrodeless discharge lamp and a graphite-tube furnace. They reported that the smoke produced during combustion of organic samples, inadequate correction of the background and interference by chloride are the main factors that affect the sensitivity. Addition of $\sim$ 1% lanthanum nitrate or organolanthanum compounds has thus been recommended for matrix modification to give an analytically useful atomic-absorption signal and to minimize the interferences caused by the matrix.

Phosphorus in various materials has been satisfactorily determined by the electrothermal atomization technique, after conversion into phosphate by one of the known decomposition methods. Plant material is decomposed by oxidative digestion, and the phosphorus stabilized by addition of lanthanum and measured at 213.6 nm [28]. Alkyl phenyl phosphite and tricresyl phosphate additives in gasoline are similarly determined at levels down to < 1 ppm after conversion into phosphate [29]. The gasoline samples are diluted with dimethylformamide or acetone and an aliquot of the solution is ashed on either magnesium oxide in the presence of hydrogen peroxide or "magnesium sulphonate" (Conostan magnesium standard) to convert the organophosphorus into magnesium pyrophosphate. The pyrophosphate is then dissolved in 1:1 nitric acid and the solution injected into the furnace along with a 1% solution of lanthanum nitrate. The phosphorus absorbance is measured at 213.6 nm; a calibration graph is prepared by use of petrol containing tritolyl phosphate as standard.

Atomization of phosphorus-containing compounds without prior decomposition to form phosphate has also been described. Phosphorus in gasoline and petroleum products is determined by injecting lanthanum nitrate solution into the graphite furnace just before addition of the gasoline samples. The organic matrix is charred at 1600°C and the phosphorus atomized at 2700°C in the presence of argon as purge gas in the interrupted flow mode [30,31]. Maximal phosphorus signal is obtained when $\sim$ 1 mg of lanthanum is present in the graphite tube. The detection limit of this method is 20 ng of phosphorus in the 90 μl of gasoline injected and the relative standard deviation at the 80 ng level of triphenyl phosphate is 2%. Similarly, phosphorus in edible oil has been determined. The oil sample is heated at 60°C, shaken, diluted with a 0.5%

solution of lanthanum acetylacetonate in chloroform and atomized [32]. The detection limit is ~ 0.5 ng of phosphorus per kg of the oil and the reproducibility is 2% at the 20-ng/kg level.

Although lanthanum is commonly used to modify the absorption characteristics of phosphorus, Prévôt and Gente-Jauniaux [33] demonstrated that such a treatment is not beneficial for all samples. A level as low as 250 ng/g in soya bean and rapeseed oil has been determined without such treatment. The oil is diluted with an equal volume of methyl isobutyl ketone and atomized. In another report by these authors [34], however, addition of 1% lanthanum 4-cyclohexylbutyrate solution has been shown to improve the results for phosphorus in vegetable oil. Havezov et al. [35] reported that addition of lanthanum to organophosphorus compounds is unnecessary if the graphite-tube furnace is coated with zirconium. Persson and Frech [36] concluded from high-temperature equilibrium calculations and experimental studies that uncoated graphite tubes can provide a sufficiently low partial pressure of oxygen and therefore favour the formation of phosphorus atoms.

In all the above-mentioned electrothermal atomization methods, the samples are introduced into the atomizer and after solvent evaporation are charred and atomized. In an attempt to increase the sensitivity, L'vov and Pelieva [37] used a tungsten probe (a wire of diameter 0.8 μm) for sample evaporation. The sample solution (~ 2-20 μl) containing as little as 2 ng of phosphorus is applied to the probe and dried under a tungsten lamp. The probe with the dry residue is then introduced into the graphite furnace through the injection hole. This technique improves the detection limit by a factor of 20–30 compared to that for evaporation from the furnace walls. Zeeman background correction and the L'vov platform have also been used to improve the sensitivity and allow determination of phosphorus down to 0.002% [38].

Mixtures of organophosphorus compounds are analysed by HPLC separation followed by electrothermal atomization and AAS measurement of each component [39]. The mixture is diluted with either xylene or methyl isobutyl ketone to bring the phosphorus concentration into the range 500–2000 μg/ml and an aliquot of the solution is separated by reversed-phase HPLC, after which the individual fractions are analysed off-line for phosphorus by measurement of the atomic absorption at 213.6 nm. The use of AAS has distinct advantages over the ultraviolet detectors usually employed in HPLC. The signals obtained are dependent only on the amount of phosphorus present and are independent of the structural features of the molecules. No inteference is caused by the organic solvents used as eluents, but these commonly limit the use of detectors based on ultraviolet absorption or HPO emission. Moreover, the detection limit (0.3 μg/ml) compares well with the detection limit obtained by ICP (~ 0.3 μg/ml) [40] but is much poorer than that obtained by using HPO flame emission (~ 0.02 μg/ml) [41]. Analysis of lubricating oil containing zinc dialkyldithiophosphate (ZTP) as anticorrosive agent, tri-p-cresyl phosphate (TCP) as anti-water agent, and triphenyl phosphate (TPP), has been satisfactorily performed by HPLC and electrothermal atomization.

5.4 METHODS BASED ON ATOMIC-ABSORPTION INHIBITION-TITRATION

The depressive effect of phosphate ions on the atomic absorption of the alkaline-earth metals in the flame has been used for the determination of phosphorus [42,43]. The decrease in the absorption of strontium is linearly related to the phosphate concentration up to 6 ppm. Addition of more phosphate than this does not cause any further depression [42]. Good linearity has also been reported for calcium, without interference from low concentrations of fluoride [43]. This technique is used for the determination of phosphorus in the range 1–5 ppm either alone [44] or simultaneously with silicate and sulphate [45]. In the latter case, the analyte solution is placed in a beaker, in which injection and aspiration tubes are positioned. With aspiration into an air–propane [44] or air–hydrogen [45] flame, magnesium titrant is added and the titration is terminated at the final rise in the titration curve (Fig. 5.1). Three distinct shifts in the slope of the

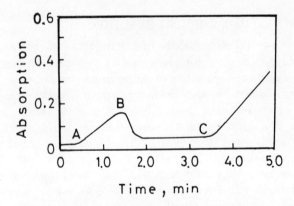

Fig. 5.1. Atomic-absorption inhibition-titration of a mixture of phosphate, silicate and sulphate, with magnesium. Solution contained 1.0 μg/ml SiO_2, 4.0 μg/ml PO_4^{3-}, 20 μg/ml SO_4^{2-}. Titrant: 50 μg/ml Mg as $MgCl_2$. [Reprinted by permission from C. I. Lin and C. O. Huber, *Anal. Chem.*, **44**, 2200 (1972). Copyright 1972 American Chemical Society.]

linear segments are obtained, which the authors suggest are probably due to the following reactions (m, n, p and q are stoichiometric factors, but not necessarily integers).

Before A:

$$SiO_3^{2-} + Mg^{2+} \xrightarrow{K_1} MgO.SiO_2 \longrightarrow \text{no reaction} \qquad (1)$$

From A to B:

$$SiO_3^{2-} + Mg^{2+} \xrightarrow{K_2} (1+m)MgO.SiO_2 \xrightarrow{K_2'} Mg \qquad (2)$$

$$PO_4^{3-} + nMg^{2+} \xrightarrow{K_3} nMgO.P_2O_5 \xrightarrow{K_3'} Mg \qquad (3)$$

After B (negative slope):

$$Mg^{2+} + (1+m)MgO.SiO_2 \xrightarrow{K_4} pMgO.SiO_2 \xrightarrow{K_4'} Mg \qquad (4)$$

$$Mg^{2+} + nMgO.P_2O_5 \xrightarrow{K_5} qMgO.P_2O_5 \xrightarrow{K_5'} Mg \qquad (5)$$

Before C (horizontal):

$$Mg^{2+} + SO_4^{2-} \xrightarrow{K_6} MgO.SO_3 \longrightarrow no\ reaction \qquad (6)$$

The end-points (A), (B) and (C) are linear functions of the amounts of silicate, silicate plus phosphate, and silicate plus phosphate plus sulphate, respectively.

5.5 METHODS BASED ON INDIRECT REACTIONS

Zaugg and Knox [46,47] described the first indirect AAS method for the determination of phosphorus. The procedure involves conversion of phosphate into 12-molybdophosphate and extraction of the heteropoly acid into 2-octanol. The molybdenum (12 atoms for each original phosphorus atom) is measured at 313.3 nm in an air–acetylene flame. The samples and standards are kept at constant temperature during aspiration since a change of $1\,^\circ C$ within the range 17–$40\,^\circ C$ significantly alters the rate of 2-octanol uptake and causes a 2% error in the absorption readings.

Quantitative heteropoly acid formation and hence maximal absorption signals are obtainable if the reaction is done at pH 0.9–1.2. Under these conditions, the molybdenum absorption is linearly related to phosphorus concentration up to 10 $\mu g/ml$. As little as 0.03 μg of phosphorus can be determined in pure solutions and biological media. This procedure is also used for the determination of the change in phosphate concentration caused by enzyme activity. Reactions involving adenosine triphosphate (ATP), and oxidative phosphorylation activity, can be monitored since such systems contain acid-labile organic phosphates which undergo hydrolysis catalysed by molybdenum and acid [48].

Determination of phosphorus in serum, blood and other biological tissues requires prior removal of protein, and digestion with a mixture of concentrated sulphuric acid with either 70% perchloric acid or 30% hydrogen peroxide or concentrated nitric acid [49]. Phosphorus in small deproteinated samples of plasma has been determined by precipitation as phosphomolybdic acid at pH 1.9 (DL-α-alanine–hydrochloric acid buffer). The heteropoly acid is extracted into methyl isobutyl ketone (MIBK), the excess of molybdic acid is stripped by shaking with citrate, then the molybdenum is determined in an air–acetylene flame [50]. This procedure is used for the determination of phosphorus in the range 5–200 $\mu g/ml$. The results of this method, however, indicate that 15 atoms of molybdenum are extracted for each atom of phosphorus. No explanation for this apparently anomalous stoichiometry was provided. It seems that the most reliable method for the determination of phosphorus in wine, organic compounds, blood serum, milk products and urine is based on direct aspiration of an isobutyl acetate or MIBK extract of the heteropoly acid into the flame [51–54].

The effect of solvent and pH on the accuracy of the method has been investigated. A pH range of 0.7–1.25 has been recommended [55,56] although slightly lower or higher pH values may also be used. Oxygen-containing solvents are found to be good extractants for heteropoly acids. Wadelin and Mellon [57] found that a 1:4 v/v n-butanol-chloroform mixture selectively extracts phosphomolybdic acid in the presence of arsenate, silicate and germanate. Isoamyl acetate [58] and isobutyl alcohol [59] may also be used. Kumamaru et al. [60] and Bernal et al. [61] recommended n-butyl acetate as an effective extractant not only for its high selectivity and extractive power for phosphomolybdic acid but also for its good combustibility in the flame (Table 5.1). On the other hand, phosphomolybdic acid can be determined without using organic extractants. Thus, after addition of excess of quinoline molybdate reagent to the phosphate, filtration of the precipitate, washing, and dissolution in aqueous ammonia, the aqueous solution can be aspirated into an air–acetylene flame for AAS determination of the molybdenum [62].

Table 5.1. AAS determination of phosphorus by complexation as phosphomolybdic acid, extraction and measurement of molybdenum

Extractant	Optimum pH	Atomization system	Limit of detection	References
n-Butyl acetate	~ 1	(Air–nitrous oxide)–acetylene	0.35 $\mu g/ml$	60
	0.3–1.7	Graphite furnace	10 pg/ml	67
Diethyl ether	0.9–1.3	Air–acetylene	0.1 $\mu g/ml$	71
Diethyl ether–pentanol	~ 1	Acetylene–nitrous oxide	0–10 $\mu g/ml$	64,66
Isobutyl acetate	0.6–1	Acetylene–nitrous oxide	0.01 $\mu g/ml$	69
	0.96M HCl	Acetylene–nitrous oxide	0.08 $\mu g/ml$	63
Methyl isobutyl ketone	1.7–2.1	Air–acetylene	0–2.5 $\mu g/ml$	50,52
2-Octanol	0.9–1.1	Air–acetylene	0.03 $\mu g/ml$	46,47

Phosphorus and silicon are simultaneously determined by conversion into phosphate and silicate, followed by formation of the heteropoly acids. Phosphomolybdic acid is then selectively extracted with either isobutyl acetate [63] or diethyl ether [64] and silicomolybdic acid is recovered from the aqueous phase by lowering the pH of the solution with $2M$ hydrochloric acid followed by extraction with butanol [63] or a 5:1 v/v diethyl ether-pentanol mixture [64]. The molybdenum content of the organic extract is measured at 313.3 nm in an air-acetylene flame [65]. These methods permit determination of 2-25 μg of phosphorus in the presence of 200 μg of silicon. Neither arsenic nor germanium interferes with the phosphorus determination but they affect the silicon results. Riddel and Turek [66] determined phosphorus and silicon by selective extraction of their heteropoly acids with a mixture of diethyl ether and pentanol (5:1) at different pH values. The distribution ratio of phosphomolybdic acid between the organic and aqueous phases is 410, compared to 0.04 for silicomolybdic acid. Thus, after extraction of phosphomolybdic acid from the solution at pH 1, the acidity of the aqueous phase is increased to $4M$ in hydrochloric acid and silicomolybdic acid is extracted with the same solvent. Molybdenum is stripped with a pH-9.3 buffer and measured in a nitrous oxide-acetylene flame. The method is applicable to samples containing not too much silicon. When the Si/P ratio exceeds 400, the phosphate results become unreliable. Arsenic interferes with phosphorus and germanium with silicon.

Phosphorus and arsenic have been sequentially determined on the micro- and picogram levels by electrothermal atomization AAS [66,68]. Phosphomolybdic acid and arsenomolybdic acid is extracted with butyl acetate and butyl acetate containing 11-19% ethanol [67] or a 1:1 butanol-ethyl acetate mixture [68]. Both extracts are decomposed with $4N$ aqueous ammonia and the molybdenum content is measured. The method is used for the determination of phosphorus and arsenic at levels as low as 10 and 25 pg, respectively. Phosphorus, arsenic and silicon are simultaneously determined by successive extraction of their heteropoly acids [69]. Isobutyl acetate selectively extracts phospho-'molybdic acid in the presence of the other two heteropoly acids, with 100% efficiency over the pH range 0.6-1 [70]. Arsenomolybdic acid is extracted from the heteropoly acid mixture with a solvent mixture containing ethyl acetate, butanol and isoamyl acetate. Silicomolybdic acid is separated from considerable amounts of the other two heteropoly acids by extraction with MIBK in the presence of citrate (which destroys the other two acids rapidly, but silicomolybdic acid only very slowly [70a]). By the separation scheme shown in Fig. 5.2, phosphorus, arsenic and silicon can be sequentially determined at levels of 1-10, 2-20 and 1-10 μg/ml, respectively.

A detailed AAS study of the distribution behaviour of heteropolymolybdic acids in various solvents as a function of solution acidity has been reported [71]. With diethyl ether as solvent, phosphomolybdic acid is quantitatively extracted over the acidity range 0.9-1.3M hydrochloric acid, whereas silicomolybdic and

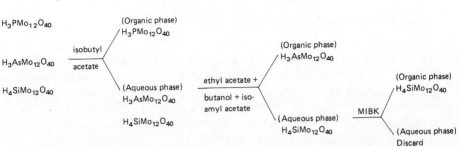

Fig. 5.2. The successive AAS determination of phosphorus, arsenic, and silicon.

germanomolybdic acid remain unextracted over the entire acidity range 0.08–4M hydrochloric acid. With diethyl ether–pentanol (5:1) mixture, all three heteropoly acids are quantitatively extracted over the acidity range 0.08–1.5M hydrochloric acid. A mixture of chloroform and n-butanol can conveniently be used for separation of phosphomolybdic acid from the other heteropoly acids over a wide acidity range (Table 5.2).

Table 5.2 Extraction of molybdoheteropoly acids with various solvents [71] [reprinted with permission from S. J. Simon and D. F. Boltz, *Anal. Chem.*, **47**, 1758 (1975). Copyright 1975 American Chemical Society]

Heteropoly acid	Extractant ($V/V_0 = 1$)	Optimum acidity (HCl, M)	Distribution ratio
Arsenomolybdic	n-Butyl acetate	2.0	<0.07
	Chloroform–n–butanol (4:1)	0.1–2.0	<0.01
	Diethyl ether	0.5	1.1
	Diethyl ether–pentanol (5:1)[a]	0.1–1.5	410
	MIBK[b]	0.1–1.5	620
Germanomolybdic	n-Butyl acetate	2.0	<0.08
	Chloroform–n–butanol (4:1)	0.1–2.0	<0.01
	Diethyl ether	2.0	0.06
	Diethyl ether–pentanol (5:1)	1–1.4	100
	MIBK[b]	0.1–2.0	600
Phosphomolybdic	n-Butyl acetate[b]	0.1–1.0	620
	Chloroform–n–butanol (4:1)	0.5–1.5	1.2
	Diethyl ether[c]	1.0–1.3	165
	Diethyl ether–pentanol (5:1)[a]	0.1–1.5	410
	MIBK[b]	0.1–1.5	630
Silicomolybdic	n-Butyl acetate	3.0	<0.6
	Chloroform–n–butanol (4:1)	0.1–2.0	<0.01
	Diethyl ether	0.1–4.0	<0.01
	Diethyl ether–pentanol (5:1)[a]	1.5–4.0	410
	MIBK[d]	0.1–4.0	710

$V/V_0 = $ (a) 4.2; (b) 6.3; (c) 1.7; (d) 7.1

5.6 METHODS BASED ON FLAME EMISSION

Although the subject of this book is AAS, flame emission measurement of phosphorus is briefly treated here, owing to its inherent advantages over AAS methods. These advantages are (a) direct aspiration of the phosphorus samples into the flame without prior decomposition or matrix modification; (b) low detection limit ($\sim$ 0.01 ng) with reasonable accuracy (1-2% error); (c) useful applications as a detector system in gas-liquid chromatography (GLC) and high-pressure liquid chromatography (HPLC).

Since the work of Wöhler in 1841 [1] on the flame emission of some phosphorus compounds, and later that of Salet [2-4], no development took place until 1907 when Geuter [72] and later Ludlam [73,74] and Rumpf [75] investigated the spectrum of phosphorus in the flame. The origin of the spectrum, however, remained unknown and was tentatively assigned to emission by PH, PH_2 and PO species [3,4,72-76]. In 1963, Lam Thanh and Peyron [77, 78] identified the emitter as HPO. The emission spectrum extends over the range 490-600 nm with maxima at 510, 526.2 and 560 nm. The band at 526.2 nm is commonly used for quantitative analysis (Fig. 5.3). Syty and Dean [79] studied the excitation conditions for the emission by HPO and PO species from shielded air-hydrogen and unshielded oxy-acetylene flames. The significant aspect of this study lies in the direct aspiration of phosphorus samples into the flame without the lengthy and often uncertain decomposition steps usually used for many organic and biochemical compounds. Direct nebulization of the sample in a cool nitrogen-hydrogen diffusion flame [80] or in a heated injection

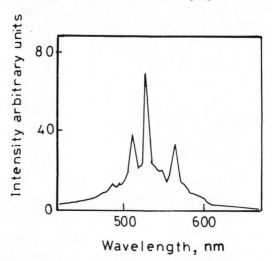

Fig. 5.3. Emission spectrum of HPO species obtained in nitrogen-hydrogen diffusion flame [reprinted from R. M. Dagnall, K. C. Thompson and T. S. West, *Analyst*, **93**, 72 (1968), by permission of the copyright holders, the Royal Society of Chemistry].

manifold system [81] in which the nebulization chamber is maintained at 90°C and the sample is vaporized, passed through the side-arm of the burner, premixed with hydrogen, and carried into the diffusion flame, have been described. Various configurations for the burners and sample introduction systems have been reported [82-87]. These types of burners are used for the determination of phosphorus in insecticides [88], detergents [84,89] and lubricating oil [90]. The detergents are heated for 15 min at 500-520°C in a muffle furnace and passed through a cation-exchanger before measurement of the HPO emission at 526.2 or 528 nm. Lubricating oil cannot be analysed by direct aspiration into the flame, owing to the strong C_2 emission at 516 nm. Thus, oil samples are ashed in the presence of potassium hydroxide, the aqueous extract of the melt is treated with ion-exchange resin and the effluent is aspirated into the flame. Phosphorus in the range 0.009-0.2% can be determined by these methods with a precision of ±5%.

Brody and Chaney [91] devised a burner for use as a detector in gas–liquid chromatography (GLC). The phosphorus compounds, at levels as low as 6.3 ng/g, are separated by GLC, the effluent is aspirated into the flame and the HPO emission at 526 nm is measured. The selectivity of the detector is enhanced by using an opaque shield to prevent the non-HPO emission from reaching the photomultiplier tube. This detector can be used for simultaneous detection of phosphorus and sulphur compounds if it is fitted with twin photomultipliers, one on each side of the flame, to receive the phosphorus and sulphur emissions at 526 and 394 nm, respectively [92]. Problems related to this system have been reported [91,93,94].

A dual-flame burner has been used advantageously, in which one flame is used to decompose the sample, and a second flame, longitudinally separated from the first, is used to produce the desired optical emission. An optically transparent Smithells-type chimney [95] is used to achieve two longitudinally separated hydrogen–air flames. One flame burns in a hydrogen-rich environment at a jet located inside the lower end of the chimney, and the second flame burns at the chimney exit, where the excess of hydrogen combines with oxygen from the surrounding air. This burner can be used for the simultaneous measurement of phosphorus and sulphur. However, the two emissions occur at different locations within the glow region and thus the optical viewing axis has to be relocated to switch from one mode of detection to the other. A convenient metallic dual-flame burner for simultaneous determination of phosphorus and sulphur compounds in gas–liquid chromatography effluents has been described [96]. A schematic illustration of this burner is shown in Fig. 5.4. This design gives a more uniform response with respect to sample concentration and composition, than the single-flame detectors do.

Julin *et al.* [41] devised a burner for use as a phosphorus detector in HPLC. The burner assembly is designed to handle the total liquid effluent from the HPLC column. Hydrogen and nitrogen enter a mixing chamber at the bottom of

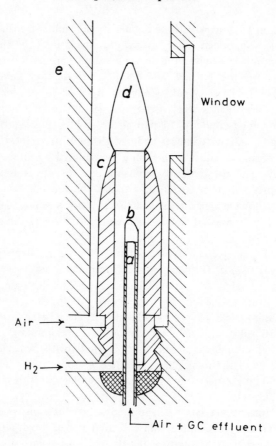

Fig. 5.4. Dual-flame burner: a, flame tip 1; b, flame 1; c, flame tip 2; d, flame 2; e, tower. [Reprinted with permission from P. L. Patterson, R. L. Howe and A. Abu-Shumays, *Anal. Chem.*, **50**, 339 (1978). Copyright 1978 American Chemical Society.]

the burner, and then aspirate the column effluent into the nebulizer, which converts much of the liquid into small droplets. This detector permits detection of about 2×10^{-8} g of phosphorus per ml of column effluent and can satisfactorily be used for the analysis of mixtures of phosphorus compounds. Figure 5.5 shows a chromatogram of a test mixture of four $5'$-monophosphate nucleotides. The detector appears best suited for ion-exchange and perhaps reversed-phase liquid chromatography. However, application of such types of detectors has been limited to systems with aqueous mobile phases. Co-elution of organic materials often leads to high loss of sensitivity and their fragmentation into CO and C_2 may cause spectral interference.

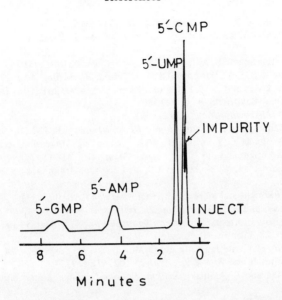

Fig. 5.5. Separation of 5′-monophosphate nucleotides (phosphorus detector): 100 µl of solution containing 0.4 mg/ml of each compound; 5′-GMP (guanosine 5′-monophosphate), 5′-AMP (adenosine 5′-monophosphate), 5′-UMP (uridine 5′-monophosphate), 5′-CMP (cytidine 5′-monophosphate). [Reprinted from B. G. Julin, H. W. Vandenborn and J. J. Kirkland, *J. Chromatog.*, 112, 443 (1975) by permission of the copyright holders, Elsevier Scientific Publishing Co.]

Determination of phosphorus in the flame by methods based on the measurement of the continuum and PO emissions has been described [97–110]. The depressive effect of phosphorus on the emission of magnesium or calcium in the flame has also been utilized for the determination of low levels of phosphorus [111–113].

REFERENCES

[1] F. Wöhler, *Ann. Chem. Pharm.*, 39, 252 (1841).

[2] G. Salet, *Bull. Soc. Chim. France*, 11, 302 (1869); 13, 289 (1870); 16, 195 (1871).

[3] G. Salet, *Compt. Rend.*, 73, 1056 (1871).

[4] G. Salet, *Ann. Chim. Phys.*, 28, 5 (1873).

[5] W. Simons and J. Robertson, *Anal. Chem.*, 22, 294 (1950).

[6] T. S. Ma and J. McKinley, *Mikrochim. Acta*, 4 (1954).

[7] C. J. F. Böttcher, C. M. van Gent and C. Pries, *Anal. Chim. Acta*, 24, 203 (1961).

[8] B. Luskina, A. Terentev and S. Syavtsillo, *Zh. Analit. Khim.*, 17, 639 (1962).

[9] M. Jacobson and S. Hall, *Anal. Chem.*, 20, 736 (1948).

[9a] J. V. Timeno Adelantrado, F. Bosch Reig, A. Pastor Garcia and V. Peris Martinez, *Talanta*, 30, 974 (1983).

[10] F. Eggertsen and F. Weiss, *Anal. Chem.*, 29, 453 (1957).

[11] M. Jureček and J. Jeník, Chem. Listy, 51, 1312 (1957).
[12] K. Fleischer, B. Southworth, J. Hodecker and M. Tuckerman, Anal. Chem., 30, 152 (1958).
[13] S. S. M. Hassan and S. A. I. Thoria, Z. Anal. Chem., 276, 74 (1975).
[14] W. I. Awad, S. S. M. Hassan and S. A. I. Thoria, Mikrochim. Acta, 111 (1976 II).
[15] S. S. M. Hassan and M. H. Eldesouki, Mikrochim. Acta, 261 (1981 II).
[16] W. Kirsten, Microchem. J., 4, 3 (1960).
[17] W. Merz, Mikrochim. Acta, 456 (1959).
[18] D. C. Manning and S. Slavin, Atom. Absorp. Newslett., 8, 132 (1969).
[19] G. C. Toralballa, G. I. Spielholtz and R. J. Steinberg, Mikrochim. Acta, 484 (1972).
[20] G. F. Kirkbright and P. J. Wilson, Anal. Chem., 46, 1414 (1974).
[21] W. B. Barnett, J. W. Vollmer and S. M. DeNuzzo, Atom. Absorp. Newslett., 15, 33 (1976).
[22] G. F. Kirkbright and M. Marshall, Anal. Chem., 45, 1610 (1973).
[23] B. V. L'vov and A. D. Khartsyzov, Zh. Prikl. Spectrosk., 11, 413 (1969).
[24] B. V. L'vov, in Atomic Absorption Spectroscopy, R. M. Dagnall and G. F. Kirkbright (eds.), p. 28. Butterworth, London, 1980.
[25] A. Prévôt, M. Gente-Jauniaux and O. Morin, Rev. Fr. Corps Gras, 24, 409 (1977).
[26] P. J. Whiteside and W. J. Price, Analyst, 102, 618 (1977).
[27] Y. Azuma and S. Aramaki, Kenkyu Hokoku-Nara-ken Kogyo Shikenjo, No. 4, 51 (1978).
[28] J. D. Kerber, D. C. Manning and S. Slavin, 7th ICAS, Prague, 1977.
[29] M. S. Vigler, A. Strecker and A. Varnes, Appl. Spectrosc., 32, 60 (1978).
[30] D. J. Driscott, D. A. Clay and R. H. Jungers, Pittsburgh Conference on Analytical Chemistry and Applied Spectroscopy, Cleveland, Ohio, 1977.
[31] D. J. Driscott, D. A. Clay, C. H. Rogers, R. H. Jungers and F. E. Butler, Anal. Chem., 50, 767 (1978).
[32] F. J. Slikkerveer, A. A. Braad and P. W. Hendrikse, Atom. Spectrosc., 1, 30 (1980).
[33] A. Prévôt and M. Gente-Jauniaux, Atom. Absorp. Newslett., 17, 1 (1978).
[34] M. Gente-Jauniaux and A. Prévôt, Rev. Fr. Corps Gras, 26, 325 (1979).
[35] I. Havezov, E. Russeva and N. Jordanov, Z. Anal. Chem., 296, 125 (1979).
[36] J.-Å. Persson and W. Frech, Anal. Chim. Acta, 119, 75 (1980).
[37] B. V. L'vov and L. A. Pelieva, Zh. Analit. Khim., 33, 1572 (1978).
[38] B. Welz, U. Voellkopf and Z. Grobenski, Anal. Chim. Acta, 136, 201 (1982).
[39] P. Tittarelli and A. Mascherpa, Anal. Chem., 53, 1466 (1981).
[40] D. M. Fraley, D. Yates and S. E. Manahan, Anal. Chem., 51, 2225 (1979).
[41] B. G. Julin, H. W. Vanderborn and J. J. Kirkland, J. Chromatog., 112, 443 (1975).
[42] K. C. Singhal, A. C. Banerji and B. K. Banerjee, Technology (Sindri, India), 5, 117 (1968).
[43] K. C. Singhal and B. K. Banerjee, Technology (Sindri, India), 5, 239 (1968).
[44] V. J. Vajgand, D. D. Stojanović and M. D. Durdević, Glas. Hem. Drus. Beograd, 39, 613 (1975).
[45] C. I. Lin and C. O. Huber, Anal. Chem., 44, 2200 (1972).
[46] W. S. Zaugg and R. J. Knox, Anal. Chem., 38, 1759 (1966).
[47] W. S. Zaugg, Atom. Absorp. Newslett., 6, 63 (1967).
[48] H. Weil-Malherbe and R. H. Green, Biochem. J., 49, 286 (1951).
[49] W. S. Zaugg and R. J. Knox, Anal. Biochem., 20, 282 (1967).
[50] J. A. Parsons, B. Dawson, E. Callahan and J. T. Pottos, Jr., Biochem. J., 119, 791 (1970).
[51] G. Devoto, Boll. Soc. Ital. Biol. Sper., 44, 424 (1968).
[52] G. Linden, S. Turk and B. Torodo de la Fuente, Chem. Anal., 53, 244 (1971).

[53] M. L. Fernandez-Feal, C. Baluja Santos and F. Bermejo Martinez, *Acta Cient. Compostelana*, 9, 35 (1972).

[54] Y. Kidani, H. Takemura and H. Koike, *Bunseki Kagaku*, 23, 212 (1974).

[55] D. F. Boltz and M. G. Mellon, *Bull. Chem. Soc. Japan*, 20, 749 (1948).

[56] M. Jean, *Chim. Anal. (Paris)*, 44, 195 (1962).

[57] C. Wadelin and M. G. Mellon, *Anal. Chem.*, 25, 1668 (1953).

[58] M. A. DeSesa and L. B. Rogers, *Anal. Chem.*, 26, 1381 (1954).

[59] C. H. Lueck and D. F. Boltz, *Anal. Chem.*, 28, 1168 (1956).

[60] T. Kumamaru, Y. Otani and Y. Yamamoto, *Bull. Chem. Soc. Japan*, 20, 429 (1967).

[61] J. L. Bernal, M. J. Del Nozal, L. Deban and A. J. Allen, *Talanta*, 28, 469 (1981).

[62] J. R. Melton, W. L. Hoover, P. A. Howard and V. S. Green, *J. Assoc. Off. Anal. Chem.*, 54, 373 (1971).

[63] G. F. Kirkbright, A. M. Smith and T. S. West, *Analyst*, 92, 411 (1967).

[64] T. R. Hurford and D. F. Boltz, *Anal. Chem.*, 40, 379 (1968).

[65] G. F. Kirkbright, A. M. Smith and T. S. West, *Analyst*, 91, 700 (1966).

[66] C. Riddle and A. Turek, *Anal. Chim. Acta*, 92, 49 (1977).

[67] V. Rozenblum, *Anal. Lett.*, 8, 549 (1975).

[68] P. Tekula-Buxbaum, *Mikrochim. Acta*, 183 (1981 II).

[69] T. V. Ramakrishna, J. W. Robinson and P. W. West, *Anal. Chim. Acta*, 45, 43 (1969).

[70] J. Paul, *Mikrochim. Acta*, 830 (1965).

[70a] R. A. Chalmers and A. G. Sinclair, *Anal. Chim. Acta*, 34, 412 (1966).

[71] S. J. Simon and D. F. Boltz, *Anal. Chem.*, 47, 1758 (1975).

[72] P. Geuter, *Z. Wiss. Photog.*, 5, 1 (1907).

[73] E. B. Ludlam, *Nature*, 128, 271 (1931).

[74] E. B. Ludlam, *J. Chem. Phys.*, 3, 617 (1935).

[75] K. Rumpf, *Z. Phys. Chem.*, B38, 469 (1938).

[76] M. Lam Thanh and M. Peyron, *J. Chim. Phys.*, 59, 688 (1962).

[77] M. Lam Thanh and M. Peyron, *J. Chim. Phys.*, 60, 1289 (1963).

[78] M. Lam Thanh and M. Peyron, *J. Chim. Phys.*, 61, 1531 (1964).

[79] A. Syty and J. A. Dean, *Appl. Opt.*, 7, 1331 (1968).

[80] R. M. Dagnall, K. C. Thompson and T. S. West, *Analyst*, 93, 72 (1968).

[81] K. M. Aldous, R. M. Dagnall and T. S. West, *Analyst*, 95, 417 (1970).

[82] H. Drägerwerk and B. Dräger, *German Patent*, 1,133,913 (26 July 1962).

[83] C. E. Vander Smissen, *U.S. Patent*, 3,213,747 (26 October 1965).

[84] A. Syty, *Anal. Lett.*, 4, 531 (1971).

[85] C. Veillon and J. Y. Park, *Anal. Chim. Acta*, 60, 293 (1972).

[86] C. Veillon and M. Margoshes, *Spectrochim. Acta*, 23B, 553 (1968).

[87] M. J. Prager and W. R. Seitz, *Anal. Chem.*, 47, 148 (1975).

[88] B. Gutsche and R. Herrmann, *Deut. Lebensm.-Rundsch.*, 67, 243 (1971).

[89] W. N. Elliott and R. A. Mostyn, *Analyst*, 96, 452 (1971).

[90] W. N. Elliott, C. Heathcote and R. A. Mostyn, *Talanta*, 19, 359 (1972).

[91] S. S. Brody and J. E. Chaney, *J. Gas Chromatogr.*, 4, 42 (1966).

[92] M. C. Bowman and M. Berze, *Anal. Chem.*, 40, 1448 (1968).

[93] S. O. Farwell and R. A. Rasmussen, *J. Chromatog. Sci.*, 14, 224 (1976).

[94] C. A. Burgett and L. E. Green, *J. Chromatog. Sci.*, 12, 356 (1974).

[95] A. Smithells and H. Ingle, *Trans. Chem. Soc.*, 61, 204 (1892).

[96] P. L. Patterson, R. L. Howe and A. Abu-Shumays, *Anal. Chem.*, 50, 339 (1978).

[97] W. A. Miller, *Phil. Mag.*, 27, 81 (1845).

[98] R. T. Simmler, *Ann. Physik*, 116, 499 (1862).

[99] A. Mitscherlich, *Ann. Physik*, 121, 459 (1864).

[100] E. Mulder, *J. Prakt. Chem.*, 91, 111 (1864).

[101] W. N. Hartley, *Phil. Trans. Roy. Soc.*, **A185**, 161 (1894).

[102] W. N. Hartley, *Proc. Roy. Soc. London*, **54**, 5 (1893).

[103] C. J. Lundström, *Proc. Roy. Soc. London*, **59**, 76 (1895).

[104] H. J. Emeléus and W. E. Downey, *J. Chem. Soc.*, **125**, 2491 (1924).

[105] D. W. Brite, *Anal. Chem.*, **27**, 1815 (1955).

[106] W. A. Dippel, C. E. Bricker and N. H. Furman, *Anal. Chem.*, **26**, 553 (1954).

[107] C. De Watteville, *Z. Wiss. Phot.*, **7**, 279 (1909).

[108] A. De Gramont and C. De Watteville, *Compt. Rend.*, **149**, 263 (1909).

[109] R. K. Skogerboe, A. S. Gravatt and G. H. Morrison, *Anal. Chem.*, **39**, 1602 (1967).

[110] K. Fuwa, H. Haraguchi, K. Okamoto and T. Nagata, *Bunseki Kagaku*, **21**, 945 (1972).

[111] L. Erdey, E. Györi and G. Svehla, *Proc. Conf. Appl. Phys. Chem. Methods Chem. Anal., Budapest*, **3**, 243 (1966).

[112] W. A. Dippel, C. E. Bricker and N. H. Furman, *Anal. Chem.*, **26**, 553 (1954).

[113] D. N. Bernhart, W. B. Chess and D. Roy, *Anal. Chem.*, **33**, 395 (1961).

6

Arsenic compounds

Arsenic is a very toxic element that represents a significant health hazard. It has been implicated in the development of hyperkeratosis, skin cancer, lung cancer and arteriosclerosis. However, many arsenic compounds are widely used in agriculture as herbicides, pesticides, insecticides and rodenticides and in industry as wood preservatives and in medicine for the treatment of various parasitic diseases. It is now firmly established that thousands of tons of organoarsenic compounds enter the environment annually either directly as pesticides or by biological transformation of the inorganic species. On the other hand, arsenic is not a normal constituent in body tissues and fluids. The estimated levels of arsenic in human blood vary greatly and range from 0.1 μg/ml or less to 0.64 μg/ml [1,2].

Most foods for human consumption contain less than 0.5 μg of arsenic per g and rarely more than 1 μg/g, but many seafoods have the capacity to concentrate arsenic up to 100 μg/g or more [3,4]. The U.K. total diet survey suggests that at least 75% of the total arsenic ingested originates from seafood. Thus, precise determination of trace levels of arsenic compounds in drinking water, foodstuffs, biological materials and organic compounds of different nature is very important. Atomic-absorption spectrometry is a successful approach for this purpose. Several review articles have been published during the last decade dealing with the development and applications of this technique for the determination of arsenic in various inorganic and organic substances [5-23].

6.1 DECOMPOSITION OF ORGANOARSENIC COMPOUNDS

Decomposition of solid organoarsenic compounds requires special precautions because arsenic and many of its compounds are quite volatile. Dry ashing of biological or organic substances at 500°C leads to considerable loss [24-27]. This loss is said to be ameliorated by ashing in the presence of sufficient magnesium nitrate [25] but others differ [28,29]. Ashing at low temperature in the presence of electronically excited oxygen has also been reported [30-32]. At a radiofrequency power level of 100 W, arsenic is converted into hydrated arsenic trioxide (arsenious acid). The degree of decomposition offered by this technique depends, however, on the composition of the sample matrix, the radiofrequency power and the chemical form of the arsenic.

Wet digestion with strong acids gives complete arsenic recovery from structurally different organic and biological compounds under suitable conditions [33,34]. Mixtures of nitric, sulphuric and perchloric acids are conveniently used. Molybdenum(VI) may be added to catalyse the digestion of chlorine-containing compounds by maintaining strongly oxidizing conditions to prevent loss of the volatile arsenic trichloride [35,36]. Raptis *et al.* [37] examined three different decomposition methods in the determination of arsenic in organic and biological materials at the nanogram and microgram levels by electrothermal atomization AAS. The samples were decomposed by heating at 140°C for 150 min with a mixture of nitric, perchloric and sulphuric acids, by ignition in oxygen for 30 min, or by heating in a bomb at 150°C for 3-8 hr. Trace studies with inorganic arsenic and addition of ^{74}As to the standard reference material confirmed the suitability of these decomposition methods for quantitative recovery of arsenic with high reproducibility. Four methods of digestion have also been evaluated for the decomposition of biological arsenic compounds [38]. The results demonstrate that though digestion with concentrated nitric acid or dilute sulphuric acid for 3 hr at 150°C under reflux or in a PTFE capsule inside a stainless-steel bomb is adequate, Kjeldahl digestion with a 4:1 v/v mixture of concentrated nitric acid and $4M$ sulphuric acid is superior.

6.2 METHODS BASED ON DIRECT ATOMIZATION IN THE FLAME

Determination of trace quantities of arsenic by direct atomization in the flame and measurement of the absorption at the three principle atomic lines at 188.99, 193.7 and 197.3 nm is often unreliable, probably for three reasons. First, the signals obtained at these wavelengths are weak and noisy [39]; secondly the air–acetylene flame, which produces a temperature sufficient to atomize most elements with few chemical interferences and burns steadily with minimal fluctuation, shows high background absorption and noise at these wavelengths; finally, many arsenic hollow-cathode lamps emit radiation of relatively low intensity and poor stability. However, determination has been reported of arsenic levels as low as 1 μg/ml, at 193.7 nm, with an air–acetylene flame in conjunction with an arsenic hollow-cathode lamp [40-45]. Under similar conditions, arsenic values higher than those obtained by chemical methods of analysis have been obtained and attributed to improper correction for light-scattering [46]. However, these difficulties have been at least partly overcome by using (a) a cooler flame to minimize the background interferences; (b) radiation sources with intensities higher than those obtainable with hollow-cathode lamps; (c) modified atomization devices.

6.2.1 Types of Flames and Burners

It has been reported that the type of the flame used for atomization of arsenic is extremely critical. Welz [41] discussed the use of flame and electrothermal atomization systems for AAS determination. Allan [40] showed that with an air-acetylene flame about 64, 72 and 88% of the maximum arsenic signal is obtained at 197.2, 193.7 and 188.99 nm, respectively. Separation of the outer mantle of the air-acetylene flame from the primary reaction zone by means of a laminar flow of nitrogen reduces the interconal background absorption and the noise level, resulting in a threefold improvement in the detection of arsenic at 193.7 and 196.3 nm [44]. The interconal zone, however, still contains molecular species (most probably oxides) at high temperatures, which results in a significant absorption of the radiation at wavelengths shorter than 200 nm, even when a shielded flame is used.

Amos [47] compared the absorption of arsenic in air-acetylene and air-hydrogen flames, using different burners. He found that an air-hydrogen flame on a wide-slot burner (0.04 in.) is to be preferred (Table 6.1). With this system, the noise level is less than 0.5% and as low an arsenic level as 0.4 μg/ml can readily be measured, compared with a noise level of 1-2% and a detection limit of 3 μg/ml with the air-acetylene flame.

Table 6.1. Flame absorption of arsenic: lines and sensitivity in different flames [47]

Flame	Burner slot (in.)	Light absorbed by flame (%)		Sensitivity for arsenic (ng/ml)	
		197.2 nm	193.7 nm	197.2 nm	193.7 nm
Air-acetylene	0.02	70	75	2.1	1.5
Air-hydrogen	0.02	55	80	2.0	1.5
Air-hydrogen	0.03	42	70	1.5	1.1
Air-hydrogen	0.04	38	61	1.2	0.9

Cold flames have also been used to provide (a) greater transparency in the ultraviolet region than the air-acetylene flame; (b) improvement of the signal: noise ratio; (c) lower fluctuation in the signal intensity. The use of an argon-hydrogen diffusion flame brings about a decrease in the background absorption and an increase in sensitivity. Argon is used to nebulize the analyte and carry it to the burner slot and combustion takes place between the hydrogen fuel and entrained or ambient air. The detection limits are 0.1 μg/ml at 188.99 and 193.7 nm and 0.2 μg/ml at 197.2 nm [48]. The flame background absorption at these wavelengths is one fourth of that obtained by using the air-acetylene flame [43,49,50].

Various burners with premixed and turbulent flames have been investigated.

The argon-entrained air-hydrogen flame proved to be suitable not only for atomic absorption but also for flame emission and atomic fluorescence measurements [50,51]. The performance of the Autolan burner II for AAS was tested for arsenic in aqueous solutions and the best results were obtained with a lean hydrogen-argon-entrained air flame and a single-slot burner [52].

Because of the low temperature of the argon-hydrogen flame compared with the air-acetylene flame, spectral and chemical interferences are much higher. Elements such as aluminium, calcium, magnesium, manganese, nickel and cobalt form arsenides, and thus interfere. Molecular absorption and incomplete salt dissociation are also inevitable. Most of these interferences are not observed with the nitrogen-hydrogen-entrained air flame [53]. It has been noticed that the degree of dissociation and atomization of arsenic strongly depends on the temperature of the relatively cold lower region. The signal:noise ratio tends to decrease with increasing hydrogen flow-rate [54].

The nitrous oxide-acetylene flame has also been demonstrated [55,56] to be an effective medium for the determination of arsenic. Although the sensitivity is lower than that obtained with the air-acetylene and argon-hydrogen flames, the low background absorption and noise of this flame offer a good detection limit for arsenic ($\sim$ 3 μg/ml compared to 0.1 μg/ml with the nitrogen-hydrogen flame). A nitrogen- or argon-shielded nitrous oxide-acetylene flame significantly reduces the background absorption at 193.7 nm and shows a limit of detection a factor or two lower than that of unseparated flames. Kirkbright *et al.* [57] used an argon-shielded nitrous oxide-acetylene flame with instrumentation incorporating a triple-pass optical arrangement, for the determination of arsenic. This technique improves the sensitivity without deterioration of the signal intensity.

Introduction of arsenic samples into the flame with an electrically heated platinum loop has been suggested [58]. The limit of detection ($\sim$ 0.9 ng/ml) is better by one or two orders of magnitude than those obtained with conventional flames. However, the stability and intensity of the atomic line given by the hollow-cathode lamp affect both the sensitivity and detection limit.

6.2.2 Radiation Sources

The hollow-cathode lamps first developed are not suitable for arsenic determination, owing to their short working life, low intensity and relative instability. Demountable hollow-cathode lamps, however, have proved superior. In one of the proposed designs, the cathode assembly and lamp body are cooled with water, and the lamp is filled with argon gas under a pressure of 1 mmHg, adjusted by a needle valve fitted to the gas control unit [59]. The cathode is made of a porous-graphite cup electrode (25.4 mm length, 6.15 mm outer diameter and 3.96 mm inner diameter) filled with $\sim$ 1 mg of elemental arsenic, and the anode is made of brass. The lamp produces an arsenic spectrum with intense emission lines at 193.7, 197.1, 200.3 and 228.8 nm. The spectrum in this range also exhibits a weak copper line at 194.2 nm from the anode. A carbon line at 193.0

nm may also be noticed if the filler gas pressure exceeds 4 mmHg. The background at the 193.7 nm line is very low and the signal-to-background intensity ratio is greater than 200. At this wavelength, arsenic in aqueous solutions has been satisfactorily determined with any of the previously described flame systems. With a premixed air–acetylene flame and a long-path (100 mm) burner, the limit of detection is 0.45 $\mu g/ml$.

The use of electrodeless discharge lamps offers the greatest potential for the development of suitable methods for the determination of arsenic. Although the usefulness of these sources depends on the selection of the cavity, lamp size and filler gas pressure [60], the intensity of arsenic radiation given by these lamps is 2–10 orders of magnitude greater than that given by the hollow-cathode lamp. These sources have a higher output intensity, and a narrower emission line, free from self-absorption or self-reversal. Moreover, these lamps allow operation at low amplifier gain and spectral band-pass settings [61–63].

Bashov [61] used an electrodeless discharge lamp with a propane–butane burner for the determination of arsenic. The lamp (20 mm in diameter) was made of fused quartz and filled with argon at 2 mmHg pressure, and had an inner wall covered with arsenic. The line at 197.262 nm was the analytically useful line, compared to the other two lines obtained (189.042 and 193.759 nm). Nakahara *et al.* [64] used an electrodeless discharge lamp and a premixed inert gas–entrained air–hydrogen flame with a multiflame burner [65] for the determination of arsenic at 193.7 nm.

An electrodeless discharge lamp has been designed consisting of a modified helical resonator mounted within the resonator coil in a quartz bulb containing arsenic and filled with argon at a pressure of 0.5–3 mmHg [66]. The lamp is powered at a frequency of 27 MHz by a solid state supply. The lamp and cavity are constructed as an integrated unit. The external dimensions of the cavity are identical to those of a standard hollow-cathode lamp, making the two types readily interchangeable in the atomic-absorption spectrometer. The emission of the 193.7 nm line is found to be 5–10 times more intense than that of the typical hollow-cathode lamp. The detection limit for arsenic is 0.1 $\mu g/ml$ with the air–acetylene flame, and slightly higher with the nitrous oxide–acetylene flame [66,67]. Although better detection limits would certainly be obtained with cooler flames such as argon–hydrogen, direct nebulization of arsenic solutions into flames, in general, has still not gained wide acceptance.

6.3 METHODS BASED ON DIRECT ELECTROTHERMAL ATOMIZATION

The use of the graphite furnace to atomize arsenic solutions, in conjunction with either a hollow-cathode or an electrodeless discharge lamp as radiation source, has greatly improved the limit of detection for this element [68–72]. The electrically heated graphite-tube furnace [73,74] and carbon rod [12,71] have been used. Massmann [75] determined as little of 0.1 μg of arsenic per ml by measuring its absorbance at 189.04 and 197.26 nm, using an electrically heated graphite tube and an electrodeless discharge lamp. The cell was heated with a current of 40–350 A to a temperature of 2400°C, and copper(II) was added to improve the sensitivity. Chu *et al.* [76] have similarly used an electrically heated absorption cell for the determination of arsenic. Pulse evaporation from a pyrolytic graphite microprobe [77] or a hollow graphite capsule [78] has also been used. These methods are used for the determination of arsenic in organic and biological materials after wet digestion with acids or dry ashing. As little as 87 ng of arsenic per g of wet tissue can be satisfactorily determined [79]. A variety of compounds and materials have been analysed for arsenic by similar procedures [70,71,80–84].

Arsenic in plant and animal tissue samples is determined by digestion with nitric–perchloric–sulphuric acid mixture followed by extraction with diethyl-ammonium diethyldithiocarbamate in chloroform and atomization in a graphite furnace [85]. The average recovery is 99% and the coefficient of variation is 3–8.5%. In an interesting development, the analyte is placed in a graphite electrode which forms the cathode of a demountable hollow-cathode cell with quartz end-windows. Upon discharge, arsenic is atomized and can be measured by using a conventional hollow-cathode lamp as a source of radiation [86].

However, formation of volatile compounds during the charring cycle, interferences by some ions and interaction of arsenic compounds with the graphite material of the furnace have been reported as the main sources of difficulties commonly encountered with the electrothermal atomization technique. Volatilization loss of arsenic by the formation of molecular species such as As_4, As_2 [87,88] and AsO [87], particularly at low temperatures, and formation of gaseous molecules in the presence of chloride and sulphate [90,91] have been reported. Since chlorine and sulphur are frequently present in many organic and biological compounds, determination of arsenic in such compounds is rather problematic, owing to the loss of arsenic [92–95]. To overcome these problems, the influence of the matrix and the interaction with the graphite tube should be circumvented.

6.3.1 Matrix Modification

Addition of nickel, lanthanum or some other inorganic salts to arsenic-containing materials increases the sensitivity of measurement and circumvents the volatility of arsenic [92,96,97]. Both nickel and lanthanum stabilize arsenic by formation of stable metal–arsenic bonds, through different mechanisms. It seems that nickel forms an intermetallic compound of composition $Ni(AsO_3)_2.NiO$ [98] and lanthanum stabilizes the oxidized centres of the atomizer and creates an interlamellar compound. Nickel is commonly used over a wide range of temperatures to provide about 50% enhancement of sensitivity, whereas with lanthanum loss of arsenic begins at temperatures above 900°C. Nickel chloride, nitrate or sulphate is used at a concentration of 1 mg/ml or more [82,97,99]. It has been reported in a recent study, however, that such a high concentration of nickel causes rapid deterioration of the graphite tube, resulting in irreproducible results. Nickel in the concentration range 100–200 μg/ml is thus recommended. Under these conditions, the life-time of the atomizer exceeds 70 injections instead of only 25 injections when a 1 mg/ml nickel concentration is used [93]. Some other matrix modifiers such as alkali metal, alkaline-earth metal, cobalt and iron nitrates, as well as perchloric acid, nitric acid and some bases, have been used to enhance the absorption and to improve the sensitivity for arsenic in some organic compounds [100].

6.3.2 Interaction of Arsenic with Graphite Furnaces

Although Nickel [101] showed that arsenic does not form stable compounds with graphite, it appears from other studies [102,103] that interaction with graphite does take place. Formation of non-stoichiometric compounds of the type AsC_n [104] and $C_mH_2AsO_4.2H_3AsO_4$ [25,26], where m is the degree of the lamellar compound (i.e. the ratio of carbon layers to reacted layers), has been suggested. This effect can be eliminated by coating the graphite tube with tantalum carbide [94], tungsten carbide [107], or tantalum foil [108] or by using molybdenum microtubes [109]. Korečková et al. [96] studied the role of the graphite furnace in the determination of arsenic and monitored the loss of arsenic during thermal pretreatment of the tube, as well as its degree of penetration into the graphite tube, using radioactive arsenic compounds. These authors used the L'vov platform technique [104] to study only the effect of the surface in contact with the sample, independent of the atomizer material. The study showed that both the atomizer material and the nature and size of the contact area with the sample are critical parameters in the determination of arsenic, and the authors recommended addition of nickel or lanthanum.

The transient signal of arsenic in pure aqueous solutions and in the presence of hydrogen peroxide has also been investigated [94]. From aqueous solution, arsenic atoms are formed at 1200°C, whereas in the presence of hydrogen peroxide, the signal does not commence until a tube-wall temperature of about 1800°C is reached, which means that almost ideal conditions for atom formation are attained. This observation is in good agreement with the results obtained from radioactive measurements. Other oxidizing agents (e.g. nitric acid, potassium permanganate and oxygen) show a similar effect.

6.3.3 Interferences

Interferences by metal ions are more common with the graphite furnace than with the flame techniques. Some metals cause chemical interferences, enhancement and suppression, which affect the production rate and population of gaseous arsenic. High concentrations of sodium, sulphate, halides [92], nitrate, iron, cobalt, nickel, perchlorate, protons [100], phosphate, bismuth, lithium, antimony, selenium, titanium [110] and different atomic and molecular species of arsenic [87] are known to interfere. Determination of arsenic at low concentrations is almost impossible when aluminium, sodium, potassium and sulphate ions are present together in the sample at concentration levels exceeding a few μg/ml [93]. Spectral interferences from phosphate matrices, due to the formation of PO and P_2 species, have also been reported [111]. However, the magnitude of most interferences depends upon the time and temperature of the cycle which in turn depend on the physical and chemical properties of both the matrix and graphite tube [80].

Although most of these interferences have been satisfactorily circumvented by solvent extraction [48,60,81,94,109,112-115], ion-exchange [126,127], flotation [128] and co-precipitation [129-132], addition of nickel as matrix modifier has proved to be rather simple and more effective in many cases [97]. Organic and biological compounds are digested with a mixture of nitric and sulphuric acid with either perchloric acid or hydrogen peroxide, and treated with nickel nitrate before injection into the graphite furnace [82-84,119, 133-146]. Arsenic levels as low as 50 pg/ml can be measured under these conditions with a precision of 5% when an electrodeless discharge lamp is used as radiation source [147].

Arsenic in feedstocks for catalytic reforming has been determined with magnesium as matrix modifier [148]. A portion of the sample ($\sim$ 10 ml) containing at least 10 ng of arsenic is treated with 0.5 ml of 1% iodine solution in toluene, followed by extraction with two 10-ml portions of 1% nitric acid. The aqueous extract is then treated with 0.1 ml of 1% magnesium nitrate solution and evaporated to dryness; the residue is dissolved in a known volume of 1% nitric acid. Aliquots (5-50 μl) are injected into the graphite furnace for arsenic measurement at 193.7 nm, by the standard addition technique. Arsenic in synthetic fuel [149], soil [150] and water [151] has been determined by similar methods. Automated systems using the graphite furnace and with matrix modification and autosampling controlled by microprocessors have been described [152].

6.3.4 Speciation of Arsenic Compounds

Mixtures of arsenic compounds can be analysed for each individual component by chromatographic separation followed by AAS measurement. Thin-layer chromatography in conjunction with AAS [153,154] has been used for separation and analysis of some mixtures. High-pressure liquid chromatography has also been used in conjunction with the graphite furnace for separation and detection of arsenical pesticide residues and some of their metabolites [155]. This involves separation of methylarsonic acid, cacodylic acid, arsenite and arsenate on a low-capacity anion-exchange column. The effluent is passed to a sampling cup from which 20-μl aliquots are automatically injected into the graphite furnace. As little as 2 ng of arsenic can be detected. The capability of this system to speciate trace quantities of arsenite, arsenate, methylarsonic acid (MAA) and dimethylarsinic acid (DMAA) has also been demonstrated by other workers [156]. Arsenite, MAA and DMAA are successfully separated and quantitatively determined on a strong anion-exchanger column, with an aqueous acetate buffer as mobile phase. Arsenite, DMAA and arsenate can be separated on a strong cation-exchanger column with an aqueous solution of ammonium acetate as mobile phase. All four arsenic compounds can be separated on a C_{18} reversed-phase column with methanol–water mixtures saturated with tetra-heptylammonium nitrate. These compounds can be identified and determined at arsenic levels of 10 ng/ml.

Ion-exchange chromatography combined with AAS has also been employed for separation of mixtures containing arsenite, arsenate, MMA and DMAA [157]. Anion-exchange (Dowex Al-X8) and cation-exchange (Dowex 50W-X8) columns are used for the separation. DMAA is isolated from one aliquot of the mixture by retention on the cation-exchanger, probably as $(CH_3)_2AsO_2H\overset{+}{N}H_4$. A second aliquot of the mixture is treated with $1.75M$ acetic acid, loaded onto the cation-exchange column and eluted with $0.02M$ acetic acid. DMAA is then stripped by passing $1M$ ammonia solution through the column. To isolate MAA, another aliquot of the mixture is mixed with ammonium acetate–acetic acid buffer of pH 4.7, allowed to pass through the anion-exchange column, which is then eluted with $0.01M$ acetate buffer (pH 4.7). The arsenite and DMAA are eluted while arsenate and MAA, probably as $CH_3AsO_3^-$ species, are retained on the column. MAA is collected, followed by arsenate, by stripping with $0.5M$ acetate buffer (pH 4.7). Each of these species, after the separation, is atomized in the graphite furnace and arsenic is measured at 193.7 nm, with a hollow-cathode lamp as light-source. The average recovery is 101%, the standard deviation being 3.5%.

A modified version of this method has been suggested by Grabinski [158]. A single column containing both exchangers and consecutive elution of the four species from the column are used. The sample solution, containing 80–4000 ng of As in a total volume not exceeding 2 ml is loaded onto the resin and trichloro-acetic acid (6mM, pH 2.5) is used as the initial mobile phase. The pH of the eluent is low enough to release the arsenite, and $0.2M$ trichloroacetic acid is added next to provide sufficient protons for elution of arsenate. Finally, $1.5M$ ammonia solution is added to strip the strongly retained DMAA. Fractions of the effluents are collected, treated with 1% nitric acid and 0.1% nickel nitrate solution, then atomized in a tantalum-treated graphite tube. The results show recoveries in the range 96–107%.

Dimethylarsinate has been preconcentrated by cation-exchange and then determined by graphite-furnace AAS, with a detection limit of 0.02 ng/ml and recovery of 100±6% [158a].

6.4 METHODS BASED ON ARSINE GENERATION

In the conventional flame AAS determination of arsenic, the analyte solution is introduced into the flame as an aerosol by means of a pneumatic nebulizer which is only about 2-5% efficient in forming the fine droplets suitable for atomization [159]. Even ultrasonic nebulization seldom has an efficiency greater than 35% and this obviously affects the lower limit of detection offered by this technique. However, flame AAS can be used for the determination of arsenic at near or above the $\mu g/ml$ level, whereas electrothermal AAS methods are useful at the ng/ml level in the absence of many extraneous metals. Hydride generation with subsequent AAS measurement is now a well known method for the determination of arsenic. The high sensitivity relative to that of direct electrothermal AAS methods, and the relative freedom from many interferences, have attracted increasing attention to the application of this method.

Chemical conversion of arsenic into arsine (Marsh reaction) dates back nearly 150 years and has been successfully used since then for detection and determination of arsenic in tissues, food, medicated animals, feeds, biological material and petroleum products [160-165]. However, determination of arsenic by hydride generation coupled with AAS began only in the late 1960s. The major AAS instrument manufacturers and several accessories firms now offer equipment for hydride generation. In principle, the arsenic solution is first acidified and treated with a suitable reducing agent such as zinc, magnesium, aluminium, or sodium borohydride or borocyanohydride, to form arsine. This volatile gas is collected and allowed to pass through a heated zone (the atomizer) on the optical axis of the AAS instrument, where it dissociates to give atomic arsenic, the absorption of which is measured at 193.7 nm. It should be noted that even at a dilution of 1:20 000 arsine is injurious and may cause death by anoxia or pulmonary oedema. The maximum allowable concentration (TLV–TWA) in the human environment is 0.05 ppm. Thus, the reaction should be done with very dilute arsenic solutions in a well ventilated or closed system.

Evolution and absorption of the hydride from the reaction medium is governed by a number of parameters, among which are the chemical nature, form and concentration of the reductant and the acid, and the carrier-gas flow. To obtain high sensitivity, it is necessary to (a) strip the hydride quickly from the solution with minimum dilution by other gases and to get as large a fraction as possible into the atomizer at one time; (b) minimize the reagent blank during sample preparation and reaction; (c) ensure that all the arsenic contained in the sample solution is in the tervalent state before reduction, since arsenic(V) is more slowly reduced than arsenic(III); (d) use an efficient atomization system for arsine decomposition.

Two main methods for generation of arsine are known. These are the earlier techniques which involved reduction by a metal–acid system, and the more effective reduction technique using borohydride, which has virtually replaced the older ones. There are two fundamentally different modes of implementing arsine generation–AAS. These are (a) prior collection of the arsine by freezing, storage or absorption in a suitable reagent before atomization; (b) direct introduction of the arsine into the atomizer as soon as the gas is produced. Prior collection of arsine is advantageous when the analytical signal is measured by peak height and slow reduction (e.g. with zinc and acid) is used. Different methods of atomization have been used, including the flame, and electrically or flame-heated graphite or silica tubes.

6.4.1 Reduction with Metals

Conversion of arsenic into arsine requires a source of nascent hydrogen, which can be smoothly provided by metal-acid reaction. Zinc metal in the form of powder, granules, tablets and columns, is the metal most frequently used for arsine generation and the first to gain acceptance for use with AAS. In 1963 Holak [166] described the first AAS method for the determination of arsenic by hydride generation with zinc metal. The arsenic was first reduced to the tervalent state with tin(II) chloride and potassium iodide, before its reduction with zinc, and the arsine was collected by freezing in a U-tube immersed in liquid nitrogen. When reaction was complete, the collection tube was allowed to warm up to room temperature and the arsine was swept with a constant flow of nitrogen (50 ml/min) to an air-acetylene flame. This method has also been applied for determination of ng-amounts of arsenic in blood, milk, hair, tissue, fat and urine after digestion at 170-200°C with a mixture of concentrated sulphuric and nitric acids; a hydrogen–argon flame was used [167]. Arsenic in tobacco has been determined by similar procedures, the arsine being atomized in a nitrous oxide-acetylene-entrained air flame [168]. Arsenic levels as low as 50 ng/g can be determined, but the relative standard deviation can be as high as 18% at the 0.3 μg/g level.

Reduction with metals has proved suitable for a variety of inorganic and organic compounds [169,170].

Fernandez and Manning [39] devised an apparatus for hydride generation in which arsine plus the excess of hydrogen gas was collected in a balloon reservoir (wall thickness 0.007 in.) attached to a reduction flask (Fig. 6.1) fitted with a 4-way stopcock. By rotation of the stopcock, an auxiliary argon flow can be set to by-pass the generation flask or flow through it to carry the collected arsine to the argon–hydrogen–entrained air flame. This arrangement provides a detection limit of about 0.02 μg of arsenic and can be used for the analysis of organoarsenic compounds after prior digestion. One of the limitations of this design, however, is that the zinc metal for hydrogen generation is introduced through a port that must then be immediately sealed to prevent loss of arsine. Also, it is not possible to purge the apparatus before initiating the arsine generation. Manning [171] overcame these drawbacks by using a closed system (Fig. 6.2). The balloon reservoir (wall thickness 0.012-0.014 in.) was attached to a hose nipple, which did not have to be replaced for each sample. The use of

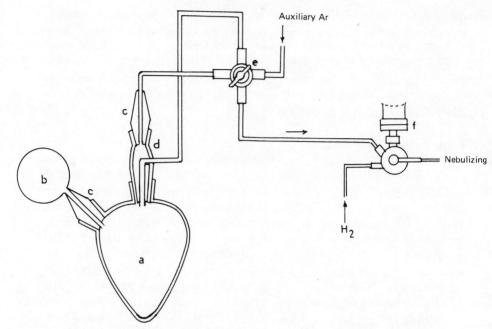

Fig. 6.1. Apparatus for arsine generation: a, generation flask (100-ml); b, collection balloon (0.007 in. wall thickness); c, adapter; d, adapter; e, 4-way stopcock (2-mm bore); f, burner. [Reprinted from F. J. Fernandez and D. C. Manning, *Atom. Absorp. Newslett.*, **10**, 86 (1971) by permission of the copyright holders, the Perkin-Elmer Corporation.]

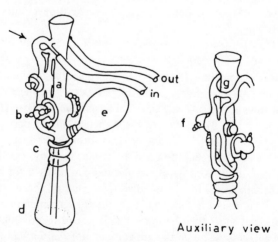

Fig. 6.2. A closed system for arsine generation: a, dosing column; b, dosing stopcock; c, 29/42 ground-glass joint; d, flask; e, balloon reservoir; f, 4-way stopcock; g, argon by-pass. [After D. C. Manning, *Atom. Absorp. Newslett.*, **10**, 123 (1971), by permission of the copyright owners, the Perkin-Elmer Corporation.]

this design appreciably reduces the blank signal and considerably lowers the detection limit. Much higher sensitivity can be achieved by using an electrodeless discharge lamp [66].

An apparatus basically the same as that devised by Manning has been developed except that the dosing stopcock is replaced by a dosing column (1-cm side-arm) at an angle of 45° to the vertical axis of the reaction flask. Zinc is added to a glass bulb reservoir joined to the side-arm with rubber tubing kept closed by a pinch-clamp [172]. A similar apparatus has also been described by Manning and Fernandez [173]. These arsine generation devices are used for the determination of arsenic in various substances [174-178]. Organic compounds are decomposed either by dry ashing or wet decomposition with a mixture of sulphuric and nitric acids (1:5 v/v) at 120-140°C, followed by reduction with tin(II) chloride and zinc. The arsine gas is collected in a balloon before being swept to the absorption tube. Introduction of arsine into an electrically heated tube (as atom cell) by means of argon carrier gas not only significantly lowered the background absorption compared with that of the argon–hydrogen flame but also increased the sensitivity and improved the reproducibility [76].

The use of a balloon to collect arsine, followed by subsequent rapid expulsion to the atomizer, was never widespread, however, probably because of the rapid degradation of the balloon by the acid vapour. Collection of arsine by condensation has proved to be more valuable, in a number of recent studies. The use of a liquid-air or liquid-nitrogen trap to condense arsine and organoarsine compounds [179] eliminates the problems arising from the different reducibilities of arsenic(III) and arsenic(V), in highly acidic media [180]. Arsine has also been collected and stored in a gasometer containing saturated aqueous sodium chloride solution [181]. In general, prior collection of arsine, if correctly performed, should free the overall analysis method from the errors caused by variations in experimental parameters affecting reduction or gas stripping kinetics. However, because of the use of inefficient gas transfer systems or ineffective atomizers the actual results obtained are much poorer than expected.

For general analytical work, a simple continuous flow system may be adequate, although the dead volume of the generator and connecting tubes is generally much larger than the actual atomizer volume and only a small fraction of the hydride can actually be atomized. Continuous introduction of directly generated arsine gas into the flame with simultaneous recording of the absorption signal of atomic arsenic has been suggested, to avoid the time-consuming freezing step [54,121,182-184]. Although some loss in sensitivity is expected due to dilution of arsine with the carrier gas, a detection limit of better than 0.1 μg and a relative standard deviation of 2-3% are obtainable with argon-hydrogen flames.

Besides the use of zinc in the form of powder [121] or tablets [54] for arsine generation, a zinc reductor column (Fig. 6.3) has also been utilized [185-

187]. A small volume of acidified solution containing methylarsonic acid or dimethylarsinic acid is reduced to give arsenic(III) and injected through an 8 × 1.5 cm glass column packed with untreated 2.5-mm mesh zinc metal and fitted with an injection port at the top and a stopcock at the bottom [188]. The gas inlet and outlet holes are positioned so that a continuous stream of gas flows through the column when it is connected to the inlet of the furnace. A small tube containing calcium chloride is inserted between the column and the furnace. Arsine may also be collected till the gas pressure reaches 0.5 kg/cm² before atomization [189].

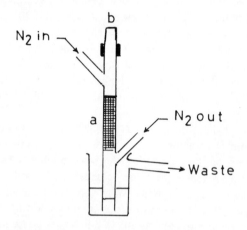

Fig. 6.3. Zinc column for arsine generation: a, zinc granules (30 mesh); b, septum cap. [Reprinted from K. C. Thompson and D. R. Thomerson, *Analyst*, 99, 595 (1974), by permission of the copyright holders, the Royal Society of Chemistry.]

It should be noted that reduction with metals has several drawbacks which somewhat offset the advantages offered by the hydride generation technique. These are (a) the reduction step needed to convert arsenic(V) into arsenic(III) prior to the hydride generation; (b) the long reaction time (~ 20 min); (c) the inapplicability in automated systems. Some of these problems have been solved by using magnesium metal [190], aluminium slurry [191] and titanium(III) chloride [190] instead of zinc. An apparatus with a solenoid valve at the gas inlet and outlet and a pressurized purging system (Fig. 6.4) has been used for arsine generation by reduction with titanium(III) or magnesium metal. The solenoid valve is opened and the system is flushed with argon, then a magnesium rod is dropped into the reaction vessel through the reagent orifice. The arsine and hydrogen gases evolved are collected, a compressed-air toggle valve (5 psig) is opened to pressurize the chamber and the solenoid valve is opened to expel the arsine into the burner.

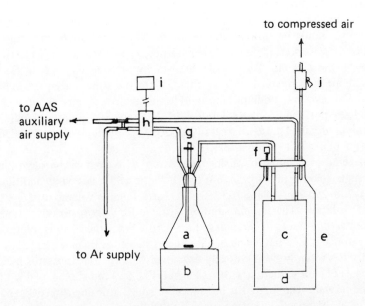

to compressed air

to AAS
auxiliary
air supply

to Ar supply

Fig. 6.4. Hydride generator and sampling system: a, 125-ml reaction flask; b, magnetic stirrer; c, collection chamber made of flexible plastic; d, pressure chamber; e, blood-pack unit (450-ml, Fenwal Laboratories); f, pressure release; g, pinch clamp; h, solenoid valve; i, electric switch; j, 5-psig toggle valve. [Reprinted from E. N. Pollock and S. J. West, *Atom. Absorp. Newslett.*, **12**, 6 (1973), by permission of the copyright holders, the Perkin-Elmer Corporation.]

6.4.2 Reduction with Borohydride

Sodium borohydride (sodium tetrahydroborate) has been used for synthetic purposes for a number of years [192,193]. The reagent was used first for hydride generation in conjunction with AAS in 1973 [194] and since then it has virtually replaced the metals. The advantages offered by this reagent are (a) efficient hydride transfer over a wide range of acidity; (b) relative ease of use in automated systems; (c) fast hydride formation ($\sim$ 10-30 sec); (d) low blank signal; (e) possible applications with some other elements such as germanium, selenium, tin, antimony, tellurium, lead and bismuth [195-197]. Recently, sodium borocyanohydride has been advocated as a reducing and masking agent [198]. It reduces arsenic to arsine and masks many metal ions that may be present in the reaction solution. However, extreme care should be taken to avoid exposure to the extremely toxic hydrogen cyanide vapour produced on acidification of the test solution, and to prevent any leak in the system.

The development of procedures utilizing the sodium borohydride reaction has progressed in much the same manner as for the metal-acid reaction, and the arsine generated is manipulated in many of the same ways. Some workers [199] have reported that the oxidation state of arsenic has a direct bearing on the choice of method. It seems from this study and other investigations [180,200] that the arsenic should be entirely present in one oxidation state before reduction. Most of these workers prefer the lower valency, as conversion of arsenic(III) into arsine is more rapid. The signal obtained by starting with arsenic(III) solution is about 30% greater than that obtained for the same concentration of arsenic(V). With excess of sodium borohydride the increase is only about 10%. The use of 1–2% potassium iodide solution [201,202] and 10% potassium iodide solution stabilized with 1% ascorbic acid [203-206] is commonly recommended for conversion of arsenic(V) into arsenic(III).

The general procedure for arsine generation with sodium borohydride involves sample acidification with 1–6M hydrochloric acid followed by addition of either a 200-250 mg pellet [196,197,207] or a suitable volume of 0.5-8% borohydride solution. The reagent solution is stable for a short period of time ranging from several hours to three days [208-210], but if the alkaline solution is filtered through a 0.45-μm membrane it remains stable for three weeks [211]. The stability has recently been critically examined [211a]. The apparatus used by Manning [171] for reduction with zinc metal (Fig. 6.2) can also be used with borohydride reagent. The reagent is introduced through a dosing stopcock, and the arsine and hydrogen produced are collected in a balloon reservoir and subsequently purged with a flow of argon into an argon–hydrogen flame [196]. Another method uses a horizontal glass tube in which a pellet of sodium borohydride is placed, and a 10-25 μl portion of the acidified arsenic solution is dropped on the pellet through a sample injection hole by means of an Eppendorf pipette (Fig. 6.5) [212].

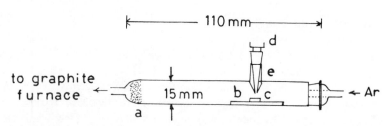

Fig. 6.5. Apparatus for arsine generation: a, silica wool; b, pulp cotton. c, sodium borohydride; d, pipette; e, sample injection port. [Reprinted from T. Inui, S. Terada and H. Tamura, *Z. Anal. Chem.*, **305**, 189 (1981), by permission of the copyright holders, Springer-Verlag.]

Kokot [213] used a hydride generation vessel sealed with a large rubber stopper with 3 glass tubes fitted into it. Two of these tubes are connected by Tygon tubing to a by-pass stopcock, and the third tube is fitted with a rubber septum through which sodium borohydride solution is injected. The by-pass stopcock is connected to a Perspex plug fitted to the atomizer. After reduction, the stopcock is opened, the balloon reservoir squeezed and the arsine passed into a nitrogen–hydrogen–entrained air flame (Fig. 6.6). Other systems for arsine generation have been suggested [214-216]. Several workers, however, have used systems that do not require prior collection of the arsine [197,208,217-220]. Alkylated arsenicals such as methylarsonic and dimethylarsinic acids have been determined directly without prior digestion [110]. Such compounds, on reduction with sodium borohydride, give methylarsine and dimethylarsine, respectively. These volatile products are directly atomized into a hydrogen–nitrogen–entrained air flame or a graphite furnace [221,222]. Peacock and Singh [223] used a 6 × 1 in. boiling tube fitted with a rubber bung with two holes for arsine generation. The first hole was connected to the nebulizer inlet and the second hole was used for borohydride injection (Fig. 6.7). The gas generated was directly nebulized into a nitrogen-supported hydrogen flame.

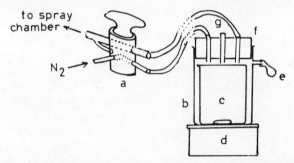

Fig. 6.6. Hydride generation vessel: a, by-pass stopcock; b, hydride generation vessel; c, 100-ml beaker; d, magnetic stirrer; e, balloon reservoir; f, rubber stopper; g, septum. [Reprinted from M. L. Kokot, *Atom. Absorp. Newslett.*, **15**, 105 (1967), by permission of the copyright holders, the Perkin-Elmer Corporation.]

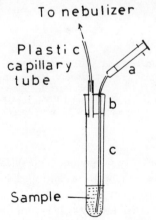

Fig. 6.7. Schematic diagram of hydride generation apparatus: a, 2-ml plastic syringe and capillary tube; b, rubber bung; c, 23 × 150 mm boiling tube. [Reprinted from C. J. Peacock and S. C. Singh, *Analyst,* **106**, 931 (1981), by permission of the copyright holders, the Royal Society of Chemistry.]

6.4.3 Methods of Arsine Atomization

Arsine is dissociated at about 800°C. This temperature can easily be attained by different types of atomizers. An air–acetylene flame has been used [166,208]. The detection limit is 0.45 ng. One of the primary difficulties encountered in the use of this flame system is the significant absorption (~ 60%) by the background at 193.7 nm [43]. The use of an argon–hydrogen flame (Fig. 6.8) provides higher sensitivity and reduces the background absorption [39,177,178, 180,196,214,215,224-226]. This flame, however, is invisible to the naked eye, so it may go out without warning to the operator. Moreover, the stoichiometry of the flame is critical, and affects the base-line. A nitrogen–hydrogen–entrained air flame with nitrogen carrier gas gives performance equal to that of the argon–hydrogen–entrained air flame [213,216,223,227-230].

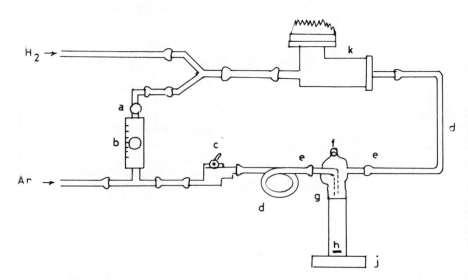

Fig. 6.8. Hydride generation cell and associated gas lines: a, regulating valve; b, flow-meter; c, on-off valve; d, flexible tubing; e, restriction 2 mm i.d. × 20 mm; f, cork stopper; g, hydride generation cell; h, magnetic follower; j, magnetic stirrer; k, burner. [Reprinted from H. D. Fleming and R. G. Ide, *Anal. Chim. Acta*, **83**, 67 (1976), by permission of the copyright holders, Elsevier Scientific Publishing Co.]

Electrically and thermally heated graphite and silica tubes have been used for arsine atomization. These temperature-controlled devices accommodate a high fraction of arsine at any given instant and give a long residence time of arsenic atoms. Electrically heated [204,206,231-238] and flame heated [202, 239-244] silica tubes have been suggested (Fig. 6.9). The length of these tubes is determined by the size of the burner chamber of the spectrometer. A high-volume T-piece quartz tube (15-17 cm long, 5-12 mm internal diameter) is

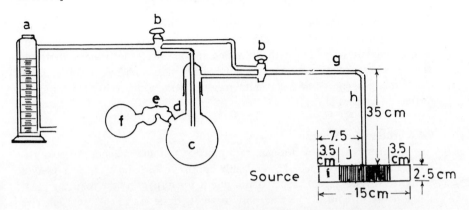

Fig. 6.9. Diagram of arsine generation apparatus and electrically heated absorption tube: a, flow-meter; b, 3-way stopcocks; c, generation flask, 100-ml; d, side neck; e, drying tube as adapter; f, collection balloon; g, Tygon tube; h, Vycor glass tubing, 8 mm o.d. and 4.5 mm i.d.; i, absorption tube; j, chromel A asbestos-covered wire, size 26. [Reprinted by permission from R. C. Chu, G. P. Barron and P. A. W. Baumgarner, *Anal. Chem.*, **44**, 1476 (1972). Copyright 1972 American Chemical Society.]

commonly used (Fig. 6.10) [242]. Similar devices [87,219,220] and graphite-tube furnaces [82,211,245–249] as well as the carbon rod atomizer [250] have also been used. These devices are satisfactorily utilized for the nanogram determination of arsenic in organic compounds [244], seaweeds [203,248], urine [195,243], biological materials [246], plant leaves [211] and fish products [202]. However, determination of arsenic in some biological materials, (e.g.

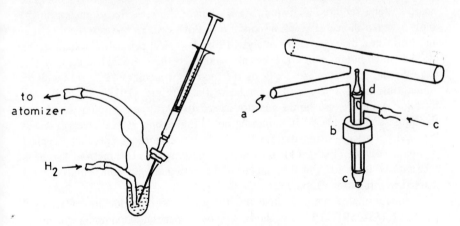

Fig. 6.10. Hydride generator and quartz-tube atomizer: a, hydrogen–arsine inlet from the generator; b, aluminium adapter to allow mounting in conventional burner base; c, rubber sleeve; d, O-ring; e, oxidant inlet. [Reprinted by permission from D. D. Siemer, P. Koteel and V. Jariwala, *Anal. Chem.*, **48**, 836 (1976). Copyright 1976 American Chemical Society.]

urine) requires some precautions due to the formation of a dense voluminous foam upon the addition of borohydride to the acidified sample solution. Some of this foam is swept out with the arsine and penetrates into the heated tube, causing severe contamination for subsequent measurement. Addition of an anti-foaming agent permits accurate measurement without prior digestion or further difficulties [195].

6.4.4 Automated Systems

An automated accessory for generating arsine (Atomic Vapour Accessory or AVA) has been developed. The assembly consists of a sample vessel, a reservoir for borohydride reagent, a peristaltic pump for dispensing the reducing agent, components to control the transport of the hydride gas to the atomizer, electronic circuitry to control the system, and remote triggering of the atomic-absorption spectrometer and the strip-chart recorder. For operation, the acidified arsenic solution is manually transferred to the sample vessel, and the AVA automatically purges the system with nitrogen or argon, adds alkaline borohydride solution and stirs the reagent–sample mixture. When the reaction is completed, the evolved arsine is swept through the atomizer and the atomic absorption measurement is automatically triggered and the peak area and/or peak height recorded. It is also possible to generate arsine in a continuous flow system and to measure the steady-state response.

Automation and semiautomation of the entire procedure for arsenic determination, with a hydrogen–argon–entrained air flame or a tube furnace have been described [232,251-254]. The semiautomatic systems use either a peristaltic pump [191,218,232,251,255] or a pressurized reagent system. The pump systems require less manipulation whereas the pressure system appears to allow a wider choice of reaction conditions. Both systems increase the precision and result in a decrease in the analysis time. A typical peristaltic system with a multichannel pump mixes the arsenic sample with hydrochloric acid, tin(II) chloride and potassium iodide, heats the mixture in a coil at 45°C, then adds an aluminium metal slurry and again heats the mixture [191]. Argon is then introduced and the gas-liquid mixture passed to a stripping column to separate the liquid and gas phases. It should be noted that arsine, even at levels as low as 0.1 μg/l., is easily stripped from the reaction solutions. A specially designed stripping column (Fig. 6.11), is used for this purpose. The method has been successfully used for the determination of organoarsenic compounds after a manual wet digestion (Table 6.2).

Semiautomated methods involving the use of borohydride are widely used [232,251,255-259]. These methods employ a peristaltic pumping system to provide continuous mixing of the reactants, separation of arsine by either a cooled or heated stripping column, and transport of the gas to the atomizer. A schematic diagram of one of these systems is shown in Fig. 6.12. Excellent precision and detection and short analysis time are offered by these systems.

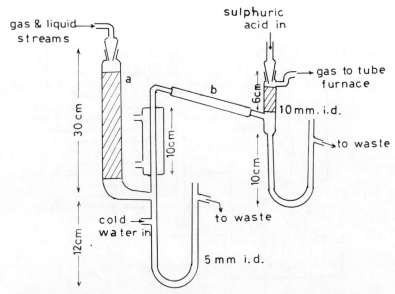

Fig. 6.11. Stripping column and arsine washer system: a, packed column (18 mm i.d.) wrapped in asbestos and heating wire; b, acidflex tubing. [Reprinted by permission from P. D. Goulden and P. Brooksbank, *Anal. Chem.*, **46**, 1431 (1974). Copyright 1974 American Chemical Society.]

Table 6.2. Automatic AAS determination of arsenic in some organic compounds by methods based on arsine generation (by permission of the copyright holders, the American Chemical Society)

Compound	Arsenic (ng/ml)		Recovery (%)
	Taken	Found	
Cacodylic acid (dimethylarsinic acid)	10	10.2	102.0[a]
Disodium methane arsonate (DSMA)	15	14.4	96.0[a]
Phenylarsonic acid	10	7.0	70.0[a]
Thorin [o-(2-hydroxy-3,6-disulpho-1-naphthylazo)benzene arsonic acid]	10	9.8	98.0[a]
p-Arsanilic acid	10	9.9	99.0[b]
p-Arsenic triphenyl	10	10.1	101.0[b]
p-Arsonophenyl urea	10	9.7	97.0[b]
Cacodylic acid	10	9.8	98.0[b]
p-Aminophenylarsonate	20	20.14	100.7[c]
Dimethylarsinate	20	19.58	97.9[c]
Methylarsonate	20	19.99	99.9[c]

[a] Decomposition with persulphate–acid mixture, reduction with borohydride, AAS with electrodeless discharge lamp [260].
[b] Wet digestion with sulphuric acid and reduction with aluminium slurry [191].
[c] Reduction with sodium borohydride, AAS with electrodeless discharge lamp [261].

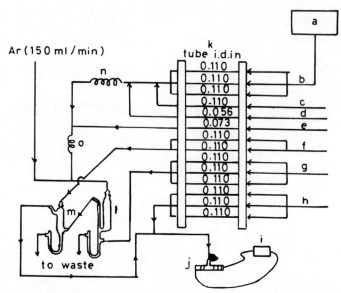

Fig. 6.12. Automated manifold for arsenic: a, automatic sampler; b, sample; c, argon (150 ml/min); d, conc. HCl; e, NaBH$_4$, 2%; f, conc. H$_2$SO$_4$; g, water; h, air; i, Variac transformer; j, quartz-tube atomizer; k, proportioning pump; l, stripping column; m, wash column; n, mixing coil (13 turns); o, mixing coil (27 turns). [Reprinted from H. Agemian and R. Thomson, *Analyst,* **105**, 902 (1980), by permission of the copyright holders, the Royal Society of Chemistry.]

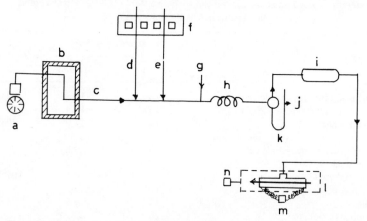

Fig. 6.13. Dionex ion chromatograph coupled with automated arsine generation and AAS detector: a, sampler; b, Dionex ion chromatograph; c, column effluent (2.65 ml/min); d, 15% HCl saturated with persulphate (0.8 ml/min); e, NaBH$_4$ (2 ml/min); f, proportioning pump; g, argon (0.4 l./min); h, mixing coil; i, expansion chamber; j, liquid waste; k, gas–liquid separator; l, heated quartz cell; m, Variac transformer; n, recorder. [Reprinted by permission from G. R. Ricci, L. S. Shepard, G. Colovos and N. E. Hester, *Anal. Chem.,* **53**, 610 (1981). Copyright 1981 American Chemical Society.]

Moreover, most trace elements (in concentrations below 300 μg/l.) do not interfere with most of the automated procedures [233]. Fundamental conditions for the determination of arsenic by these techniques have also been discussed [238].

A fully automated method for the determination of arsenic in organic compounds, that incorporates both the decomposition and measurement steps, has been described [260]. Two automated digestion procedures using either ultraviolet radiation or acid–persulphate decomposition are used. Both procedures degrade the organic arsenic compounds to about the same degree. The arsenic released is reduced to arsine with borohydride, stripped from the solution and atomized. Three samples containing arsenic at levels as low as 1 μg/l. can be analysed with good precision in 1 hr (Table 6.2).

An automated chromatography–AAS method has been used for the simultaneous determination of arsenite, arsenate, methylarsonate, dimethylarsinate and p-aminophenylarsonate [261]. A diagram of the complete system is shown in Fig. 6.13. Dionex anion-exchanger is used with a mixture of 2.4mM sodium bicarbonate–1.9mM sodium carbonate–1mM borax as eluent. All arsenic components except arsenite and dimethylarsinate are effectively separated (Fig. 6.14).

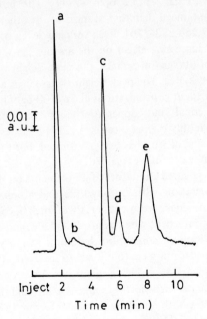

Fig. 6.14. IC-AAS chromatogram of trace inorganic and organic arsenic species separated by gradient elution: a, 20 ng/ml dimethyl arsenite; b, 20 ng/ml arsenite; c, 20 ng/ml monomethyl arsonate; d, 20 ng/ml p-aminophenylarsonate; e, 60 ng/ml arsenate. [Reprinted by permission from G. R. Ricci, L. S. Shepard, G. Colovos and N. E. Hester, *Anal. Chem.*, **53**, 610 (1981). Copyright 1981 American Chemical Society.]

The precision of the analysis is better than 8% for most of these species in the concentration range 20-80 ng/ml (Table 6.2). Mixtures containing 5-12 ng of trimethylarsine, dimethyl selenide and tetramethyl stannate are isolated by gas-liquid chromatography and the effluent is atomized on a graphite rod [262].

6.4.5 Interferences

Since arsine generation involves separation and preconcentration of arsenic from the associated substances, it is expected that AAS methods based on this reaction are less subject to interferences than those dealing with the whole test solution. However, other interferences are commonly experienced. These are (a) signal suppression caused by some anions and cations of groups Ib, VI and VIII; (b) reduction of other hydride-forming elements; (c) the initial oxidation state of the arsenic.

6.4.5.1 Effect of cations. Silver(I), nickel(II), cobalt(II), barium(II), copper(II), lead(II), aluminium(III), iron(III) and platinum(IV) cause signal suppression, depending on their concentration, order of reagent addition and the pH of the test solution [7,204,215,222,233,236,263–271]. The interference of nickel, however, is more pronounced than that of most of the other elements [180,204, 215,233,263,265,267,268,272–276]. Even very low levels of nickel (> 10 μg/ml) may have a strongly depressing effect on the arsenic signal [204]. Iron(III) at a concentration level of 900 μg/ml reduces the peak height of arsenic by 4% but does not affect the peak area. The peak height of arsenic is also decreased by the presence of copper or tin at concentrations above 100 μg/ml. Some workers have reported that nickel, cobalt and copper interfere in the absence of iron, whereas in the presence of iron only copper interferes [215]. The less pronounced effect of nickel in the presence of iron has been confirmed, but the mechanism of this antagonistic effect is not yet fully explained.

Smith [265] investigated the effect of 48 elements on the determination of arsenic by arsine generation with borohydride and atomization in an argon-hydrogen flame. Interferences from copper, silver, nickel, cobalt, gold, platinum and ruthenium were confirmed, whereas alkali and alkaline-earth metals, boron, aluminium, gallium, titanium, zirconium, hafnium, mercury, lanthanum, manganese, vanadium and yttrium were found not to interfere. The effect of various cations in the presence of some reducing agents has also been investigated [187]. However, masking agents such as thiosemicarbazide and 1,10-phenanthroline [278] or their mixtures with pyrogallol and cupferron [271] as well as thiocyanate [279] and citrate [180], minimize the depressive effect of nickel and the platinum group metals. The interfering metals may be extracted with 8-hydroxyquinoline immobilized on glass beads [280], or co-precipitated with lanthanum [267,269] before the hydride generation. Thiourea can be used to prevent the interference of several metals. Addition of 0.7 g of thiourea nullifies the effect of 100 mg of copper, silver, cobalt and nickel [223]. The interference of some cations in the automated determination of arsenic with a graphite-tube furnace depends on the order of reagent addition [263]. No interference from cationic species has been reported for a fully automated system using a heated silica tube atomizer [263].

The mechanism of cationic interference is still tentative. Meyer *et al.* [281] proposed that arsenic is reduced to arsine, which then reacts with the interfering ion to form insoluble products. Preferential reduction of the interferent to lower oxidation states or to the free metal, and adsorption of arsine on these products followed by catalytic decomposition and/or formation of insoluble products, is another route for the same effect [265]. The finding that an increased amount of borohydride tends to reduce the magnitude of interference agrees with this proposed mechanism [277].

6.4.5.2 Effect of anions. The effect of anions on the determination of arsenic by arsine generation has been widely reported [187,189,233,263]. This effect is significant in the analysis of organoarsenic compounds, because these compounds are usually decomposed by wet digestion with acids, resulting in a high concentration of anions. Most of the acids interfere at high concentration. The pattern is found to be mainly dependent on the concentration of acid present. The peak area remains relatively constant irrespective of the acid concentration in the solution up to 24% (v/v) [282]. By control of the amount of residual acid before reduction, over 95% of the arsenic can be recovered [70,282]. Brown *et al.* [198] have reported that nitrate is not the interfering species *per se*, but that digestion of samples with nitric acid results in the formation of nitrite, which on disproportionation in acid gives volatile nitrogen oxides, which lead to arsine oxidation. These authors recommended evaporation of the acid and addition of sulphamic acid before the borohydride reduction.

6.4.5.3 Effect of hydride-forming elements. It has been reported that under suitable reaction conditions arsenic, bismuth, germanium, lead, selenium, antimony, tin and tellurium are quantitatively converted into their corresponding hydrides with borohydride. Reduction with zinc is, however, limited to arsenic, selenium, antimony and bismuth. Antimony(III) and selenium(IV), in concentrations greater than 60 ng/ml and 10 ng/ml respectively, interfere with determination of arsenic by the borohydride method [283]. With zinc as reductant, antimony at a concentration as high as 130 μg/ml reduces the arsenic signal by only 10%. Germanium and tin are likely to inhibit the formation of arsine when present at trace levels. It is difficult to determine arsenic unless these interfering elements are separated by a prior chemical treatment. Recent work has indicated that addition of EDTA and the use of ion-exchangers will remove many of these hydride interferences [284,285].

6.4.5.4 Effect of the oxidation state. Some workers have demonstrated that arsenic must be present in the tervalent form for quantitative arsine generation [39,76,221]. However, arsenic(V) can be more than 90% reduced if the concentration of borohydride is increased [180,217,286]. Selective determination of arsenic(III) and arsenic(V) by careful adjustment of the pH of the analyte solution before borohydride reduction has been described [180,287].

Table 6.3 summarizes the analytical features of some methods used for the determination of arsenic by arsine generation.

Table 6.3. AAS determination of arsenic at 193.7 nm by methods
based on arsine generation

Reductant	Radiation source	Atomizer	Limit of detection		Reference
Zinc-HCl	Hollow-cathode lamp	Air-acetylene	0.04	μg	166
			1.5	ng/ml	197
		Argon–hydrogen	5	ng	54,167
			0.04	μg	184
		Acetylene–N$_2$O	0.2	μg	168
		Silica tube (15 × 2.5 cm)	0.05	μg	76
Sodium borohydride		Air-acetylene	1	μg	215
		Argon–hydrogen	0.56	ng/ml	272,278
			0.03	μg	265
			25	ng	180
		Hydrogen–nitrogen	1	μg	110,216
			6	μg	223
	Electrodeless discharge lamp	Graphite furnace	1	μg/ml	99
	Hollow-cathode lamp	Silica tube (12 × 1.1 cm)	0.3	ng	242
		(17 × 0.8 cm)	0.8	ng/ml	197
		(17 × 0.8 cm)	1	ng/ml	234
		(20.5 × 1.0 cm)	10	ng	220
		(14 × 0.8 cm)	90	ng	271
	Electrodeless discharge lamp	Silica T-tube (15 × 0.5 cm)	1.1	μg/kg	241
			20	μg/kg	202

6.5 METHODS BASED ON INDIRECT REACTIONS

The general procedure most frequently used for the indirect determination of arsenic is based on the addition of ammonium molybdate to the arsenate solution and the subsequent extraction of arsenomolybdic acid into an organic solvent. The molybdenum in the complex is then measured at 313.2 nm. These methods are sensitive enough to permit determination of as little as 10 ng of arsenic per ml. Danchik and Boltz [288] described a method based on this principle and involving a reaction with molybdate at pH 1.8 after treatment of the arsenic solution with bromine water to ensure that the arsenic was present in the quin-quevalent state. The arsenomolybdic acid was extracted into methyl isobutyl ketone (MIBK), and the organic layer washed with dilute hydrochloric acid and water to remove the excess of molybdate, followed by stripping of the hetero-poly acid into an alkaline aqueous solution. The aqueous phase containing the equivalent amount of molybdenum was then measured [21,289].

Atomization of the MIBK extract of the arsenomolybdic acid, after washing, gives recoveries better than 99.5% at the 50 μg arsenic level [289]. A mixture of ethyl acetate–butanol–isopentyl acetate (1:1:2) may also be used for extraction and direct aspiration of arsenomolybdic acid into a nitrous oxide–acetylene flame. This improves the detection limit for arsenic to 25 ng/ml [290]. Arsenic and phosphorus have been determined in urine by reaction with molybdate at pH 1, extraction of phosphomolybdic acid into isobutyl acetate, increase of the pH to 2, and extraction of arsenomolybdic acid into cyclohexane [291]. Deter-mination of arsenic and phosphorus or silicon by successive extraction of their heteropoly acids has similarly been suggested [290,292–294] (see also p. 138).

Prior separation of arsenic from high concentrations of interfering ions, such as sulphur, antimony, silicon and phosphorus, can also be utilized before indirect measurements. This can be performed by one of the following methods.

(a) Generation of arsine, absorption in iodine, formation of arsenomolybdic acid, extraction into MIBK and aspiration into the flame [289,295].
(b) Generation of arsine, absorption in silver nitrate and nebulization of the resulting solution into the flame [296].
(c) Generation of arsine, absorption in mercury(II) chloride, reduction with hydroxylamine to give arsenious acid, and atomization [297].
(d) Generation of arsine, absorption in a chloroform solution of silver diethyl-dithiocarbamate (AgDDC) in the presence [298] or absence [113] of ephedrine, followed by flame or electrothermal atomization.
(e) Reduction of arsenate to arsenite, reaction with diethylammonium diethyl-dithiocarbamate, extraction into diethyl ether, decomposition of the complex with concentrated nitric acid and formation of arsenomolybdic acid [299].

These methods are used for the determination of organoarsenic compounds and total arsenic in feed-grade wheat [299] and biological materials [113,297] after a digestion step. The sensitivity of these methods is in the range from 4 μg [297] to 0.2 ng [298] per ml.

Hassan and Eldesouky [300] have described a method for the indirect AAS determination of arsenic in organic compounds. The method is based on heating to dryness with a mixture of 70% nitric acid and 30% hydrogen peroxide at 250°C, followed by digestion with 70% perchloric acid. The arsenate ions formed are then precipitated with excess of magnesium as magnesium ammon-ium arsenate, followed by AAS measurement of the excess of magnesium at

285.2 nm. The results obtained (Table 6.4) show a mean recovery of 100.7% and a precision of ±0.4%.

Table 6.4 Indirect AAS determination of arsenic in organic compounds [300] (reprinted by permission of the copyright holders, Springer-Verlag)

Compound	Arsenic (%)	
	Expected	Found
o-Aminophenylarsonic acid	34.51	35.0
3,4-Difluorophenylarsonic acid	31.85	31.7
p-Hydroxybenzene arsonic acid	34.51	37.7
4-Hydroxy-3-nitrophenylarsonic acid	28.48	28.3
p-Toluene arsonic acid	21.40	21.6
p-Toluene-3-nitroarsonic acid	28.71	28.4

REFERENCES

[1] K. Thoma, A. Schöntag and E. Kuchinke, *Naunyn-Schmiedebergs Arch. Exptl. Pathol. Pharmakol.*, **226**, 255 (1955).
[2] H. Guthman and H. Grass, *Arch. Gynäkol.*, **152**, 127 (1932).
[3] H. E. Cox, *Analyst*, **50**, 3 (1925); **51**, 132 (1926).
[4] A. C. Chapman, *Analyst*, **51**, 548 (1926).
[5] R. G. Godden and D. R. Thomerson, *Analyst*, **105**, 1137 (1980).
[6] F. Vajda, *Magy. Kem. Lapja*, **35**, 503 (1980).
[7] W. B. Robbins and J. A. Caruso, *Anal. Chem.*, **51**, 889A (1979).
[8] R. G. Lewis, *Residue Rev.*, **68**, 123 (1977).
[9] B. Welz, *Chim. Ind. (Milan)*, **59**, 771 (1977).
[10] R. R. Lauwerys, J. P. Buchet and H. Roels, *Arch. Toxicol.*, **41**, 239 (1979).
[11] I. G. Shafran, K. Z. Zonenberg and V. A. Oboznenko, *Tr. Vses. Nauchno-Issled. Inst. Khim. Reakt.*, **31**, 183 (1969).
[12] D. R. Thomerson and K. C. Thompson, *Am. Lab.*, **6**, No. 3, 53 (1974).
[13] I. Paralescu and A. Tasca-Danescu, *Rev. Chim. (Bucharest)*, **26**, 507 (1975).
[14] I. A. Blyum and Yu. A. Zolotov, *Zh. Analit. Khim.*, **31**, 159 (1976).
[15] D. R. Thomerson and K. C. Thompson, *Chem. Brit.*, **11**, 316 (1975).
[16] N. T. Crosby, *Analyst*, **102**, 225 (1977).
[17] P. Mushak, *J. Anal. Toxicol.*, **1**, 286 (1977).
[18] L. E. Smythe and R. J. Finlayson, *Aust. Water Resour. Tech.*, *Paper*, No. 32, 1978.
[19] R. V. Slates, *U.S. Energy Res. Dev. Adm. Rept.*, No. DP-1421 (1976).
[20] W. J. Price, *Paint Oil Colour J.*, **158**, 282 (1970).
[21] M. Pinta, *Methods Phys. Anal.*, **6**, 268 (1970).
[22] G. F. Kirkbright and H. N. Johnson, *Talanta*, **20**, 433 (1973).
[23] R. R. Brooks, D. E. Ryan and H. Zhang, *Anal. Chim. Acta*, **131**, 1 (1981).
[24] T. T. Gorsuch, *Analyst*, **84**, 135 (1959).
[25] R. J. Evans and S. L. Bandemer, *Anal. Chem.*, **26**, 595 (1954).
[26] J. Pijck, J. Gidlis and J. Hoste, *Intern. J. Appl. Radiat. Isotopes*, **10**, 149 (1961).
[27] E. I. Hamilton, M. J. Minski and J. J. Cleary, *Analyst*, **92**, 257 (1967).

[28] J. E. Portmann and J. P. Riley, *Anal. Chim. Acta*, **31**, 509 (1964).

[29] W. F. Carey, *J. Assoc. Off. Anal. Chem.*, **51**, 1300 (1968).

[30] C. E. Gleit and W. D. Holland, *Anal. Chem.*, **34**, 1454 (1962).

[31] H. A. Schroeder and J. J. Balassa, *J. Chronic Dis.*, **19**, 85 (1966).

[32] C. E. Mulford, *Atom. Absorp. Newslett.*, **5**, 135 (1966).

[33] W. R. Penrose, *Crit. Rev. Environ. Control*, **5**, 465 (1974).

[34] Y. Talmi and C. Feldman, in *Arsenical Pesticides*, E. Woolson (ed.), p. 13. Am. Chem. Soc., Washington D.C. (1975).

[35] H. C. Leifheit, in *Standard Methods of Clinical Chemistry*, Vol. III, D. Seligson (ed.), Academic Press, New York, 1961.

[36] R. K. Simon, G. D. Christian and W. C. Purdy, *Am. J. Clin. Pathol.*, **49**, 207 (1968).

[37] S. E. Raptis, W. Wegscheider and G. Knapp, *Mikrochim. Acta*, **93** (1981 I).

[38] K. Yanagi and M. Ambe, *Bunseki Kagaku*, **30**, 209 (1981).

[39] F. J. Fernandez and D. C. Manning, *Atom. Absorp. Newslett.*, **10**, 86 (1971).

[40] J. E. Allan, *4th Australian Spectroscopy Conference*, Canberra, Australia, August 1963.

[41] B. Welz, *Chim. Ind. (Milan)*, **59**, 771 (1977).

[42] W. Slavin and S. Sprague, *Atom. Absorp. Newslett.*, **3**, 1 (1964).

[43] H. L. Kahn and J. E. Schallis, *Atom. Absorp. Newslett.*, **7**, 5 (1968).

[44] G. F. Kirkbright, M. Sargent and T. S. West, *Atom. Absorp. Newslett.*, **8**, 34 (1969).

[45] C. S. Rann and A. N. Hambly, *Anal. Chim. Acta*, **32**, 346 (1965).

[46] W. Slavin, C. Sebens and S. Sprague, *Atom. Absorp. Newslett.*, **4**, 341 (1965).

[47] M. D. Amos, *The Element*, **16**, 1 (1966).

[48] O. Menis and T. C. Rains, *Am. Chem. Soc. Natl. Meeting*, Chicago, September 1967.

[49] P. Johns, *Spectrovision*, **24**, 6 (1970).

[50] C. Veillon, J. M. Mansfield, M. L. Parsons and J. D. Winefordner, *Anal. Chem.*, **38**, 204 (1966).

[51] K. Zacha and J. D. Winefordner, *Anal. Chem.*, **38**, 1537 (1966).

[52] J. Ramírez-Muñoz, *Flame Notes*, **7**, 41 (1975).

[53] A. Ando, M. Suzuki, F. Fuwa and B. I. Vallee, *Anal. Chem.*, **41**, 1974 (1969).

[54] T. Maruta and G. Sudoh, *Anal. Chim. Acta*, **77**, 37 (1975).

[55] G. F. Kirkbright and L. Ranson, *Anal. Chem.*, **43**, 1238 (1971).

[56] K. C. Thompson and R. G. Godden, *Analyst*, **101**, 96 (1976).

[57] G. F. Kirkbright, L. Ranson and T. S. West, *Spectrosc. Lett.*, **5**, 25 (1972).

[58] H. Berndt and J. Messerschmidt, *Spectrochim. Acta*, **36B**, 809 (1981).

[59] G. F. Kirkbright and P. J. Wilson, *Anal. Chem.*, **46**, 1414 (1974).

[60] O. Menis and T. C. Rains, *Anal. Chem.*, **41**, 952 (1969).

[61] A. S. Bazhov, *Zavodsk. Lab.*, **33**, 1096 (1967).

[62] W. B. Barnett, J. W. Vollmer and S. M. DeNuzzo, *Atom. Absorp. Newslett.*, **15**, 33 (1976).

[63] E. R. Likaits, R. F. Farrell and A. J. Mackie, *Atom. Absorp. Newslett.*, **18**, 53 (1979).

[64] T. Nakahara, H. Nishino, M. Munemori and S. Musha, *Bull. Chem. Soc. Japan*, **46**, 1706 (1973).

[65] W. B. Barnett and J. D. Kerber, *Atom. Absorp. Newslett.*, **13**, 56 (1974).

[66] W. B. Barnett, *Atom. Absorp. Newslett.*, **12**, 142 (1973).

[67] A. G. Coedo and M. T. D. Lopez, *Rev. Met. (Madrid)*, **10**, 355 (1974).

[68] R. Wagemann, *Tech. Rept. Fish. Mar. Serv. (Can.)*, 555 (1975).

[69] A. Croce and C. Tonini, *Ind. Carta*, **14**, 259 (1976).

[70] G. C. Kunselman and E. A. Huff, *Atom. Absorp. Newslett.*, **15**, 29 (1976).

[71] D. B. Ratcliffe, C. S. Byford and P. B. Osman, *Anal. Chim. Acta*, **75**, 457 (1975).

[72] R. B. Baird, S. Pourian and S. M. Gabrielian, *Preparat. Paper Natl. Meeting, Div. Environ. Chem.*, *Am. Chem. Soc.*, **13**, 67 (1973).

[73] P. Aruscavage, *J. Res. U.S. Geol. Surv.,* **5**, 405 (1977).

[74] R. R. Brooks, D. E. Ryan and H. F. Zhang, *Atom. Spectrosc.,* **2**, 161 (1981).

[75] H. Massmann, *Z. Anal. Chem.,* **255**, 203 (1967).

[76] R. C. Chu, G. P. Barron and P. A. W. Baumgarner, *Anal. Chem.,* **44**, 1476 (1972).

[77] E. D. Prudnikov, *Izv. Vyssh. Vchebn. Zaved., Khim. Tekhnol.,* **18**, 186 (1975).

[78] D. A. Katskov, L. R. Kruglikova and B. V. L'vov, *Zh. Analit. Khim.,* **30**, 238 (1975).

[79] D. B. Lo and R. L. Coleman, *Atom. Absorp. Newslett.,* **18**, 10 (1979).

[80] D. B. Lo and G. D. Christian, *Can. J. Spectrosc.,* **22**, 45 (1977).

[81] A. W. Fitchett, E. H. Daughtrey, Jr. and P. Mushak, *Anal. Chim. Acta,* **79**, 93 (1975).

[82] H. Freeman, J. F. Uthe and B. Flemming, *Atom. Absorp. Newslett.,* **15**, 49 (1976).

[83] J. W. Owens and E. S. Gladney, *Atom. Absorp. Newslett.,* **15**, 47 (1976).

[84] R. M. Hamner, D. L. Lechak and P. Greenberg, *Atom. Absorp. Newslett.,* **15**, 122 (1976).

[85] K. S. Subramanian and J. C. Meranger, *Anal. Chim. Acta,* **124**, 131 (1981).

[86] B. W. Gandrud and R. K. Skogerboe, *Appl. Spectrosc.,* **25**, 243 (1971).

[87] J. W. Robinson, R. Garcia, G. Hindman and P. Slevin, *Anal. Chim. Acta,* **69**, 203 (1974).

[88] I. Rubeška and J. Korečková, *Chem. Listy,* **73**, 1009 (1979).

[89] K. Dittrich, *Talanta,* **24**, 725 (1977).

[90] I. Barin and O. Knacke, *Thermochemical Properties of Inorganic Substances,* Springer, Berlin (1973).

[91] K. Fujiwara, J. N. Bower and J. D. Winefordner, *Anal. Chim. Acta,* **109**, 229 (1979).

[92] P. R. Walsh, R. A. Duce and J. L. Fashing, *Anal. Chem.,* **48**, 1014 (1976).

[93] D. Chakrabarti, W. de Jonghe and F. Adams, *Anal. Chim. Acta,* **119**, 331 (1980).

[94] P. Hocquellet, *Analusis,* **6**, 426 (1978).

[95] J. F. Alder and D. A. Hickman, *Atom. Absorp. Newslett.,* **4**, 110 (1977).

[96] J. Korečková, W. Frech, E. Lundberg, J. Persson and A. Cedergren, *Anal. Chim. Acta,* **130**, 267 (1981).

[97] R. D. Ediger, *Atom. Absorp. Newslett.,* **14**, 127 (1975).

[98] K. M. Zhumanova, A. Z. Beilina and Z. M. Muldakhmetov, *Deposited Document,* VINITI 3713-77 (1977); *Chem. Abstr.,* **91**, 203456x (1979).

[99] R. B. Denyszyn, P. M. Groshe and D. E. Wagoner, *Anal. Chem.,* **50**, 1094 (1978).

[100] Y. Odanaka, O. Matano and S. Goto, *Bunseki Kagaku,* **28**, 517 (1979).

[101] H. Nickel, *Spectrochim. Acta,* **23B**, 323 (1968).

[102] B. V. L'vov and L. A. Pelieva, *Can. J. Spectrosc.,* **23**, 1 (1978).

[103] B. V. L'vov and L. A. Pelieva, *Zavodsk. Lab.,* **44**, 173 (1978).

[104] B. V. L'vov, *Spectrochim. Acta,* **33B**, 153 (1978).

[105] W. Rüdorf, *Z. Anorg. Allg. Chem.,* **254**, 319 (1947).

[106] G. R. Henning, in *Progress in Inorganic Chemistry,* F. A. Cotton (ed.), Vol. I, Interscience, New York, 1959.

[107] H. M. Ortner and E. Lassner, *Mikrochim. Acta,* Suppl. 7, 41 (1977).

[108] R. Baird and S. M. Gabrielian, *Appl. Spectrosc.,* **28**, 273 (1974).

[109] K. Ohta and M. Suzuki, *Talanta,* **25**, 160 (1978).

[110] J. S. Edmonds and K. A. Francesconi, *Anal. Chem.,* **48**, 2019 (1976).

[111] K. Saeed and Y. Thomassen, *Anal. Chim. Acta,* **130**, 281 (1981).

[112] K. C. Tam, *Environ. Sci. Technol.,* **8**, 734 (1974).

[113] E. Chreneková and N. Rusinová, *Chem. Listy,* **72**, 990 (1978).

[114] N. Thieux, *J. Assoc. Off. Anal. Chem.,* **63**, 496 (1980).

[115] J. C. Chambers and B. E. McLellan, *Anal. Chem.,* **48**, 2061 (1976).

[116] W. Lautenschlaeger and J. Maasen, *GIT Fachz. Lab.,* **23**, 176 (1979).

[117] T. J. Forehand, A. E. Dupuy, Jr. and H. Tai, *Anal. Chem.,* **48**, 999 (1976).

[118] K. Ebato, K. Ito and H. Harada, *Tokyo Toritsu Eisei Kenkyusho Kenkyu Nempo*, 29, 129 (1978).
[119] S. Dupire and M. Hoenig, *Analusis*, 8, 153 (1980).
[120] M. Ishazaki, *Bunseki Kagaku*, 26, 667 (1977).
[121] T. Korenaga, *Mikrochim. Acta*, 435 (1979 I).
[122] A. Yasui, C. Tsutsumi and S. Toda, *Agric. Biol. Chem.*, 42, 2139 (1978).
[123] A. Heres, O. Girard-Devasson, J. Gaudet and J. C. Spuig, *Analusis*, 1, 408 (1972).
[124] M. T. Friend, C. A. Smith and D. Wishart, *Atom. Absorp. Newslett.*, 16, 46 (1977).
[125] T. Kamada, *Talanta*, 23, 835 (1976).
[126] K. Dixon, R. C. Mallett and R. C. Kocaba, *S. Afr. Natl. Inst. Metall. Rept.*, No. 1689 (1975).
[127] S. Tagawa and Y. Kojima, *Bunseki Kagaku*, 29, 216 (1980).
[128] S. Nakashima, *Bunseki Kagaku*, 28, 561 (1979).
[129] W. Reichel and B. G. Bleakley, *Anal. Chem.*, 46, 59 (1974).
[130] E. A. Jones, *S. Afr. Natl. Inst. Metall.*, Rept. No. 1787 (1976).
[131] A. Sato and N. Saitoh, *Bunseki Kagaku*, 25, 663 (1976).
[132] J. D. Mullen, *Talanta*, 24, 657 (1977).
[133] D. G. Iverson, M. A. Anderson, R. R. Holm and R. S. Stanforth, *Environ. Sci. Technol.*, 13, 1491 (1979).
[134] P. Aruscavage, *J. Res. U.S. Geol. Surv.*, 5, 405 (1977).
[135] B. W. Haynes, *Atom. Absorp. Newslett.*, 17, 49 (1978).
[136] R. C. Mallett, *S. Afr. Natl. Inst. Metall.*, Rept. No. 1809 (1976).
[137] C. J. Molnar, R. D. Reeves, J. Winefordner, M. T. Glenn, J. R. Ahlstrom and J. Savory, *Appl. Spectrosc.*, 26, 606 (1972).
[138] T. Kamada, T. Kumamaru and Y. Yamamoto, *Bunseki Kagaku*, 24, 89 (1975).
[139] K. Korn, *Z. Anal. Chem.*, 279, 288 (1976).
[140] W. B. Barnett and E. A. McLaughlin, Jr., *Anal. Chim. Acta*, 80, 285 (1975).
[141] J. E. Forrester, V. Lehecka, J. R. Johnston and W. L. Ott, *Atom. Absorp. Newslett.*, 18, 73 (1979).
[142] I. G. Yudelevich, L. V. Zelentsova, N. F. Beisel, T. A. Chanysheva and L. Vechernish, *Anal. Chim. Acta*, 108, 45 (1979).
[143] L. Pozzoli and C. Minoia, *Ann. Ist. Super. Sanita*, 13, 377 (1977).
[144] B. W. Haynes, *Atom. Absorp. Newslett.*, 18, 46 (1979).
[145] E. Sefzik, *Vom Wasser*, 50, 285 (1978).
[146] C. A. Johnson and J. F. Lewin, *Anal. Chim. Acta*, 82, 79 (1976).
[147] J. Futhe, H. C. Freeman, J. R. Johnson and P. Michalik, *J. Assoc. Off. Anal. Chem.*, 57, 1363 (1974).
[148] F. Lavilla and F. Queraud, *Rev. Inst. Fr. Pet.*, 32, 413 (1977).
[149] D. C. Manning, R. D. Ediger and D. W. Hoult, *Atom. Spectrosc.*, 1, 52 (1980).
[150] S. Costantini, R. Giordano and P. Ravagnan, *Ann. Ist. Super. Sanita*, 16, 287 (1980).
[151] D. Chakrabarti, W. De Jonghe and F. Adams, *Anal. Chim. Acta*, 120, 121 (1980).
[152] M. Cooksey and W. B. Barnett, *Atom. Absorp. Newslett.*, 18, 101 (1979).
[153] Y. O. Danard, O. Matano and S. Goto, *Bunseki Kagaku*, 28, 517 (1979).
[154] R. R. Stanforth, *Env. Sci. Tech.*, 12, 1491 (1979).
[155] E. A. Woolson and N. Aharonson, *J. Assoc. Off. Anal. Chem.*, 63, 523 (1980).
[156] F. E. Brinkman, K. L. Jewett, W. D. Iverson, K. J. Irogolic, K. C. Ehrhardt and R. A. Stockton, *J. Chromatog.*, 191, 31 (1980).
[157] G. E. Pacey and J. A. Ford, *Talanta*, 28, 935 (1981).
[158] A. A. Grabinski, *Anal. Chem.*, 53, 966 (1981).
[158a] J.-A. Persson and K. Irgum, *Anal. Chim. Acta*, 138, 111 (1982).
[159] H. L. Kahn, Jr., *J. Chem. Educ.*, 43, A7 (1966).
[160] G. R. Kingsley and P. R. Schaffert, *Anal. Chem.*, 23, 914 (1951).

[161] G. W. Powers, Jr., R. L. Martin and F. J. Piehl, *Anal. Chem.*, **31**, 1589 (1959).
[162] D. Liederman, J. E. Bowen and O. I. Milner, *Anal. Chem.*, **31**, 2052 (1959).
[163] J. L. Morrison, *J. Assoc. Off. Anal. Chem.*, **44**, 740 (1961).
[164] J. L. Morrison and G. M. George, *J. Assoc. Off. Anal. Chem.*, **52**, 930 (1969).
[165] H. K. Hundley and J. C. Underwood, *J. Assoc. Off. Anal. Chem.*, **53**, 1176 (1970).
[166] W. Holak, *Anal. Chem.*, **41**, 1712 (1969).
[167] R. M. Orheim and H. H. Bovee, *Anal. Chem.*, **46**, 921 (1974).
[168] H. R. Griffin, M. B. Hocking and D. G. Lowery, *Anal. Chem.*, **47**, 229 (1975).
[169] J. Ramírez-Muñoz, *Flame Notes*, **7**, 45 (1975).
[170] S. Musha, *Nippon Kaisui Gakkai-Shi*, **27**, 255 (1974).
[171] D. C. Manning, *Atom. Absorp. Newslett.*, **10**, 123 (1971).
[172] H. C. Freeman and J. F. Uthe, *Atom. Absorp. Newslett.*, **13**, 75 (1975).
[173] D. C. Manning and F. J. Fernandez, *U.S. Patent*, 3,801,282 (2.2.1974).
[174] Y. Yamamoto, T. Kumamaru and Y. Hayashi, *Anal. Lett.*, **5**, 419 (1972).
[175] Y. Yamamoto, T. Kumamaru, Y. Hayashi and M. Manke, *Anal. Lett.*, **5**, 717 (1972).
[176] R. D. Snook, *Anal. Proc.*, **18**, 342 (1981).
[177] R. Tsuijino, H. Yamamoto, S. Ueda, T. Sudo and Y. Sawasaki, *Bunseki Kagaku*, **23**, 1378 (1974).
[178] S. Terashima, *Bunseki Kagaku*, **23**, 1331 (1974).
[179] E. J. Knudsen and G. D. Christian, *Anal. Lett.*, **6**, 1039 (1973).
[180] J. Aggett and A. C. Aspell, *Analyst*, **101**, 341 (1974).
[181] J. F. Chapman and L. S. Dale, *Anal. Chim. Acta*, **111**, 137 (1979).
[182] E. F. Dalton and A. J. Malanoski, *Atom. Absorp. Newslett.*, **10**, 92 (1971).
[183] A.-M. de Kersabiec, B. Giraud and J. Nicolas, *Compt. Rend. D*, **282**, 2025 (1976).
[184] S. Terashima, *Anal. Chim. Acta*, **86**, 43 (1976).
[185] M. Murokowa, M. Kaneko, N. Nishiyama, S. Fukui and S. Kanno, *Eisei Kagaku*, **21**, 77 (1975).
[186] F. E. Lichte and R. K. Skogerboe, *Anal. Chem.*, **44**, 1480 (1972).
[187] K. C. Thompson and D. R. Thomerson, *Analyst*, **99**, 595 (1974).
[188] W. A. Maher, *Anal. Chim. Acta*, **126**, 157 (1981).
[189] K. Ebato, E. Amakawa, H. Yamanobe, S. Suzuki and T. Totani, *Shokuhin Eiseigaku Zasshi*, **15**, 469 (1974).
[190] E. N. Pollock and S. J. West, *Atom. Absorp. Newslett.*, **12**, 6 (1973).
[191] P. D. Goulden and P. Brooksbank, *Anal. Chem.*, **46**, 1431 (1974).
[192] G. W. Schaeffer and M. Emilius, *J. Am. Chem. Soc.*, **76**, 1203 (1954).
[193] E. D. Macklen, *J. Chem. Soc.*, 1989 (1959).
[194] F. J. Schmidt and J. L. Royer, *Anal. Lett.*, **6**, 17 (1973).
[195] B. Welz and M. Melcher, *Atom. Absorp. Newslett.*, **18**, 121 (1979).
[196] F. J. Fernandez, *Atom. Absorp. Newslett.*, **12**, 93 (1973).
[197] K. C. Thompson and D. R. Thomerson, *Analyst*, **99**, 595 (1974).
[198] R. M. Brown, Jr., R. C. Fry, J. L. Moyers, S. J. Northway, M. B. Denton and G. S. Wilson, *Anal. Chem.*, **53**, 1560 (1981).
[199] H. W. Sinemus, M. Melcher and B. Welz, *Atom. Spectrosc.*, **2**, 81 (1981).
[200] R. Kaszerman and K. Theurer, *Atom. Absorp. Newslett.*, **15**, 129 (1976).
[201] R. D. Wauchope, *Atom. Absorp. Newslett.*, **15**, 64 (1976).
[202] P. J. Brooke and W. H. Evans, *Analyst*, **106**, 514 (1981).
[203] S. Peats, *Atom. Absorp. Newslett.*, **18**, 118 (1979).
[204] I. Rubeška and V. Hlavinková, *Atom. Absorp. Newslett.*, **18**, 5 (1979).
[205] A. M. Dekersabiec, *Analusis*, **8**, 97 (1980).
[206] T. Yamashige, M. Yamamoto and Y. Yamamoto, *Bunseki Kagaku*, **30**, 324 (1981).
[207] K. G. Brodie, *Intern. Lab.*, July/Aug., 40 (1979).
[208] D. D. Siemer and L. Hagemann, *Anal. Lett.*, **8**, 323 (1975).

[209] L. P. Greenland and E. Y. Campbell, *Anal. Chim. Acta*, **87**, 323 (1976).
[210] R. C. Rooney, *Analyst*, **101**, 749 (1976).
[211] J. R. Knechtel and J. L. Fraser, *Analyst*, **103**, 104 (1978).
[211a] R. Bye, *Talanta*, **29**, 797 (1982).
[212] T. Inui, S. Terada and H. Tamura, *Z. Anal. Chem.*, **305**, 189 (1981).
[213] M. L. Kokot, *Atom. Absorp. Newslett.*, **15**, 105 (1976).
[214] H. G. King and R. W. Morrow, *Rept.*, Y-1956 (1974); *Nuclear Sci. Abstr.*, 1975, **31**, No. 8040.
[215] H. D. Fleming and R. G. Ide, *Anal. Chim. Acta*, **83**, 67 (1976).
[216] I. May and L. P. Greenland, *Anal. Chem.*, **49**, 2376 (1977).
[217] R. S. Braman, L. L. Justen and C. C. Foreback, *Anal. Chem.*, **44**, 2195 (1972).
[218] P. N. Vijan and G. R. Wood, *Atom. Absorp. Newslett.*, **13**, 33 (1974).
[219] D. L. Collett, D. E. Fleming and G. A. Taylor, *Analyst*, **103**, 1074 (1978).
[220] D. E. Fleming and G. A. Taylor, *Analyst*, **103**, 101 (1978).
[221] R. S. Braman and C. C. Foreback, *Science*, **182**, 1247 (1973).
[222] A. U. Shaikh and D. E. Tallman, *Anal. Chim. Acta*, **98**, 251 (1978).
[223] C. J. Peacock and S. C. Singh, *Analyst*, **106**, 931 (1981).
[224] Y. Yamamoto, T. Kumamaru, Y. Hayashi and T. Kamada, *Bunseki Kagaku*, **22**, 876 (1973).
[225] Y. Nakamura, H. Nagai, D. Kuboto and S. Himeno, *Bunseki Kagaku*, **22**, 1543 (1973).
[226] M. Ihnat and J. H. Miller, *J. Assoc. Off. Anal. Chem.*, **60**, 813 (1977).
[227] G. De Groot, A. Van Dijk and R. A. A. Maes, *Pharm. Weekbl.*, **112**, 949 (1977).
[228] D. I. Rees, *J. Assoc. Public Anal.*, **16**, 71 (1978).
[229] D. R. Corbin and W. M. Barnard, *Atom. Absorp. Newslett.*, **15**, 116 (1976).
[230] S. Nakashima, *Bunseki Kagaku*, **28**, 561 (1979).
[231] R. V. D. Robért and G. Balaes, *S. Afr. Natl. Inst. Metall., Rept.*, No. 2033 (1979).
[232] F. D. Pierce, T. C. Lamoreaux, H. R. Brown and R. S. Frazer, *Appl. Spectrosc.*, **30**, 38 (1976).
[233] F. D. Pierce and H. R. Brown, *Anal. Chem.*, **49**, 1417 (1977).
[234] A. J. Thompson and P. A. Thoresby, *Analyst*, **102**, 9 (1977).
[235] G. Ivanov and G. Kurchatova, *Khig. Zdraveopaz.*, **23**, 266 (1980).
[236] R. G. Smith, J. C. Van Loon, J. R. Knechtel, J. L. Fraser, A. E. Pitts and A. E. Hodges, *Anal. Chim. Acta*, **93**, 61 (1977).
[237] H. Agemian and V. Cheam, *Anal. Chim. Acta*, **101**, 193 (1978).
[238] M. Ikeda, J. Nishibe and T. Nakahara, *Bunseki Kagaku*, **30**, 368 (1981).
[239] M. Dujmović, *GIT Fachz. Lab.*, **20**, 336 (1976).
[240] W. Lautenschlaeger, J. Maassen and W. Oelschlaeger, *Landwirtsch. Forsch.*, **29**, 59 (1976).
[241] W. H. Evans, F. J. Jackson and D. Dellar, *Analyst*, **104**, 16 (1979).
[242] D. D. Siemer, P. Koteel and V. Jariwala, *Anal. Chem.*, **48**, 836 (1976).
[243] D. H. Cox, *J. Anal. Toxicol.*, **4**, 207 (1980).
[244] J. T. Kinard and M. Gales, Jr., *J. Environ. Sci. Health*, **16A**, 27 (1981).
[245] M. McDaniel, A. D. Shendrikar, K. D. Reiszner and P. W. West, *Anal. Chem.*, **48**, 2240 (1976).
[246] G. Drasch, L. von Meyer and G. Kauert, *Z. Anal. Chem.*, **304**, 141 (1980).
[247] K. Tsujii and K. Kuga, *Anal. Chim. Acta*, **72**, 51 (1978).
[248] A. Kuldvere, *Atom. Spectrosc.*, **1**, 138 (1980).
[249] M. Ishizaki, *Bunseki Kagaku*, **26**, 667 (1977).
[250] K. G. Brodie, *Intern. Lab.*, Sept./Oct., 65 (1977).
[251] J. Y. Hwang, P. A. Ullucci, C. J. Mokeler and S. B. Smith, *Am. Lab.*, **5**, No. 3, 43 (1973).
[252] K. T. Kan, *Anal. Lett.*, **6**, 603 (1973).

[253] P. N. Vijan, A. C. Rayner, D. Sturgis and G. R. Wood, *Anal. Chim. Acta*, **82**, 329 (1976).
[254] P. N. Vijan and G. R. Wood, *Talanta*, **23**, 89 (1976).
[255] F. J. Schmidt, J. L. Royer and S. M. Muir, *Anal. Lett.*, **8**, 123 (1975).
[256] H. Agemian and E. Bedek, *Anal. Chim. Acta*, **119**, 323 (1980).
[257] H. Agemian and R. Thomson, *Analyst*, **105**, 902 (1980).
[258] J. A. Fiorino, J. W. Jones and S. G. Capar, *Anal. Chem.*, **48**, 120 (1976).
[259] L. Ebdon, J. R. Wilkinson and K. W. Jackson, *Anal. Chim. Acta*, **136**, 191 (1982).
[260] M. Fishman and R. Spencer, *Anal. Chem.*, **49**, 1599 (1977).
[261] G. R. Ricci, L. S. Shepard, G. Colovos and N. E. Hester, *Anal. Chem.*, **53**, 610 (1981).
[262] G. E. Parris, W. R. Blair and F. E. Brinckman, *Anal. Chem.*, **49**, 379 (1977).
[263] F. D. Pierce and H. R. Brown, *Anal. Chem.*, **48**, 693 (1976).
[264] S. Terashima, *Chishitsu Chosasho Geppo*, **26**, 153 (1975).
[265] A. E. Smith, *Analyst*, **100**, 300 (1975).
[266] Y. Talmi and D. T. Bostick, *Anal. Chem.*, **47**, 2145 (1975).
[267] J. Azad, G. F. Kirkbright and R. D. Snook, *Analyst*, **104**, 232 (1979).
[268] J. Azad, G. F. Kirkbright and R. D. Snook, *Analyst*, **105**, 79 (1980).
[269] M. Thompson, B. Pahlavanpour, S. J. Walton and G. F. Kirkbright, *Analyst*, **103**, 705 (1978).
[270] M. Thompson and B. Pahlavanpour, *Anal. Chim. Acta*, **109**, 251 (1979).
[271] M. H. Arbab-Zavar and A. G. Howard, *Analyst*, **105**, 744 (1980).
[272] Y. Yamamoto and T. Kumamaru, *Z. Anal. Chem.*, **281**, 353 (1976).
[273] A. Miyazaki, A. Kimura and Y. Umezaki, *Anal. Chim. Acta*, **90**, 119 (1977).
[274] R. S. Braman, D. L. Johnson, C. C. Foreback, J. M. Ammons and J. L. Bricker, *Anal. Chem.*, **49**, 621 (1977).
[275] S. S. Sandhu and P. Nelson, *Anal. Chem.*, **50**, 322 (1978).
[276] A. G. Howard and M. H. Arbab-Zavar, *Analyst*, **105**, 338 (1980).
[277] M. McDaniel, A. D. Shendrikar, A. D. Reizner and P. W. West, *Anal. Chem.*, **48**, 836 (1976).
[278] G. F. Kirkbright and M. Taddia, *Anal. Chim. Acta*, **100**, 145 (1978).
[279] J. Guimont, M. Pitchette and N. Rhéaume, *Atom. Absorp. Newslett.*, **16**, 53 (1977).
[280] B. L. Dennis, J. L. Moyers and G. S. Wilson, *Anal. Chem.*, **48**, 1611 (1976).
[281] A. Meyer, C. Hofer, G. Tölg, S. Raptis and G. Knapp, *Z. Anal. Chem.*, **296**, 337 (1979).
[282] H. K. Kang and J. L. Valentine, *Anal. Chem.*, **49**, 1829 (1977).
[283] S. Nakashima, *Analyst*, **103**, 1031 (1978).
[284] R. Belcher, S. L. Bogdanski, E. Henden and A. Townshend, *Analyst*, **100**, 522 (1975).
[285] G. Mausbach, *GIT Fachz. Lab.*, **23**, 898 (1979).
[286] E. J. Knudson and G. D. Christian, *Atom. Absorp. Newslett.*, **13**, 74 (1974).
[287] S. Nakashima, *Analyst*, **104**, 172 (1979).
[288] R. S. Danchik and D. F. Boltz, *Anal. Lett.*, **1**, 90 (1968).
[289] Y. Yamamoto, T. Kumamaru, Y. Hayashi, M. Kanke and A. Matsui, *Talanta*, **19**, 1633 (1972).
[290] T. V. Ramakrishna, J. W. Robinson and P. W. West, *Anal. Chim. Acta*, **45**, 43 (1969).
[291] G. Devoto, *Boll. Soc. Ital. Biol. Sper.*, **44**, 425 (1968).
[292] V. Rozenblum, *Anal. Lett.*, **8**, 549 (1975).
[293] J. Paul. *Mikrochim. Acta*, 830 (1965).
[294] S. J. Simon and D. F. Boltz, *Anal. Chem.*, **47**, 1758 (1975).
[295] S. S. Michael, *Anal. Chem.*, **49**, 451 (1977).
[296] R. E. Madsen, Jr., *Atom. Absorp. Newslett.*, **10**, 57 (1971).
[297] S. Tamura and S. Nazaki, *Shika Gakuho*, **27**, 62 (1972).

[298] A. U. Shaikh and D. E. Tallman, *Anal. Chem.*, **49**, 1093 (1977).
[299] S. P. Singhal, *Mikrochim. J.*, **18**, 178 (1973).
[300] S. S. M. Hassan and M. H. Eldesouky, *Z. Anal. Chem.*, **259**, 346 (1972).

7

Oxygen compounds

7.1 CARBOXYLIC ACIDS

Solvent extraction of many of the first-row transition metal elements can readily be achieved at room temperature by use of a variety of carboxylic acids. The extraction involves formation of binary and ternary metal complexes and constitutes the main principle upon which AAS determinations of carboxylic acids are based.

A method for determining thiosalicylic acid has been described, based on its reaction with phthalic acid and chromium(III) at pH 3. The reactants are boiled in n-butanol for 20–60 min and the chromium content in the organic phase is measured by AAS, as a function of thiosalicylic acid concentration. Sodium chloride is added to the reaction mixture as a salting-out agent to increase the efficiency of extraction to > 95%, and a fuel-rich air–acetylene flame is used to inhibit the formation of refractory chromium oxides [1].

Phthalic acid has been determined by measuring the copper content of its ion-association complex with copper(I)–neocuproine (2,9-dimethyl-1,10-phenanthroline. The complex is $[Cu(neocuproine)_2^+]$ $[C_6H_4(COOH)COO^-]$ and can be easily extracted into either chloroform or MIBK. Relatively non-polar solvents such as carbon tetrachloride, cyclohexane and benzene do not extract the complex either with or without phthalic acid present. MIBK (methyl isobutyl ketone) also provides a stable flame during combustion and gives high sensitivity. About 80% of the phthalic acid is recovered in a single extraction step and quantitative extraction is easily achieved under optimal conditions. The use of phosphate buffer of pH 8 to keep the final pH in the aqueous phase at about 5, and use of at least tenfold molar ratio of copper to phthalic acid are required for full extraction of the complex.

The calibration graph is prepared by treating 1–10 ml portions of $10^{-4}M$ phthalic acid with 1 ml of $10^{-2}M$ copper sulphate, 2 ml of 5% hydroxylammonium sulphate, and 5 ml of $0.25M$ phosphate buffer. The mixture is diluted to 2 ml, shaken for 2 min with 10 ml of $2 \times 10^{-2}M$ neocuproine solution in MIBK, and the copper absorbance in the extract is measured at 324.7 nm in an air–

acetylene flame. The air-fuel ratio has a significant effect on the copper absorbance. Flow-rates of 1 and 7 l./min for the acetylene and air respectively, are used. Under these conditions, a linear calibration graph in the range 4×10^{-6}–$4 \times 10^{-5}M$ phthalic acid is obtained without interference from equimolar quantities of isophthalic and terephthalic acids. The presence of these two isomers or of benzoic acid in tenfold molar ratio to phthalic acid causes a 10% positive bias of the phthalic acid recovery.

Anthranilic acid has been determined by its reaction with cobalt(II) and bathophenanthroline in a phosphate buffer of pH 6.3 to form an insoluble cobalt complex, followed by extraction of the complex into methyl isobutyl ketone (MIBK) and measurement of the cobalt content [4]. According to Chemical Abstracts, one mole of cobalt(II) and four moles of bathophenanthroline are combined with two moles of anthranilic acid to give a mixed-ligand complex. The structure of the complex was determined by paper chromatography and a continuous-variations spectrophotometric method. The organic extract is aspirated into an air-acetylene flame and the cobalt absorbance is measured at 240.7 nm, as a function of anthranilic acid concentration in the range 3-22 μg/ml. The average recovery is 99.5%.

Anthranilic acid has also been determined by reaction with iron(III) in acetate buffer of pH 5.8-6 to give a metal complex extractable into MIBK [5]. The optimal concentration of the acetate buffer is 0.7-0.9M and that of the iron(III) chloride is $> 5 \times 10^{-2}M$. Under these conditions, the atomic absorption of iron at 248.3 nm in the organic extract is linearly related to the concentration of anthranilic acid in the range 3-6 $\times 10^{-3}M$.

The reaction of p-aminobenzoic acid with copper in the presence of bathophenanthroline to give a ternary complex has been utilized for the determination of concentrations of p-aminobenzoic acid as low as 5 μg/ml [6]. A methanolic solution of the acid is treated with a methanolic $10^{-3}M$ solution of copper and this mixture is heated to dryness on a water-bath. After cooling, the residue is washed twice with ethyl acetate to remove the unreacted copper, and the copper aminobenzoate salt is extracted with a solution of bathophenanthroline in MIBK. The organic layer is then aspirated into an air-acetylene flame and the absorbance of copper is measured at 324.7 nm. A linear calibration graph is obtained for the range 13.7-38 μg/ml and the average recovery of the acid is 100.6% [6].

β-Hydroxynaphthoic acid is determined by a method based on extraction of the tris(1,10-phenanthroline)nickel-β-hydroxynaphthoate ion-association complex from a solution at pH 5, into nitrobenzene, followed by AAS measurement of the nickel [7]. The cation should be at fivefold molar ratio to the anion to ensure quantitative extraction, the optimum concentrations being $> 1.6 \times 10^{-3}M$ and $10^{-4}M$, respectively. Interference by β-naphthol (the main contaminant) is avoided if the raction is done at pH > 9.

Oxalic acid forms an insoluble calcium salt, and this reaction has been used for its AAS determination. Oxalic acid in urine is determined by reaction with calcium at pH 5 followed by measurement of the residual calcium, and then the total calcium added plus that already present in urine is determined at pH 2. The oxalate present in the precipitate is calculated from the difference between the two calcium results [8]. This method has been modified by Koehl and Abecassis [9]. The urine is collected and acidified with hydrochloric acid to pH < 1 to ensure good stability before the reaction. Then two 8-ml portions of the acidified sample are transferred to two 10-ml graduated tubes, one marked 'reference' and the other 'test'. To each of these tubes, 0.4 ml of 'overloading' oxalic acid solution (4 g/l.), and 0.4 ml of 'overloading' calcium chloride (8 g/l.) are added. The pH of the solution in the tube marked 'test' is adjusted with aqueous ammonia to 5, and both solutions are diluted to 10 ml with water. The tubes are left for 24 hr at 4°C for precipitation of calcium oxalate. The tubes are centrifuged, and aliquots of the supernatant liquids are recentrifuged, diluted 100-fold with lanthanum chloride solution, and aspirated into the flame. The standard deviations found for 17 and 186 mg/l. oxalate concentrations were 3 and 6 mg/l., respectively and no interference was caused by phosphate, ammonium, magnesium and urea.

Ethylenediaminetetra-acetic acid (EDTA) in the antibiotic streptomycin has been determined by its complexation reaction with nickel(II) followed by pH adjustment to release the nickel, which is directly proportional to the EDTA present [10]. The limit of detection obtained by AAS measurement is about 4 μg/g and the reproducibility is within ± 1.3 μg/g without significant interference from phosphate ions. Another method for determining EDTA is based on formation of its complex with copper, and consequent masking of copper in extraction of the copper–oxinate complex into methyl isobutyl ketone (MIBK) at pH 6.5 [11]. In this method, exactly 4 ml of $10^{-5}M$ aqueous copper solution and 1 ml of $10^{-4}M$ oxine solution in MIBK are shaken with the EDTA solution (containing < 40 nmole of EDTA), and the copper content of the organic layer is then measured by AAS. The decrease in copper absorbance is linearly related to EDTA concentration.

Flufenamic acid in the presence of copper(II) and 2-(2-hydroxyethyl)-pyridine forms a 1:1:1 complex extractable into propyl acetate. Measurement of the copper in the organic layer by AAS provides a reliable method for determining the acid at low levels [12]. The method is applicable to the measurement of the acid in biological systems (e.g. small intestine of rats) and pharmaceutical preparations. The feasibility of determining various carboxylic acids by precipitation of their insoluble silver salts, followed by measurement of the excess of silver by AAS has been demonstrated [13]. Estimations of equivalent weights by this method showed a positive bias of 0.5–3.5%.

7.2 ESTERS AND ANHYDRIDES

Organic esters are known to form hydroxamic acids by reaction with hydroxylamine. Most hydroxamic acids react with iron(III) to form characteristic complexes.

$$RCOOR' + NH_2OH \xrightarrow{OH^-} R\overset{\overset{\displaystyle O}{\|}}{C} - NHOH + R'-OH$$

$$3R\overset{\overset{\displaystyle O}{\|}}{C}-NHOH + Fe^{3+} \longrightarrow \left(\begin{array}{c} HN-O \\ | \quad \diagdown \\ | \quad \nearrow \\ R-C=O \end{array}\right)_3 Fe + 3H^+$$

This reaction has been adapted for AAS determination of esters of some carboxylic acids [14]. An aliquot of the sample solution containing 0.035-3.5 mg of the ester per ml is heated with 4 ml of 20% hydroxylammonium chloride in 12% sodium hydroxide solution for 5 min at 70°C in a water-bath. The solution is cooled and treated with 2 ml each of $3M$ hydrochloric acid and ferric chloride solution (0.33% Fe). The mixture is filtered after 15 min and iron(III) in the filtrate is measured at 248.3 nm (air-acetylene flame). Recoveries of aliphatic esters are between 92.4 and 105% without interference from ethers, acids, ketones, phenols and aldehydes. The same procedure has also been satisfactorily used for determining some acid anhydrides (recoveries 97.3-102.8%) [15].

7.3 CARBONYL COMPOUNDS

Oxidation of the carbonyl group by silver(I) has formed the basis of several procedures for the determination of structurally different aldehydes. Successful titrimetric methods based on the use of silver-tert. butylamine, ammoniacal silver nitrate (Tollen's reagent) and solid silver oxide have been developed for determining many aldehydes at millimolar levels. The usefulness of some of these reagents on the micromolar level, however, is limited by the possible formation of non-oxidizable imines. The reaction of the ammonia in Tollen's reagent with the carbonyl group to form such imines is apparently insignificant compared with that of the amine moiety of the silver-tert. butylamine reagent.

$$H_2CO + H_2NC(CH_3)_3 \rightleftharpoons H_2C=NC(CH_3)_3 + H_2O$$

For this reason, Tollen's reagent has been recommended for AAS determination of low levels of aldehydes [16,17]. The reagent is simply prepared by addition of 1 ml of $6M$ sodium hydroxide and 2 ml of concentrated ammonia solution to 5 ml of $1M$ silver nitrate. The reaction of aldehydes with the reagent proceeds according to:

$$RCHO + 2Ag(NH_3)_2^+ + 2OH^- \longrightarrow 2Ag + RCOONH_4 + 3NH_3 + H_2O$$

Aldehydes at 0.25-4mM concentration levels are determined by reaction with Tollen's reagent at room temperature, followed by collection of the metallic silver on a fine fritted-glass funnel, washing with aqueous ammonia and dissolution of the silver in nitric acid $(1+1)$. The solution is then diluted with water and the silver content measured at 328.1 nm by AAS with an air–acetylene flame. The silver content of the filtrate may also be determined after acidification with $12M$ nitric acid [18]. The optimum working range for the silver determination is 2-20 μg/ml (equivalent to 0.1-1 μmole of aldehyde) in a final volume of 10 ml. The error can range up to ± 3% for 1-2 μmole of aldehyde.

The reactivity of aldehydes varies according to the nature of the substituent groups. Aliphatic aldehydes, other than formaldehyde, are much less reactive than substituted benzaldehydes, and aromatic aldehydes substituted with electron-withdrawing groups are more reactive than those substituted with electron-donating groups. It has also been reported that substituent groups that enhance imine formation affect the reaction with silver. Examples of these compounds are p-acetamidobenzaldehyde and p-methoxybenzaldehyde. The former gives p-aminobenzaldehyde on hydrolysis, which reacts with another mole of the parent aldehyde to form an imine derivative, and the latter slowly forms an imine with Tollen's reagent.

In general, the reactivities of various substituted aldehydes can be estimated from the Hammett function. The detection limit varies between about 4 μg/ml (e.g. phenylacetaldehyde) and 0.23 mg/ml (e.g. 2-chlorobenzaldehyde). The recovery is in the range 97-103% for levels of 0.025-15 mg/ml. The presence of a 50-80-fold amount of acetone, ethanol or acetic acid, and a 3.7-fold amount of phenol, aniline, benzoic acid, nitrobenzene or 2-aminophenol is completely tolerated, whereas o- and p-aminophenol, resorcinol, benzoin and benzil interfere seriously [18].

The reaction of biacetyl with hydroxylammonium chloride and nickel(II) to give nickel dimethylglyoximate has been used for determining biacetyl in the concentration range 0.05-2 mg/ml [19]. The sample solution is mixed with hydroxylammonium chloride–sodium acetate mixture (81 mg of $NH_2OH.HCl$ and 40 mg of CH_3COONa per ml) and heated at 75°C for 10 min. The mixture is allowed to cool and react with nickel solution (Ni 6.3 mg/ml), then the precipitate is filtered off, washed, dissolved in concentrated nitric acid, diluted with water and aspirated into an air–acetylene flame for nickel measurement at 232 nm. Acetone, ethyl acetate, ethanol, acetic acid, acetaldehyde, phenol, lactic acid and glucose do not interfere.

Low molecular-weight ketones and aldehydes have been determined by reaction with thiosemicarbazide solution and copper acetate at 70°C. The reaction product was extracted with benzene, and the extract dried, mixed with ethanol, and analysed for copper content by AAS, with measurement at 324.7 nm. Recoveries for acetone, butyraldehyde, cyclopentanone, ethyl methyl ketone, formaldehyde and propionaldehyde were in the range 96.7-102.2% [20]. Phenylacetaldehyde, p-dimethylbenzaldehyde, cinnamaldehyde and salicylaldehyde did not interfere.

7.4 PHENOLS

Phenols are determined by a reaction involving concomitant nitrosation and complexation with cobalt. The phenol solution is heated with $Na_3Co(NO_2)_6$ in 0.14-1M acetic acid for 5 min at 100°C followed by extraction of the cobalt complex into methyl isobutyl ketone and aspiration of the extract into the flame for cobalt measurement [21].

$$[Co(NO_2)_6]^{3-} + 6H^+ \longrightarrow Co^{3+} + 6HNO_2$$

The cobalt complexes formed are stoichiometric in composition, permitting accurate determination of various phenols, recoveries generally being in the range 97–103%.

Many other derivatives of phenols have been determined by a variety of reactions. Pentachlorophenol (PCP) in the presence of tris(1,10-phenanthroline)-iron(II) forms a complex with the composition $[Fe(phen)_3] [C_6Cl_5O]_2$ which is extractable into some organic solvents [22]. The PCP is allowed to react with 0.02M tris(1,10-phenanthroline)iron(II) sulphate in 0.3M phosphate buffer at pH 7, and the complex formed is extracted into nitrobenzene; the extract is dried over anhydrous sodium sulphate and aspirated into an air–acetylene flame. The absorbance of iron is linearly related to PCP concentration in the range 0–3 × 10⁻⁴M.

8-Hydroxyquinoline (oxine) has been determined by reaction with copper(II) in ammonium acetate buffer at pH 6.5. The copper-oxinate complex is extracted into methyl isobutyl ketone or ethyl acetate and the copper measured in the air–acetylene flame [11]. The reaction of copper(II) with catechol to form a yellow 2:1 metal:ligand chelate at pH 9.8–10.2 has been adapted for determining catechol by AAS [23]. The chelate is extracted into chloroform in the presence of capriquat (tri-n-octylmethylammonium chloride). The organic extract is diluted with methanol and its copper content is measured at the resonance line of 324.7 nm, in an air–acetylene flame. The copper and capriquat concentrations should be 5 and 10 times that of the catechol, respectively. The composition of the complex is probably $[(Catechol)_2Cu(II)(capriquat)_2]$. The relationship between copper absorbance and catechol concentration is linear in the range 11–176 μg/ml. The standard deviation and coefficient of variation are 0.3 and 0.7%, respectively and many ions cause no interference.

Catecholamines are determined by oxidation with a mild oxidant followed by complexation of the reaction product with transition metals. Concomitant oxidation and complexation of this group of biogenic amines with potassium dichromate has been reported as the basis of a sensitive AAS method for their determination [24]. The choice of potassium dichromate is based on its suitability as a staining agent for the preservation of the dense core vesicles of adrenergic nerves, through rapid oxidation of the catecholamines contained in the vesicles. Oxidation of the catechol moiety to a quinonoid structure either prior to or concomitant with complexation with chromium gives a reaction product containing one chromium atom for every molecule of catechol. The exact structure of these complexes has not yet been elucidated, but presumably the electrophilic metal co-ordinates with the electron-rich oxygen and the benzene ring.

This reaction has been used for the determination of dopamine levels in urine. The catecholamines from urine are adsorbed on alumina, eluted with acetic acid, allowed to react with dichromate, and separated on a Sephadex G-25 column. The dopamine–chromium chelate is retained on the column, and the effluent contains other catechol–chromium products and the excess of dichromate. The chromium content in the effluent is measured by either flame or carbon-rod atomization. By comparison of the concentration of chromium in the test samples and a blank and a test sample with a known amount of catecholamine added, the level of catecholamine in the test sample can be calculated. Application of the AAS method to determination of the urinary excretion of dopamine by eight normal subjects gave a value of 0.18 ± 0.04 $\mu g/mg$ of creatinine, which compares favourably with the values of 0.15 ± 0.05 and 0.19 ± 0.03 $\mu g/mg$ of creatinine obtained by gas–liquid chromatographic and fluorimetric methods, respectively. However, the extreme sensitivity of the carbon-rod atomization method permits detection of as little of 5 ng of chromium bound to catecholamine, and enables the use of as little of 1 μl of the sample test solution [24].

Coomans et al. [25] described a method for AAS determination of tannins in tea. The tannins are defined as heterogeneous mixtures of polyphenolic substances, which have the property of tanning leather. Tea tannins are often classified as a group of polyphenolic character consisting of catechin and catechin tannins, the levels of which vary from tea to tea. Determination of this group of phenolic compounds is based on the precipitation of the copper salt and measurement of the copper content of either the filtrate or the precipitate. The tea sample (~ 5 g) is boiled for 10 min with 100 ml of water and the extract is filtered and diluted fivefold. An aliquot (25 ml) of the solution is then treated with 10 ml of $0.1M$ copper acetate solution, boiled for 10 min and the precipitate removed by filtration or centrifugation. A known fraction of the precipitate or the solution is digested with 1:1 v/v nitric–sulphuric acid mixture, diluted and aspirated into an air–acetylene flame for copper measurement. The tannin content is calculated from the quantity of copper bound per gram of tea and may also be expressed as the tannic acid equivalent. The mean ratios of bound copper to tannic acid, obtained by AAS and potentiometric measurements, are in good agreement: 27.23 ± 0.34 and 27.14 ± 0.32, respectively. Significantly low results are usually obtained by gravimetric methods [25].

7.5 ALCOHOLS

Monohydric alcohols are determined by a reaction involving formation of chromium complexes [26]. The solution of alcohol in benzene is treated with a chromium reagent prepared by mixing 20 g of chromium(III) iodide and 31.6 ml of pyridine in 250 ml of acetic acid. The reaction mixture is diluted with benzene and after one hour aqueous ammonia solution is added. The benzene layer is then separated, dried over sodium sulphate, filtered, mixed with an equal volume of methanol and analysed for chromium at 359.7 nm by AAS with an air–acetylene flame. The limit of detection for various alcohols is about 10 μg, and the recovery ranges from 95.4 to 103.8%. Isopropyl alcohol cannot be determined by this method. Acetone, toluene, phenol, aniline, benzoic acid and nitrobenzene in concentrations up to 40 times that of the alcohol do not interfere. However, propionaldehyde interferes in the determination of 1-propanol [26].

Oxidation of the vicinal diol group with periodic acid (Malaprade reaction) is by far the most widely applied reaction for determining diols:

$$R.CHOH.(CHOH)_n.RCHOH + (n-1)IO_4^- \longrightarrow 2R.CHO + nHCOOH$$
$$+ H_2O + (n-1)IO_3^-$$

The reaction is highly selective, although hydroxy compounds substituted at the α-position with carbonyl or amino groups, α-dicarbonyl and α-aminocarbonyl compounds react similarly.

Addition of silver nitrate to slightly acidified diol samples treated with periodic acid results in stoichiometric precipitation of silver iodate in amounts proportional to the diol compound. Separation of excess of reagent by filtration and dissolution of the silver iodate precipitate in concentrated ammonia solution followed by measuring the silver content by AAS has been recommended [27]. The optimal concentrations of nitric acid, silver nitrate and periodic acid in a solution containing 0.1–4 μmole of diol per ml are $2.0M$, $0.34M$ and 0.34–$7.4mM$, respectively. The reaction mixture should be cooled to -10 or $-15°C$ before filtration, to suppress the solubility of silver iodate, then the precipitate is washed with 1:1 v/v acetone–water mixture containing 0.2% v/v nitric acid to prevent peptization. Under these conditions, the blank value is about 0.5–1 ppm of Ag in 50 ml of solution. The recovery is generally 97–99%.

The reactivity of various diols is in the order: ethylene glycol > 1,2-propanediol > 2,3-butanediol. It seems that the electron-releasing effect of the alkyl groups stabilizes the reaction intermediate of diols with periodate. On the other hand, the reaction time required for quantitative oxidation of various compounds, whether substituted with electron attracting or repelling groups, is not very long and 45 min are sufficient for quantitative oxidation. It should also be noted that the time required for silver iodate to precipitate from the reaction solution increases significantly when the concentrations of iodate and diols are less than $0.07 \, \mu$mole/ml.

Oxidation of the vicinal diol groups in some compounds (e.g. tartaric acid) may result in the formation of products which are susceptible to further oxidation with periodate.

$$HOOCCH(OH)CH(OH)COOH + HIO_4 \rightarrow 2OHCCOOH + H_2O + HIO_3$$

$$OHCCOOH + HIO_4 \rightarrow HCOOH + CO_2 + HIO_3$$

In this case, the reaction is done either under extremely mild conditions (i.e. low reactant concentrations and temperature) so that only the first step is quantitatively achieved, or under relatively severe conditions whereby the reaction is forced to completion.

Vicinal diols have also been determined by reaction with potassium periodate followed by precipitation of the excess of periodate with lead. The lead periodate precipitate is filtered off, dissolved in nitric acid and analysed for lead by AAS [28,29]. Although some other heavy metal ions [e.g. copper(II), iron(III), mercury(II) and silver(I)] are known to precipitate periodate, the reaction with lead gives the only stoichiometric product $[Pb_3H_4(IO_6)_2]$.

An aliquot of diol solution (0.4–$4 \, \mu$mole/ml) is treated with $0.02M$ potassium periodate followed after 2–5 min by $0.04M$ lead nitrate. After standing for not more than 2 min the lead periodate is filtered off on a 0.2-μm membrane filter under suction. The precipitate is dissolved in $1M$ nitric acid and the solution aspirated into the air–acetylene flame for lead measurement. The method is applicable to both aliphatic and aromatic diols. The relative standard deviation ranges from 2.6 to 6.1% at concentration levels of about $2 \, \mu$mole/ml, but can be as high as 14% at lower concentrations. The significant difference in the oxidation rates of some diols renders it possible to determine more than one diol in a mixture, or a particular diol in the presence of another. For example, 1-amino-2,3-propanediol and DOPA can be determined in their mixtures, and styrene glycol can be measured in the presence of styrene epoxide. This is of particular interest because epoxide hydrase enzyme catalyses the hydrolysis of the epoxide into the 1,2-diol which renders it possible to monitor this enzymatic reaction by AAS [29]. Furthermore, methanol, ethanol, 2-propanol, 1-butanol, formaldehyde and formic acid do not interfere, but glycine, 1,3-dihydroxybenzene and 1,3,5-trihydroxybenzene at concentrations $> 0.05 \, \mu$mole/ml undergo slight oxidation with periodate. This is also of particular significance, as ethanol and 1,3-dihydroxybenzene are present as impurities in 1,2-ethanediol and DOPA, respectively.

7.6 CARBOHYDRATES

Oxidation with periodate and reduction of certain metal ions are widely used for the determination of reducing sugars. Frequently, the equivalent amount of the reagent consumed or one of the reaction products is measured by a variety of techniques. These reactions have been adapted for the determination of low concentrations of carbohydrates by AAS. Sugars in ethanol are determined by reaction with periodic acid, followed by precipitation of the resultant iodate with silver nitrate. The precipitate is dissolved in aqueous ammonia and the silver content measured by flame AAS [30]. The limits of detection for lactose, sucrose and glucose are 0.82, 1.93 and 0.21 μg/ml, respectively. Tartaric acid, benzoin, glycerol and vicinal diols react similarly. The method is non-selective, of course.

A mixture of sugars has been determined by reaction with silver(I) after chromatographic separation. The sugar solution and standards (up to 45 μg of each sugar) are subjected to paper chromatography on the same sheet, with ethyl acetate/acetic acid/water (9:2:2 v/v) or 1-butanol/2-propanol/water (8:3:2 v/v) as developing solvent. After development, the paper is dried and drawn through an acetone solution of silver nitrate, dried again and then drawn through ethanolic sodium hydroxide to locate the spots of the various sugars. The spot of each sugar is fixed by drawing the paper through 10% sodium thiosulphate solution and the paper is then washed in running water. The greenish spots are then cut out and the colloidal silver is dissolved with 20% aqueous nitric acid and determined by AAS [31]. The AAS test solution may be diluted until its silver content is < 20 ppm, since the absorbance of silver is a linear function of amount of sugar up to 45 μg. This method has been used for the analysis of mixtures containing glucose, mannose, arabinose and xylose.

Reducing sugars have also been determined by reaction with copper(II) in alkaline media, followed by separation of the copper(I) oxide precipitate by centrifugation and AAS measurement (at 249 nm) of the excess of copper(II) in the supernatant liquid. The method has been applied to dried parsley and frozen strawberries [32]. The results were in good agreement with those obtained by the AOAC method.

Another approach for determining glucose, suggested by Christian and Feldman [11], is based on the effect of glucose on the atomic absorption of calcium in the flame. These authors found that glucose, at very low concentrations ($< 10^{-6}M$), causes a marked decrease in the absorption or emission by calcium chloride at the $10^{-4}M$ level. As the concentration of glucose increases above $10^{-6}M$, the absorption or emission of the calcium increases and then levels off when the glucose concentration is about $10^{-5}M$. Over this range of glucose concentration, no background emission is detected by the d.c. system. These results have been confirmed by Pinta [33] who found that glucose in the range of 10^{-6}-5 × $10^{-5}M$ linearly enhances the atomic absorption of calcium. The phenomenon, however, is not entirely understood. It seems that small amounts of glucose may complex part of the calcium or form compounds with the calcium salt, resulting in suppression of the calcium absorption, whereas at high glucose concentrations interferences by other species in the solution are alleviated. It is also known that sucrose and EDTA exhibit a physical effect in the removal of chemical interferences in the flame. The observed behaviour of glucose may originate from this effect.

REFERENCES

[1]   D. G. Sebastian and D. C. Hilderbrand, *Anal. Chem.*, 50, 488 (1978).
[2]   T. Kumamaru, Y. Hayashi, N. Okamoto, E. Tao and Y. Yamamoto, *Anal. Chim. Acta*, 35, 524 (1966).
[3]   T. Kumamaru, *Anal. Chim. Acta*, 43, 19 (1968).
[4]   Y. Kidani, N. Osugi, K. Inagaki and H. Koike, *Bunseki Kagaku*, 24, 218 (1975); *Chem. Abstr.*, 83, 84926f (1975).
[5]   Y. Kidani, K. Inagaki and H. Koike, *Ann. Rept. Fac. Pharm. Sci., Nagoya City University*, 20, 21 (1972).
[6]   Y. Kidani, T. Saotome, K. Inagaki and Y. Koike, *Bunseki Kagaku*, 24, 463 (1975).
[7]   Y. Yamamoto, T. Kumamaru, Y. Hayashi and S. Matsushita, *Bull. Chem. Soc. Japan*, 42, 1774 (1969).
[8]   R. Menache, *Clin. Chem.*, 20, 1444 (1974).
[9]   C. Koehl and J. Abecassis, *Clin. Chim. Acta*, 70, 71 (1976).
[10] R. J. Hurtubise, *J. Pharm. Sci.*, 63, 1131 (1974).
[11] G. D. Christian and F. J. Feldman, *Anal. Chim. Acta*, 40, 173 (1968).
[12] T. Fukushima, K. Sakai, N. Hashitani, E. Fukushima and, N. Yamagishi, *Chem. Pharm. Bull. Tokyo*, 21, 1632 (1973).
[13] H. K. L. Gupta and D. F. Boltz, *Mikrochim. J.*, 16, 571 (1971).
[14] Y. Minami, T. Mitsui and Y. Fujimura, *Bunseki Kagaku*, 28, 513 (1979).
[15] Y. Minami, T. Mitsui and Y. Fujimura, *Bunseki Kagaku*, 29, 491 (1980).
[16] P. J. Oles and S. Siggia, *Anal. Chem.*, 46, 911 (1974).
[17] T. Mitsui and Y. Fujimura, *Bunseki Kagaku*, 25, 429 (1976).
[18] T. Mitsui and T. Kojima, *Bunseki Kagaku*, 26, 182 (1977).
[19] Y. Minami, T. Mitsui and Y. Fujimura, *Bunseki Kagaku*, 28, 717 (1979).
[20] Y. Minami, T. Mitsui and Y. Fujimura, *Bunseki Kagaku*, 30, 566 (1981).
[21] T. Mitsui and Y. Fujimura, *Bunseki Kagaku*, 23, 1303 (1974).
[22] Y. Yamamoto, T. Kumamaru and Y. Hayashi, *Talanta*, 14, 611 (1967).
[23] Y. Kidani, S. Uno, Y. Kato and Y. Koike, *Bunseki Kagaku*, 23, 740 (1974).
[24] N. E. Naftchi, M. A. Becker and A. S. Akerkar, *Anal. Biochem.*, 66, 423 (1975).
[25] D. Coomans, J. Siberklang, Y. Michotte, L. Dryon and D. L. Massart, *Z. Anal. Chem.*, 294, 140 (1979).
[26] T. Mitsui and T. Kojime, *Bunseki Kagaku*, 26, 228 (1977).
[27] P. J. Oles and S. Siggia, *Anal. Chem.*, 46, 2197 (1974).
[28] B. Tan, P. Melius and M. V. Kilgore, *Anal. Chem.*, 52, 602 (1980).
[29] B. Tan and P. Melius, *Anal. Proc.*, 18, 384 (1981).
[30] T. Mitsui and Y. Fujimura, *Bunseki Kagaku*, 25, 429 (1976).
[31] A. Halonen, *Paperi Puu*, 54, 99 (1972).
[32] A. L. Potter, E. D. Ducay and R. M. McCready, *J. Assoc. Off. Anal. Chem.*, 51, 748 (1968).
[33] M. Pinta, *Method. Phys. Anal.*, 6, 268 (1970).

8

Sulphur compounds

8.1 TOTAL SULPHUR

8.1.1 Direct Methods

Sulphur has long been determined routinely by arc or spark emission spectrometry at its principle resonance line at 180.7 nm. Attempts to use AAS for direct determination of sulphur have been hampered by the difficulties encountered in the commercial manufacture and operation of hollow-cathode lamps for this element. The high rate of sputtering of sulphur from the cathode and the formation of a relatively dense atomic sulphur vapour in front of the cathode cause loss of signal by self-absorption. Moreover, the rapid deposition of the cathode material on the lamp envelope after operation shortens the operation lifetime. The first direct AAS method for the determination of sulphur and the first usage of a microwave discharge as an atomizer were described by Taylor et al. [1]. Microgram quantities of organosulphur, sulphate and sulphide are determined by measuring the absorption at 216.89 nm. Organic compounds or sulphate solutions are dissolved in acetone, and 10 μl of the solution are deposited on a platinum filament; the sample is then vaporized into a stream of argon, which is used as plasma gas. For sulphide solutions, hydrogen sulphide is evolved and passed directly into the plasma. The microwave power is adjusted to 25 W and the discharge initiated with a Tesla coil. This method has been used for the determination of sulphur in some pesticides such as parathion, ethion and guthion.

The successful production and operation of the microwave-excited electrodeless discharge lamp (EDL) for sulphur [2] has led to the use of this source in conjunction with a nitrous oxide–acetylene flame and a vacuum or inert-gas-purged monochromator for direct AAS determination of sulphur at 180.7 nm [3]. The lamp is made from a silica bulb (60 mm length, 8 mm inner diameter, 1 mm wall thickness) filled with argon at a pressure of 3 mmHg and containing about 1 mg of sublimed sulphur. The lamp is operated in a ¾-wave resonance cavity powered by a 2450-MHz microwave generator at 20 W. Glass tubes

(25 mm diameter) purged with nitrogen and fitted with optically flat fused-silica windows are placed between the source and the flame on one side, and the flame and the monochromator on the other (Fig. 8.1). This device provides a light-path appreciably more transparent than the atmosphere and has been used satisfactorily for the determination of sulphur in some organic compounds. In aqueous solution, sulphate-sulphur in the concentration range 50–700 ppm gives

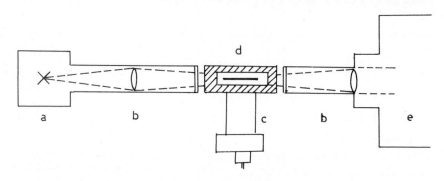

Fig. 8.1. Instrumental assembly for sulphur determination: a, EDL source; b, nitrogen-filled glass tubing with silica end-windows; c, indirect nebulizer and expansion chamber; d, 7-cm slot burner; e, vacuum monochromator. [Reprinted by permission from G. F. Kirkbright and M. Marshall, *Anal. Chem.,* **44**, 1288 (1972). Copyright 1972, American Chemical Society.]

a linear calibration graph. The sensitivity increases remarkably in the presence of organic solvents, probably owing to an improvement of the nebulization efficiency (Fig. 8.2). A nitrogen-separated fuel-rich nitrous oxide–acetylene flame has also been used for AAS determination of sulphur and non-metals which have resonance lines at wavelengths shorter than 200 nm [4,5].

A demountable sulphur hollow-cathode lamp has been used as the source of radiation [6]. Operation of this lamp at high current produces an intense and sharp emission line and minimizes self-absorption. This advantage and the good signal-to-background and signal-to-noise levels make this type of radiation source very useful for flame AAS with a nitrogen-shielded nitrous oxide–acetylene or air–acetylene flame. The lamp is made from a hollow porous graphite-cup cathode (25.4 mm in length, 6.15 mm outer diameter and 3.96 mm inner diameter) filled with elemental sulphur, and a brass anode. The cathode assembly and lamp body are cooled with water and flushed with a constant flow of argon (Fig. 8.3). The modulated lamp is far superior to the corresponding electrodeless discharge lamp for AAS work. With 1 mg of resublimed elemental sulphur in the graphite cathode, and a current of only 2–5 mA, the intensities of the lines at 180.7, 182.0 and 182.6 nm become four times greater than those obtained from the electrodeless discharge lamp under optimal conditions. The

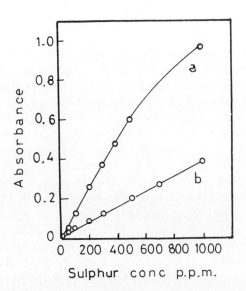

Fig. 8.2. Calibration curves for sulphur: a, thiourea in ethanol; b, potassium sulphate in water. [Reprinted by permission from G. F. Kirkbright and M. Marshall, *Anal. Chem.*, **44**, 1288 (1972). Copyright 1972 American Chemical Society.]

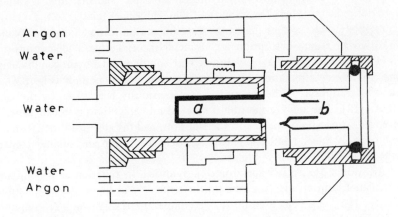

Fig. 8.3. Demountable sulphur hollow-cathode lamp assembly: a, cathode; b, anode. [Adapted by permission from G. F. Kirkbright and P. J. Wilson, *Anal. Chem.*, **46**, 1414 (1974). Copyright 1974 American Chemical Society.]

limit of detection for sulphur is 5 μg/ml, compared with 30 μg/ml with the electrodeless discharge lamp. The relative merits of electrothermal atomization and the conventional flame system have been well documented for sulphur compounds [7-9]. A graphite-rod atomizer similar to that used for the determination of iodine [10] has been used for the determination of some sulphur compounds [7]. An optical path purged with inert gas, a vacuum monochromator, and an electrodeless discharge lamp containing hydrogen sulphide at 3 mmHg pressure and an equal pressure of argon, are utilized. The radiation is focused onto the graphite atomizer by means of calcium fluoride lenses held in the furnace housing, and a second similar lens at the entrance slit of the monochromator. The whole optical path is enclosed by glass tubing which is purged of atmospheric oxygen by a constant flow of oxygen-free nitrogen at 1 l./min. A 5-μl portion of aqueous solution of the sulphur compound (e.g. thiourea or a sulphate) is introduced into the furnace, dried by application of 1 V to the tube until 100% transmission is obtained, then atomized at 8 V applied voltage. The characteristic concentrations (amounts giving 1% absorption) at 180.7, 182.0 and 182.6 nm are 0.42, 0.68 and 1.5 ng, respectively. It has been reported that the calibration curves deviate from linearity at absorbance values of approximately 0.3, probably owing to molecular absorption by SO_2, which is formed during thermal decomposition of sulphur compounds at the high atomization temperature.

Adaptation of the atomic-absorption spectrometer to measure the absorption by molecular species allows determination of sulphur dioxide and hydrogen sulphide. The first of these is formed by direct atomization of organosulphur compounds or sulphate ions in a hydrogen–air flame, and the other is formed by a prior reduction step. The stability of these molecules relative to the relatively unstable atomic species makes the experimentation easier. Fuwa and Vallee [11] used an atomic-absorption spectrometer incorporating a long absorption cell (1 × 273 cm) heated by a nichrome wire winding, and a hydrogen discharge lamp, for the determination of some organosulphur compounds (Fig. 8.4). A detailed description of the instrument had previously been given by the same authors [12]. The analyte solution is sprayed into the absorption cell either at the end or through an opening in the side-arm, and the absorption at 207 nm is monitored. A linear relationship between SO_2 absorption and sulphur content in the range 10–80 μg/ml is obtained.

An atomic-absorption spectrometer modified to function as a non-flame molecular-absorption spectrometer [in the same way as for ammonia determination (p. 115)] has been used for the determination of sulphide [13]. Sulphuric acid is introduced from a burette into the reaction vessel and nitrogen is passed through at a flow-rate of 1.63 l./min. After establishment of the base-line, an aliquot of the sulphide test solution (equivalent to 1–100 μg of sulphur per ml) is injected through the port of the reaction vessel, and the absorption signal due to the liberated hydrogen sulphide is recorded. Under optimum conditions the linear portion of the plot of peak absorbance *vs.* concentration of injected sulphide solution extends up to at least 400 μg/ml, with a lower limit of detection of 1.2 μg/ml.

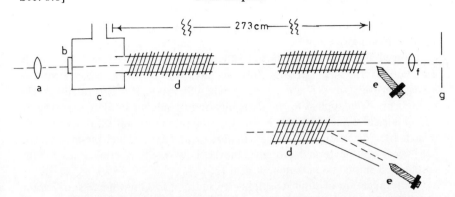

Fig. 8.4. Schematic diagram of flame absorption system for sulphur determination: a, collimating lens; b, quartz window; c, exhaust chamber; d, Vycor cell; e, burner; f, focusing lens; g, monochromator slit. [Reprinted by permission from K. Fuwa and B. L. Vallee, *Anal. Chem.*, 41, 188 (1969). Copyright 1969 American Chemical Society.]

8.1.2 Indirect Methods

The ease with which organically bound sulphur can be converted into sulphate and sulphide permits the application of indirect AAS methods to the determination of a wide variety of organic sulphur compounds. Sulphate is usually released either by wet or dry oxidation whereas sulphide is formed by reduction or fusion with metals [14-18]. However, one of the simplest and most direct methods for safe, rapid and quantitative conversion of sulphur in various compounds into sulphate is the oxygen-flask method [19-22]. In this method the samples are combusted in a ground-neck 250-ml Erlenmeyer flask with a stopper into which one end of a platinum wire (1 mm diameter) is sealed, the other end being attached to a 1.5 × 2 cm piece of platinum gauze which forms a sample holder. Solid samples are wrapped in an L-shaped 2 × 3 cm piece of filter paper cut with a 2-3 cm tail to act as a fuse. Glass capillaries or gelatine capsules may be used as containers for non-volatile liquids and wrapped in the filter paper. The fuse of the paper is ignited and the stopper is immediately inserted into the flask, which has been flushed with oxygen and contains 10 ml of water and 3-5 drops of 30% hydrogen peroxide as absorbent. The combustion is completed in 15-30 sec and the sulphur oxides produced are dissolved in the absorbent to give sulphate.

The first indirect AAS method for the determination of organosulphur compounds was based on their conversion into sulphate and precipitation of barium sulphate followed by dissolution of the precipitate and measurement of the barium content [23]. The samples (50-100 mg) are weighed into methylcellulose capsules and combusted in an oxygen-filled flask containing 10 ml of 6% aqueous ammonia/6% hydrogen peroxide mixture. Alternatively the samples are digested with Benedict sulphur reagent (200 g of copper nitrate + 50 g of sodium chlorate in 1 litre of water). The sulphate ions released are treated with 15% barium chloride solution and the barium sulphate precipitate is repeatedly centrifuged and washed to remove all the residual barium, before dissolution in disodium EDTA and aspiration into an air–acetylene flame. Results with an average recovery of 98% and a mean standard deviation of 4% are obtained. The method has been applied to the determination of sulphur in diets, urine, faeces and biological tissues [23,24]. Similarly, total sulphur in soil [25] can be determined after digestion with nitric–sulphuric acid mixture, and plasma-sulphate [26] after phosphate and protein removal. Low results were obtained for DL-cysteic acid and L-cysteinesulphinic acid [23].

Determination of sulphate-sulphur by addition of a known excess of barium to acidified analyte solutions ($SO_4^{2-} \geqslant 100$ $\mu g/ml$), followed by measurement of the excess of barium has also been suggested [27,28]. This approach may avoid the manipulation steps required for isolation of a pure precipitate and minimize the errors caused by co-precipitation of other ions. Errors due to the partial solubility and incomplete precipitation of barium sulphate can, however, be minimized by using a standard sulphate solution, treated similarly, for calibration. Variations of the solubility product of barium sulphate with the composition of the reaction solution are controlled either by carrying out the precipitation in ethanol solution [29], or by seeding the reaction mixture with a specially prepared barium sulphate precipitate [30]. Reduction of either the concentration or the volume of the barium chloride precipitant and addition of sodium ions (> 1.5 mg/ml) to suppress the ionization of barium have also been reported to improve the sensitivity of the method [31]. In general, measurement of sulphate-sulphur by precipitation with barium gives results with a recovery better than 97% and an accuracy comparing favourably with that obtained by turbidimetry and titrimetry.

AAS determination of sulphur by precipitation of either lead or strontium sulphate may also be used. The solubility of lead sulphate is minimized by addition of 25% ethanol to the solution [32], and this method gives results with a relative standard deviation of 3.6%. Gersonde [33] has described a method for determination of sulphur in protein by combusting lypophilized samples in an oxygen-filled flask containing hydrogen peroxide and ammonia, and precipitating the resulting sulphate with 25% strontium chloride solution at 65°C. An equal volume of ethanol is added and the precipitate is removed by centrifugation after 15 hr, washed with 70% ethanol and dissolved in 1.5% EDTA solution containing ammonia. The method has been applied to the determination of sulphur in cystine, cysteine, cysteine methyl ester hydrochloride, glutathione, ribonuclease, bovine serum albumin and ferric perchlorate diphenylsulphoxide.

Atomic-absorption inhibition-titration may also be used for the indirect determination of sulphur. One of these methods is based on the inhibition effect of sulphate on the absorption of magnesium in the flame [34]. An aliquot of the sulphate solution (containing 0.05–1.0 mg of SO_4^{2-}) is pipetted into the titration vessel by means of suitable polyethylene tubing (0.027 in. bore) and diluted to 50 ml. The titration vessel is placed on a magnetic stirrer and the aspiration and titrant-delivery tubes inserted. The solution is then aspirated into the flame by use of a constant-flow pump fitted with a 30-ml syringe. Addition of the magnesium titrant at constant flow-rate is started and the chart-recorder is switched on at the same time, to monitor the atomic absorption of magnesium at 285.2 nm. Since complete inhibition of the atomic absorption of magnesium usually takes place at a magnesium:sulphate ratio of less than unity, when the ratio is in excess of this there is a linear increase in the magnesium absorption, and the end-point of the titration is located at the intersection of extrapolations of the base-line and this rise in absorption (see Fig. 3.3, p. 79).

Sulphur may be determined indirectly by methods based on its conversion into sulphide followed by reaction with heavy metals and AAS measurements of the metals. Mercury(II) chloride and tin(II) chloride are warmed at 40°C with submicrogram quantities of sulphide, and the residual uncombined mercury is

Table 8.1   AAS determination of bivalent sulphur in some aliphatic compounds by desulphurization with alkali metal plumbite [37] (reprinted by permission of the copyright holders, Springer-Verlag)

Compound	Sulphur (%)			
		Found		
	Theory	By flask combustion	AAS	Pb–ISE	Visual titration
Acetazolamide	28.80	28.7	14.5[a]	14.4[a]	14.4[a]
Benzoylthiosemicarbazide	16.41	16.3	16.4	16.3	16.2
Benzylthiazonium chloride	15.84	15.4	15.6	15.6	15.4
Carbimazole	17.18	16.9	17.2	16.7	17.1
L-Cystine	26.63	26.4	26.1	26.2	26.3
2,5-Diaminobenzoyl isothiocyanate	22.93	22.6	22.6	22.7	22.8
2,4-Diaminothiazole	27.58	27.4	27.2	27.3	27.5
Dibenzyl disulphide	25.95	25.6	25.6	25.5	25.8
Dithio-oxamide	53.24	52.7	53.0	52.9	52.8
2-Phenyl-5-aminobenzoyl isocyanate	18.82	18.5	18.7	18.6	18.6
2-Phenyl-4-aminothiazole	18.07	17.7	17.8	17.7	18.1
Sodium diethyldithiocarbamate	28.40	28.1	28.0	28.2	28.5
Sodium thioglycollate	28.04	27.0	27.3	27.5	27.6
Sulphathiazole	25.06	24.7	12.4[a]	12.3[a]	12.6[a]
Thiambutosine	9.31	9.1	9.2	9.2	9.4
Thioacetamide	42.59	42.1	42.3	42.3	41.9
Thiourea	42.10	42.2	41.8	41.8	41.9

[a] Compounds containing 2 atoms of sulphur per molecule, only one being bivalent.

volatilized and determined by conventional cold-vapour AAS [35]. Results accurate to ±2% can be obtained for down to 20 ng of sulphide. Sulphide ions may also be determined by precipitation with zinc and measurement of the zinc content of either the precipitate or the excess of reagent [36].

Hassan and Eldesouki [37] found that bivalent sulphur in structurally different aliphatic compounds (i.e. thiols, sulphides, disulphides and thiocarbonyls) reacts with solid potassium hydroxide at 250–280°C with quantitative formation of lead sulphide on addition of alkali-metal plumbite. This permits use of AAS measurement of the excess of lead as a measure of the sulphur. The excess of plumbite is diluted with nitric acid to bring the final lead concentration within the linear range of the calibration graph (1–10 μg/ml), then aspirated into an air–acetylene flame, and the absorbance of lead is measured at 217 nm. The results obtained with many organic compounds show an average recovery of 98.8% and agree favourably with those obtained by measuring the excess of lead with a lead-selective electrode or by combusting the samples in an oxygen-filled flask and titrating the sulphate (Table 8.1). However, bivalent sulphur directly attached to an aromatic moiety (e.g. in Methylene Blue, promazine, thioridazine, thiosalicylic acid, toluene dithiols, diphenyl disulphide and thiophenol) is not converted into sulphide under these conditions. Sulphur(IV) and (VI) in either aliphatic or aromatic compounds will not be converted into sulphide. No significant interference is caused by chloride, bromide, iodide, arsenate and phosphate.

8.2 THIOLS (MERCAPTANS)

Determination of thiols by AAS is based on the ability of these compounds to form stable metal derivatives. Mercaptobenzothiazole is determined by heating with aqueous silver nitrate solution at 50–60°C for 1 min, followed by aspiration of a portion of the supernatant liquid into the flame [38]. Various other thiols are similarly determined on the micro scale by using the 2-propanol–water azeotrope as solvent for the compounds and the silver nitrate [39]. Then 1–2 ml of silver nitrate solution is added to 1 ml of sample solutions containing 0.8–20 μmole of thiol, at room temperature, followed by 2 ml of water. After 20–30 min the precipitate is filtered off on a fine fritted glass filter and washed with 5–10% v/v aqueous ammonia solution, then dissolved with hot concentrated nitric acid, and after dilution analysed for silver. Simple aliphatic and aromatic thiols react stoichiometrically with silver in 1:1 ratio. Substances that reduce silver nitrate or that form an insoluble silver salt that is not soluble in ammonia solution will interfere. For mercaptocarboxylic acids sodium acetate should be added to the reaction mixture to enhance dissociation of the carboxylic acid group so that both the thiol and the carboxylate group react with silver. The precipitate is then washed only with water (to avoid the removal of silver from the −COOAg grouping that would occur with an ammonia wash). Thioacetic acid and silver sulphide are produced:

$$CH_3COS^- + 2Ag^+ + H_2O \rightarrow CH_3COO^- + Ag_2S + 2H^+$$

8.3 THIOETHERS (SULPHIDES)

Aqueous solutions of sodium metaperiodate quantitatively oxidize organic sulphides at room temperature to the corresponding sulphoxides, without further oxidation of the sulphur to a higher state:

$$R-S-R + HIO_4 \rightarrow R-\underset{\downarrow}{\overset{}{S}}-R + HIO_3$$
$$O$$

Sample solution (1 ml) containing 100-500 μg of the sulphide is treated with 1 ml of 0.015M sodium periodate solution, and the iodate formed is precipitated as the silver salt under acidic conditions. The precipitate is removed by filtration and the excess of silver is measured by AAS in an air-acetylene flame [40]. The solubility of silver iodate is minimized by the use of a relatively large excess of silver ions ($\sim$ 0.65M), cooling to −15°C, and washing the precipitate with 1:1 water-acetone solution containing 0.2% of concentrated nitric acid. The reaction time for most sulphides is approximately 15 hr, but longer is required with dilute periodate solutions.

The average recovery for 1–5 μmole of sulphide was 99.3%, the precision being ±2-6%. The method was applied to the determination of total sulphides in shale oil and the naphtha cut of shale oil. It should be noted that disulphides do not react with periodate, whereas thiols are partially oxidized. Thus determination of sulphides in the presence of thiols necessitates prior assessment of the thiol content by one of the standard methods. Vicinal diol, diamine and diketone compounds react quantitatively with periodate but they cause few problems in the determination of sulphides as they rarely occur in admixture with them.

8.4 THIOCARBONYLS

8.4.1 Carbon Disulphide

The use of large quantities of carbon disulphide by the chemical industry necessitates monitoring of this compound in the industrial and ambient atmosphere. The reaction of carbon disulphide with copper(II) in the presence of aliphatic amines gives the corresponding copper dithiocarbamate which can be extracted and measured by AAS. Carbon disulphide solutions are shaken with a copper sulphate solution containing diethylamine and aqueous ammonia and the complex formed is extracted with carbon tetrachloride. A portion of the organic extract is then evaporated to dryness, the residue is dissolved in dilute nitric acid and the copper content of the solution is measured by AAS [41]. This method is suitable for the measurement of as little as 25 ng of CS_2. Carbon disulphide in coke-oven gas has been determined by absorption in alcoholic potassium hydroxide to form ethyl xanthate, followed by reaction with copper sulphate and AAS measurement of the copper content of the complex [42]. The sensitivity of the method for CS_2 is 9 mg/m^3. Kovatsis [43] described a method for the determination of the carbon disulphide liberated by acid decomposition of diethyldithiocarbamate at $\sim 100°C$. The carbon disulphide is absorbed in a methanolic solution containing zinc chloride or acetate and dibenzylamine. After a reaction period of 30 min an equal volume of water is added, the pH is adjusted to 9.1, the zinc-dibenzyldithiocarbamate complex produced is extracted into toluene, and the zinc is measured by flame AAS. The limit of detection is 0.165 μg/ml.

8.4.2 Dithiocarbamates

Ammonium pyrrolidine dithiocarbamate (APDC) is determined by the measurement of its copper or cobalt complex. These complexes are extracted into methyl isobutyl ketone (MIBK) at pH 3 and aspirated into the flame. APDC concentrations as low as $10^{-5}M$ can be determined by this method [44]. APDC has similarly been determined by nebulizing its copper complex in MIBK into an air-acetylene flame and measuring the copper at 324.7 nm [45]. Diethyldithiocarbamate as a metabolite of thiram [46] or disulphiram [47] is determined by complexation with copper and electrothermal atomization of the carbon tetrachloride extract of the copper complex. The lower limit of detection is 0.6 μg/ml.

8.4.3 Xanthates

Xanthates react with silver ions to give sparingly soluble salts. The reaction mixture is heated for 10-15 sec at 50°C, and the precipitate is immediately separated before catalytic decomposition of the product by the excess of silver:

$$C_2H_5-O-\overset{\overset{\displaystyle S}{\|}}{C}-S-Ag + 2Ag^+ + H_2O \longrightarrow C_2H_5-O-\overset{\overset{\displaystyle O}{\|}}{C}-S-Ag + Ag_2S + 2H^+$$

$$2C_2H_5-O-\overset{\overset{\displaystyle O}{\|}}{C}-S-Ag + H_2O \longrightarrow 2C_2H_5OH + Ag_2S + COS + CO_2$$

The supernatant liquid (after suitable dilution) is aspirated into the flame for silver measurement [38].

8.5 SULPHONAMIDES

Hassan and Eldesouki [48] investigated the solubility of various metal sulphonamides in aqueous solutions at various pH values. They found that both silver and copper form extremely insoluble precipitates (AgX and CuX_2) with many structurally different sulphonamides in the pH range 7.4-8.6. Application of this method to the determination of some sulphonamides gave an average recovery of 99.4% and a mean standard deviation of 0.6% (Table 8.2). The results agreed well with those obtained by direct potentiometric titration with either copper or silver, with use of the appropriate ion-selective electrode.

Table 8.2 AAS determination of some sulphonamides by reaction with silver and copper [48] (reprinted by permission of the copyright holders, The Association of Official Analytical Chemists)

| Compound | Average recovery %[a] | | | |
| | Reaction with silver | | Reaction with copper | |
	AAS	Ag–ISE	AAS	Cu–ISE
Sulphadiazine	99.4	99.7	99.0	99.7
Sulphadimidine	99.0	99.5	99.4	99.6
Sulphamerazine	99.4	99.6	99.1	49.7
Sulphamethoxazole	99.7	99.8	99.4	99.7
Sulphamethoxine	99.4	99.6	99.6	99.7
Sulphamethoxydiazine	99.5	99.7	99.4	99.7
Sulphamethoxypyridazine	99.5	99.8	99.3	99.9
Sulphapyridine	99.6	99.8	99.6	99.8
Sulphathiazole	99.3	99.7	99.6	99.8

[a] Average of 5 measurements; the mean standard deviation is ±0.5%.

REFERENCES

[1]   H. E. Taylor, J. H. Gibson and R. K. Skogerboe, *Anal. Chem.*, **42**, 1569 (1970).
[2]   B. V. L'vov, *Atomic Absorption Spectrochemical Analysis*, p. 255, Elsevier, New York (1970).
[3]   G. F. Kirkbright and M. Marshall, *Anal. Chem.*, **44**, 1288 (1972).
[4]   G. F. Kirkbright, A. F. Ward and T. S. West, *Anal. Chim. Acta*, **62**, 241 (1972).
[5]   G. F. Kirkbright, M. Marshall and T. S. West, *Anal. Chem.*, **44**, 2379 (1972).
[6]   G. F. Kirkbright and P. J. Wilson, *Anal. Chem.*, **46**, 1414 (1974).
[7]   M. J. Adams and G. F. Kirkbright, *Can. J. Spectrosc.*, **21**, 127 (1976).
[8]   G. F. Kirkbright, *Analyst*, **96**, 609 (1971).
[9]   R. Woodriff, *Appl. Spectrosc.*, **28**, 413 (1974).
[10] M. J. Adams, G. F. Kirkbright and T. S. West, *Talanta*, **21**, 573 (1974).
[11] K. Fuwa and B. L. Vallee, *Anal. Chem.*, **41**, 188 (1969).
[12] K. Fuwa and B. L. Vallee, *Anal. Chem.*, **35**, 942 (1963).
[13] A. Syty, *Anal. Chem.*, **51**, 911 (1979).
[14] L. Carius, *Ann.*, **116**, 1 (1860); **136**, 129 (1865); **146**, 301 (1868).
[15] W. Kirsten, *Anal. Chem.*, **25**, 74 (1953).
[16] J. B. Niederl, H. Baum, J. S. McCoy and J. A. Kuck, *Ind. Eng. Chem., Anal. Ed.*, **12**, 428 (1940).

[17] A. M. G. Macdonald, in *Comprehensive Analytical Chemistry*, C. Wilson and D. W. Wilson (eds.), Vol. IB, p. 510, Elsevier, Amsterdam (1960).

[18] C. J. Thompson and C. S. Allbright, in *The Analytical Chemistry of Sulfur and its Compounds*, J. H. Karchmer (ed.), p. 89, Wiley–Interscience, New York (1970).

[19] W. Schöniger, *Mikrochim. Acta*, 123 (1955); 869 (1956).

[20] I. Lysyj and J. E. Zarembo, *Anal. Chem.*, **30**, 428 (1958).

[21] A. Steyermark, *Quantitative Organic Microanalysis*, 2nd Ed. p. 276, Academic Press, New York (1961).

[22] A. M. G. Macdonald, in *Advances in Analytical Chemistry and Instrumentation*, C. N. Reilley (ed.), Vol. IV, p. 75, Interscience, New York (1965).

[23] D. A. Roe, P. S. Miller and L. Lutwak, *Anal. Biochem.*, **15**, 313 (1966).

[24] J. L. Meyer and R. T. Rundquist, *Biochem. Med.*, **12**, 398 (1975).

[25] A. Oeien, *Acta Agric. Scand.*, **29**, 71 (1979).

[26] D. Michalk and E. Manz, *Clin. Chim. Acta*, **107**, 43 (1980).

[27] R. Dunk, R. A. Mostyn and H. C. Hoare, *Atom. Absorp. Newslett.*, **8**, 79 (1969).

[28] M. I. Couto and A. J. Curtius, *Appl. Spectrosc.*, **34**, 228 (1980).

[29] N. V. Hue and F. Adams, *Commun. Soil Sci. Plant Anal.*, **10**, 841 (1979).

[30] B. Magyar and F. Santos, *Helv. Chim. Acta*, **52**, 820 (1969).

[31] G. G. Galindo, H. Applet and E. B. Schalscha, *Soil Sci. Soc. Am. Proc.*, **33**, 974 (1969).

[32] S. A. Rose and D. F. Boltz, *Anal. Chim. Acta*, **44**, 239 (1969).

[33] K. Gersonde, *Anal. Biochem.*, **25**, 459 (1968).

[34] R. Looyenga and C. O. Huber, *Anal. Chim. Acta*, **55**, 179 (1971).

[35] Z. Yoshida and M. Takahashi, *Mikrochim. Acta*, 459 (1977 I).

[36] R. C. Ray, P. K. Nayar, A. K. Misra and N. Sethunathan, *Analyst*, **105**, 984 (1980).

[37] S. S. M. Hassan and M. H. Eldesouki, *Mikrochim. Acta*, 27 (1979 II).

[38] H. K. L. Gupta and D. F. Boltz, *Microchem. J.*, **16**, 571 (1971).

[39] J. S. Marhevka and S. Siggia, *Anal. Chem.*, **51**, 1259 (1979).

[40] R. P. D'Alonzo, A. P. Carpenter, Jr., S. Siggia and P. C. Uden, *Anal. Chem.*, **50**, 326 (1978).

[41] T. Mitsui and Y. Fujimura, *Bunseki Kagaku*, **25**, 429 (1976).

[42] C. A. Coutinho, J. G. De Sousa and P. Y. Saito, *Metal. ABM*, **32**, 683 (1976).

[43] A. V. Kovatsis, *Atom. Absorp. Newslett.*, **17**, 104 (1978).

[44] G. D. Christian and F. J. Feldman, *Anal. Chim. Acta*, **40**, 173 (1968).

[45] J. E. Allan, *Spectrochim. Acta*, **17**, 459 (1961).

[46] F. K. Martens and A. Heyndrickx, *Meded. Fac. Landbouwwet., Rijksuniv. Gent*, **41**, 1393 (1976).

[47] F. K. Martens and A. Heyndrickx, *J. Anal. Toxicol.*, **2**, 269 (1978).

[48] S. S. M. Hassan and M. H. Eldesouki, *J. Assoc. Offic. Anal. Chem.*, **64**, 1158 (1981).

9

Halogen compounds

Halogen atoms are present in organic compounds in ionizable, hydrolysable and tightly bound forms. Such halogenated compounds can be determined by AAS, after conversion into halide ions either by hydrolysis with alkalis or decomposition by the Schöniger oxygen-flask method [1–4]. Since the principle atomic resonance lines of halogens occur in the vacuum ultraviolet region of the spectrum, it is not possible to use the normal atomic-absorption procedures without instrumental modification or indirect measurement. The methods now in use for the determination of halogens are based on the following approaches.

1. Suppression or enhancement of the atomic absorption of certain metals by interaction with the halides in the flame. For example, fluorine depresses the absorption of magnesium and enhances the absorption of zirconium or titanium.

2. Precipitation of the halides with excess of metal reagent and measurement of the metal in the solution or the precipitate. Fluoride can be precipitated as the calcium salt and chloride, bromide and iodide as the silver salts.

3. Formation of ion-association complexes in which the halide and cationic metal complex form an extractable species, the metal content of which can be measured and will give the halide concentration. An example is the formation of the cadmium–phenanthroline–iodide complex.

4. Reduction of ions such as chromium(VI) and selenium(IV) with iodide, followed by separation of the oxidized and reduced forms, by extraction or precipitation, and AAS measurement.

5. Displacement of metals from their complexes [e.g. iron(III) from its thiocyanate complex by fluoride] and extraction and measurement of the excess of metal complex.

6. Modification of the instrument to permit measurement in the ultraviolet region. Modified radiation sources and atomization devices have been suggested for the determination of iodine.

7. Oxidation of the halides (e.g. bromide and iodide) and measurement of the molecular absorption of the halogens. In another approach, the halides are allowed to form molecular metal halide species (e.g. AgX, InX, CuX and HgX_2)

the metal content of which is monitored by recording the emission spectra by use of the atomic-absorption spectrometer accessories.

9.1 FLUORINE COMPOUNDS

The first genuine study of the spectra of fluorides was reported by Mitscherlich in 1862 [5,6] who observed band spectra of what are now known to be BaF and CaF, when mixtures of ammonium fluoride and barium or calcium fluoride are supplied to the flame. Since then, many methods have been described for flame spectrometric determination of fluorine. With the development of AAS, indirect methods have also been proposed.

Fluoride ions have a depressive effect on the atomic absorption of some cations and an enhancement effect on the absorption of others. These phenomena are used for determining fluoride at low concentration levels [7]. Absorption by magnesium, calcium, strontium, barium, lead and iron in an air–coal gas flame is known to be depressed to various extents by fluoride ions. Of these metals, magnesium displays a significant depression, the magnitude of which depends on the concentration of both the magnesium and the fluoride ions. The absorption of magnesium at 285.2 nm decreases linearly with increase of fluoride concentration, to reach a constant value when the Mg:F ratio becomes 1:2. A similar effect has been reported for barium in a turbulent-flow oxygen–hydrogen flame [8]. In a premixed air–acetylene flame, however, neither magnesium nor barium absorption is affected by fluoride.

On the other hand, fluoride ions enhance the atomic absorption of zirconium, titanium and hafnium in a luminous nitrous oxide–acetylene flame [7,9]. The absorption of zirconium at 360.1 nm increases linearly with increase in fluoride concentration, then deviates from linearity at fluoride concentration slightly in excess of that of zirconium. A 0.1–$0.3M$ hydrochloric acid solution containing 500 μg of zirconium per ml, and 500 μg of sodium chloride per ml is used for the determination of fluoride at concentrations $< 10^{-3}M$. Sodium chloride is added to the standard and sample solutions to suppress ionization of zirconium in the flame. The presence of phosphate interferes seriously, owing to precipitation of zirconium phosphate. It has also been reported that fluoride similarly enhances the absorption of titanium at 364.3 nm with a sensitivity comparable to that for the zirconium system. The major advantage of titanium over zirconium is that phosphate in concentrations $< 0.1M$ does not interfere. The use of magnesium offers the advantages of high sensitivity, applicability to low fluoride concentrations, and the use of safe flames such as air–coal gas.

It should be noted that the depression and enhancement effects of fluoride on the atomic absorption of some metals are mainly due to formation of difficultly and easily dissociated species in the flame, respectively. The depression of magnesium absorption is probably due to the formation of magnesium fluoride, which has a higher boiling point (~ 2600 K) than any other magnesium salt [10], and hence resists complete dissociation in the low-temperature air–coal gas flame. The fact that fluoride does not depress magnesium absorption in the high-temperature nitrous oxide–acetylene flame, because there is complete dissociation of magnesium fluoride, is in good agreement with this assumption. Metals which have their absorption enhanced by the presence of fluoride are all known to exist in aqueous solutions as oxy or hydroxy species. For example, zirconium cations are present in solutions as trimeric and tetrameric hydroxy species [11]. Atomization of these solutions in the absence of fluoride ions gives a stable zirconium oxide aggregate, whereas in the presence of fluoride, easily atomized oxyfluoro complexes such as $ZrOF_2$ are formed [12].

Menis and House [13] described a method for the determination of fluoride by precipitation with calcium and measurement of the excess of calcium by flame-emission spectrometry. Campanella et al. [14] precipitated the fluoride with calcium acetate, bismuth nitrate, or potassium bromide plus lead nitrate followed by separation of the precipitates, dissolution in acid and measurement of the calcium, lead or bismuth at 422.7, 283.3 and 223.0 nm, respectively. Of these methods, the calcium fluoride method gives the best results. At the 15-200 µg/ml level, the precision is about 10%. In another indirect method for the determination of fluorine in organic compounds, Kidani and Ito [15] decomposed the samples (ca. 30-60 µg) with sodium biphenyl, adjusted the pH of the decomposition solution to 3, and treated an aliquot of the solution with excess of iron(III) chloride and ammonium thiocyanate. The excess of the iron(III) thiocyanate complex was extracted with MIBK and atomized in an air-acetylene flame. The calibration graph was linear in the fluoride range 0.5-6.0 µg/ml and the recoveries for fluorine in various organic compounds were in the range 94.8-100.5%.

The flame band spectra of fluoro compounds of magnesium, calcium, barium, strontium and copper in both hydrogen–oxygen and carbon monoxide flames have long been used for the determination of fluorine [16]. Simultaneous nebulization of strontium nitrate and fluoride solutions into the flame gives an SrF band, which has an intensity linearly related to the fluoride concentration; the detection limit is 9.2 µg and the precision ±1.6%. A hydrogen-oxygen flame is the best for this method [17]. A more sensitive method is based on aspiration of fluoride solution in the presence of an aluminium salt solution into the flame (or injection into a carbon rod atomizer) to give the AlF radical [18]. This radical displays a strong absorption at 227.5 nm and can be used for subnanogram determination of fluorine in organic and inorganic compounds.

The formation and the absorption of AlF are both influenced by the type of flame. In a nitrous oxide-acetylene flame, aluminium atoms are generated and display absorption bands near to that of AlF, at 226.9 and 226.3 nm. Prior formation of aluminium atoms seems to be an essential step in the formation of the AlF radical under these conditions. The absorption of AlF and Al increases with increase of the acetylene level in the flame, reaching a maximum at an acetylene–nitrous oxide flow-rate ratio of about 0.75 and at a height of 8 mm above the burner-head. In the air–acetylene flame, where the temperature is about 300°C lower, the band due to Al at 226.3 nm disappears and the sensitivity of the AlF band becomes independent of the fuel–oxidant ratio. Formation of AlF in a carbon rod atomizer is independent of the ashing current up to 70 A (1000°C). When the atomization current reaches 200 A (2400°C), the absorption signal becomes maximal and levels off at higher current (temperature).

In the actual measurement, fluoride (at concentrations up to 1 μg/ml) is mixed with 0.02M aluminium nitrate and atomized. This gives results with a precision of ±3% for 0.2 ng of fluoride. As little as 21 pg of fluoride can be measured by use of electrothermal atomization. This level is the lowest so far reported for spectrochemical determination of this element. The method has been used for the determination of fluorine, chlorine and bromine in biological and organic compounds, at the nanogram level (see Table 9.1). The samples are ashed with 0.1 g of sodium carbonate in a porcelain crucible at 550°C for 10 hr, then the ash is dissolved in hot water and neutralized with 1M nitric acid, and the solution is diluted. An aliquot of solution is mixed with 0.02M aluminium nitrate and 0.005M strontium or nickel nitrate, and the AlF absorption is measured, with a deuterium lamp as the source of radiation.

Table 9.1 Determination of halogens in some organic compounds by measuring the molecular absorption of aluminium halides [18,44] [Reprinted by permission from K. Tsunoda, K. Fujiwara and K. Fuwa, *Anal. Chem.*, **49**, 2035 (1977); **50**, 861 (1978). Copyright 1978 and 1979 American Chemical Society.]

Compound	Weight (ng)		Standard deviation (ng)
	Taken	Found	
o-Fluorobenzoic acid	0.59	0.63	0.03
Sodium fluoroacetate	0.62	0.68	0.03
Trifluoroacetic acid	0.52	0.56	0.03
o-Chlorobenzoic acid	5.1	5.7	0.3
Sodium chloroacetate	5.0	4.9	0.4
p-Bromobenzoic acid	25.5	28.2	1.3
Sodium 2-bromoethane sulphonate	25.6	25.8	0.9

The main problems with use of the deuterium lamp for this purpose are the needs to use a narrow spectral band-pass and to correct for the background. These can be overcome, however, by using an irradiation source giving lines close to that of AlF. Since the atomic lines of platinum are located at 227.44 and 227.48 nm, very close to the AlF line at 227.5 nm, a platinum hollow-cathode lamp can be used, with spectral band wider than that of the deuterium lamp [19]. It is known that the sensitivity of the atomic-absorption technique is independent of the slit-width when a hollow-cathode lamp is used. The molecular AlF and atomic Pt signals are time-resolved in the electrothermal atomization procedure, the AlF signal being the more transient. The sensitivity of the method, in general, is about half that obtained by using the deuterium lamp. Possibilities for determining traces of fluoride by molecular absorption of the monofluorides of gallium, indium and thallium have been explored [20], but the sensitivity is less than that obtained with AlF.

9.2 CHLORINE COMPOUNDS

The interaction of chlorine-containing compounds with the flame has attracted the attention of spectroscopists since 1860 [21-25]. The greenish-blue colour of the mantle of the flame in the presence of these compounds, however, has no analytical utility, as the emitting molecules contain no chlorine. All of the practical flame-spectroscopic methods for chlorine utilize line or band spectra of metals. The Beilstein test is extensively used for detection of organochlorine compounds; a trace amount of the sample is introduced into a non-luminous flame on a copper wire [26]. Jurány [27] described a simple device for the detection of halogens, based on the Beilstein method. Application of this technique for quantitative determination of chlorine by measuring the emission band of CuCl in an oxy–hydrogen flame has been described [28-31]. Investigation of the spectrum of the green Beilstein flame reveals the presence of OH, Cu, CuH, CuCl, CuOH and CuO species. The green colour is actually due to a diffused CuOH band rather than a CuCl band. The emission flame profile of CuCl is weak, and observed only when the chloride concentration exceeds 500 μg/ml [32].

In general, the methods used for chlorine determination by the Beilstein reaction have three main limitations. The first is that some compounds containing nitrogen and sulphur, pure volatile copper compounds, and lard, butter and suet (or products containing them) give a green colour in the flame and thus interfere [33-35]. It is a generally accepted fact that only a negative Beilstein test can be taken as being conclusive. The second limitation is the formation and dissociation of CuO and CuH in the flame, which affects the background. The third is the interference caused by other halides.

A burner assembly for direct determination of organochlorine compounds by measuring the copper spectrum has been described [32]. The burner provides a contact between the chlorine and hot metallic copper (Fig. 9.1). In a fuel-rich flame, organochlorine compounds are decomposed to give HCl, which strikes the hot copper surface to form CuCl, and this on volatilization dissociates to yield free copper atoms in concentrations dependent upon the concentration of chlorine. Gilbert [36,37] used an indium-coated copper tube burner with an air–hydrogen flame for the determination of chlorine, by measuring the emission of InCl at 360 nm. Organochlorine compounds such as o-chloroaniline, trichloroethylene and pesticides such as lindane, DDT, dieldrin, heptachlor and methoxychlor can be determined at 0.35 mg/ml by using this burner [38,39].

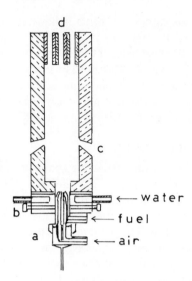

Fig. 9.1. Burner chimney assembly used for halogen determination: a, Beckman burner; b, aligning screws; c, air holes; d, copper tubes. [Reprinted with permission from D. F. Tomkins and C. W. Frank, *Anal. Chem.*, **44**, 1451 (1972). Copyright 1972 American Chemical Society.]

The detection limit of most of these methods has later been improved so that as little as 14 ng of chlorine can be measured by using a modified version of Gilbert's burner [40]. A larger indium surface is used in the modified burner, to permit complete contact between the vapour of the chlorine compounds and indium. Figure 9.2 shows the original burner used by Gilbert [36,37], and its modified version [40]. The latter burner has been used for the determination of chlorinated pesticides in baby foods (carrots, spinach and noodles) [40]. It can also be used for bromine determination [41]. The effect of indium concentration on the InCl emission is not critical, provided that the chloride-to-indium ratio is less than about 3 [42]. Absorption measurement of InCl at 267.2 nm may also be used for the determination of chlorine in the range 0–7.1 μg/ml [43].

Nanogram levels of chlorine in organic compounds have been determined by measuring the absorption of AlCl, with use of a nitrous oxide–acetylene flame or a carbon-rod atomizer [44]. The spectrum of AlCl exhibits fine structure with peaks of varying intensity, at 261.4–262 nm. Since background correction by the deuterium lamp or Zeeman effect cannot be applied to the AlCl absorption signals, a two-channel spectrometer is conveniently used for correction. The continuous light-source channel 'a' measures the absorbance at the bandhead of the molecular absorption, while channel 'b' records the absorption at a wavelength where the molecular species exhibits minimal absorption. The difference between the signals gives the true absorption. This method is applied

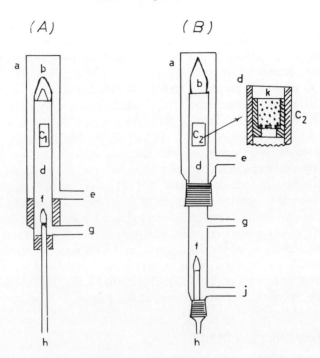

Fig. 9.2. Gilbert burner (A) for flame determination of halogen, and its modified version (B): a, external burner tube; b, secondary combustion zone (upper flame); C_1, indium-coated tube; C_2, beryllium–copper cylinder coated with indium; d, inner burner tube; e, air supply for secondary combustion; f, primary combustion zone (lower flame); g, sample vapour with nitrogen or air; h, hydrogen supply; j, air supply for secondary combustion zone; k, indium-coated copper–beryllium coils; l, indium-coated copper–beryllium perforated disc. [Reprinted by permission from P. T. Gilbert, *Anal. Chem.*, **38**, 1920 (1966), copyright 1966 American Chemical Society, and from R. Herrmann and B. Gutsche, *Analyst*, **94**, 1033 (1969) by permission of the copyright holders, the Royal Society of Chemistry.]

to the determination of halogens in organic and biological materials containing as little as 0.12 ng of Cl and 1.1 ng of Br (see Table 9.1). The samples are decomposed by heating with milk of lime in a porcelain crucible at 550°C and an aliquot of a solution of the decomposition product is mixed with 0.01M aluminium nitrate, 0.01M cobalt nitrate and 0.01M strontium nitrate followed by injection of this mixture into the carbon-rod atomizer.

Indirect determination of chlorine by AAS involves precipitation with silver and measurement of silver in the excess of reagent or in the precipitate. Atomization of silver in an air–acetylene flame permits measurements of 1-10 μg of chlorine per ml with a coefficient of variation in the range 2-3% [45,46]. This method is used for the determination of chlorine in serum [47], plant liquors [48], plant extracts [49] and citrus fruit juices [50]. The silver content of the silver chloride precipitate may also be measured. The precipitate is thoroughly washed, dissolved in aqueous ammonia and aspirated into the flame [51-53]. Chlorine may also be determined by treatment with mercury(II) solution, passage of the solution through a cation-exchanger in the hydrogen form to retain the unreacted mercury(II), reduction of $HgCl_2$ in the effluent by tin(II), and measurement of the mercury content by electrothermal AAS at 253.7 nm [54]. The limit of detection is 0.2 ng, the coefficient of variation being 6.1%. This method can also be used for the determination of bromine and iodine separately or in mixtures with chlorine.

Organic compounds containing ionizable chlorine (e.g. quaternary ammonium salts) are determined similarly [55]. The samples (equivalent to 0.6–0.7 mg of Cl per ml) are treated with excess of silver nitrate, followed by AAS measurement of the silver. Organically-bound chlorine may also be determined by a similar procedure after decomposition into chloride ions either by fusion with metallic sodium or by the Schöniger oxygen-flask method [56,57]. Sublimable compounds are decomposed by refluxing for 30 min with metallic sodium in n-butanol or isoamyl alcohol [58]. This method is used for the determination of chlorine, bromine and iodine at levels of 0.2–5 μg/ml.

9.3 BROMINE COMPOUNDS

The bands of CuBr, AgBr and InBr are as persistent and conspicuous in flames as those of the corresponding chlorides. Hence many of the methods described above for determination of chlorine can be applied to bromine. A burner assembly has been devised to provide a contact between the flame and a hot metallic silver surface [59]. The sampling system is inserted into a cooling-water jacket which is bolted directly onto the burner. When a solution containing bromine or iodide is nebulized into the flame, AgBr or AgI is formed on the surface of the silver insert at a rate dependent upon the concentration of the halide in the sample solution. In contrast to copper and indium halide, most of the silver halide remains on the insert surface at this stage. After 1–5 min of sample nebulization, the aqueous solution is replaced with a 95% ethanol solution to allow vaporization of the metal halides in the flame and measurement of the atomic emission of silver at 338.3 nm. The detection limit is 5 ng/ml for Br and 2 ng/ml for I and the useful analytical range is 5–10 000 ng/ml. Binary mixtures of bromide and iodide can be analysed by this method provided that the maximum I:Br or Br:I ratio is 12.

The band spectrum of CuBr in the flame is not as sensitive as that of CuCl. With the hydrogen–oxygen flame, a linear relation between the emission at 434.1 nm and bromide concentration is obtainable in the range $0.025–0.2M$ without interferences from chloride at concentrations up to $0.2M$ [60]. Bromine-containing compounds may be dissolved in dimethylformamide–acetone mixture and directly nebulized into the flame. Bromide and iodide have also been determined by oxidation to the free halogens and measurement of the absorption of bromine and iodine at 410 nm and 530 nm, respectively [61]. A reaction vessel similar to that used for the determination of sulphide and ammonia (see p. 115, Fig. 4.2) is used. Oxidizing mixtures consisting of $7.2M$ sulphuric acid–$0.01M$ potassium permanganate and $5.4M$ sulphuric acid–$0.05M$ vanadium pentoxide are utilized for oxidation of bromide and iodide ions, respectively. The detection limit offered by this method is 7–14 μg/ml, the average relative deviation being ±0.2–4%. Chloride ions do not interfere under these conditions.

9.4 IODINE COMPOUNDS

The main atomic resonance line of iodine lies at 183.04 nm and is due to the $5_p{}^2P_{5/2} - 6_s{}^4P_{5/2}$ transition. Even with use of a high spectral-output primary source and a nitrous oxide–acetylene flame, poor sensitivity is commonly obtained [62]. Purging the monochromator with nitrogen [63] and using a nitrogen-separated nitrous oxide–acetylene flame [64] improves the sensitivity. L'vov and Khartsyzov [65,66] determined 2 ng of iodine by pulse vaporization of the samples, by use of a 2.5 mm diameter graphite cuvette and a vacuum monochromator. The slit-width used in this measurement (~ 0.5 mm), however, allows the radiation from the furnace wall and molecular absorption and scattering to cause severe interference. Other workers have determined iodine by measuring the emission at non-resonance lines, using microwave [67] and radiofrequency [68] sources. Such methods are applicable to the determination of some organoiodine herbicides and metabolites [69]. The use of a high-power plasma source and purging of both the monochromator and the optical path between the source and entrance slit of the spectrometer with nitrogen allows measurement of the emission of iodine at its resonance (178.28 and 183.04 nm) and non-resonance (206.16 nm) lines [70]. The last-mentioned line is more sensitive, less affected by self-absorption, usable for the measurement of a wide range of iodine concentration, and useful for determining organically bound iodine.

A cathode sputtering cell similar to that devised by Gandrud and Skogerboe [71] for the determination of metals has been used for direct determination of iodine [72]. The atomic absorption or emission at 183 nm is measured in a cell made from a silica tube with an inlet and an outlet (Fig. 9.3). The analyte is introduced by evaporation of a 5-μl aliquot on the cathode, then the cell is evacuated to a pressure of 0.2 mmHg and powered from the mains by a 400-V (225 mA) transformer with a diode rectifier and smoothing capacitors. The input to the cell is set as low as 5–10 W to give high stability and sensitivity. This cell is used for the measurement of nanogram quantities of iodine. A platinum-loop technique has also been described for the AAS determination of iodine [73]. This involves the use of a Tesla-initiated iodine electrodeless discharge

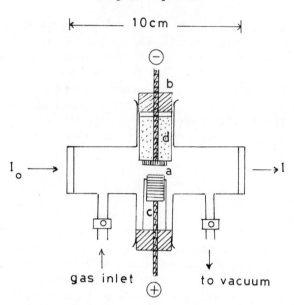

Fig. 9.3. Cathode sputtering cell: a, brass; b, rubber; c, glass; d, Teflon. [Reprinted from G. F. Kirkbright, T. S. West and P. J. Wilson, *Anal. Chim. Acta*, **68**, 462 (1974) by permission of the copyright holders, Elsevier Scientific Publishing Co.]

lamp (EDL) made from fused quartz (8 mm inner diameter, 5 cm long) and filled with argon at 5 mmHg pressure, as a primary source of radiation. The lamp is operated at a power of 15 W in a foreshortened ¾-wave resonance cavity and cooled with a flow of nitrogen. The radiation from the source is allowed to pass through the absorption tube (12.5 cm length, 7 mm inner diameter) where a platinum loop (1 cm of wire of 0.1 mm diameter) is directly suspended in the absorption path just in front of the monochromator entrance slit (Fig. 9.4).

In this system, the loop is used as an atomization device. When the loop is dipped into the analyte solution, a surface tension film of about 0.1 μl volume is picked up. Passage of a small d.c. current of about 0.1 A through the loop evaporates the water in the sample, and further increase of the current to about 1.9 A causes atomization of the dried sample. The atomic absorption and scattering signals at 183 and 206.2 nm respectively, can be measured. The former signal is not entirely due to atomic absorption when the concentration of the atomized iodine-containing compound exceeds 500 μg of iodine per ml, as some scattering may contribute to the signal. The sensitivity and detection limit of this method are similar to those obtained by using an inert-gas-separated nitrous oxide–acetylene flame [64]. The standard deviation at the 100 $\mu g/ml$ level is ±4% and the limit of detection is 18 $\mu g/ml$.

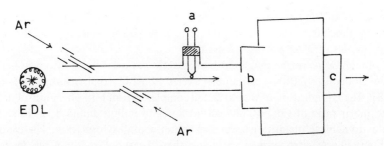

Fig. 9.4. Platinum loop atomization system: a, d.c. supply; b, monochromator entrance slit; c, photomultiplier. [Reprinted from J. M. Manfield, T. S. West and R. M. Dagnall, *Talanta*, **21**, 787 (1974) by permission of the copyright holders, Pergamon Press.]

A water-cooled demountable hollow-cathode lamp, with the cathode made from a porous graphite cup (25.4 mm length, 6.15 mm outer diameter and 3.96 inner diameter) and filled with either iodine or mercuric iodide has been used for the determination of iodine [74]. The analyte solution is atomized in a nitrogen-shielded nitrous oxide–acetylene or air–acetylene flame and the absorbance is measured at 184.4, 187.6 and 206.1 nm. The characteristic concentration (1% absorption) of the method is 14 μg/ml, which compares favourably with that obtained by using the electrodeless discharge lamp. An electrodeless discharge lamp source containing a few milligrams of iodine and powered by a microwave generator in conjunction with a small graphite-tube atomizer (6 mm outer diameter, 4 mm inner diameter, 20 mm length) and a vacuum monochromator has been used for AAS determination of iodine [75]. The atomization cell is similar to that described by Dagnall *et al.* [76] (see p. 57, Fig. 2.18). The sample solution ($\sim$ 10 μl) is injected into the cell through a 2-mm port, and the absorbance is measured. The calibration graph for the 183-nm line is linear up to 6 μg/ml with a characteristic concentration of 0.04 μg/ml. Measurement at 178.2 nm is twice as sensitive.

Iodine can also be determined by indirect methods. For example, when an iodine solution at pH 3.4–4.8 is treated with mercury(II) nitrate and injected into a graphite-tube furnace heated in a ramp mode under a flowing argon atmosphere, two absorption peaks for mercury are displayed [77]. The first peak, due to decomposition of mercury(II) nitrate, appears after 7–23 sec at about 60–180°C and is formed as a result of reduction of the mercury salt by the carbon of the graphite tube. The second peak appears after 37 sec at 440°C, and is due to decomposition of HgI_2. The absorbance of the second peak increases with increase in iodide concentration, whereas the absorbance of the first peak decreases with increase in the iodide level. With a constant flow of argon ($\sim$ 50 ml/min), $10^{-3}M$ mercury(II) nitrate and a heating rate of 4 A/sec, the height of the second peak can be measured as a function of iodide concentration. Precipitation of iodide with silver and AAS measurement of the equivalent amount of silver can also be used.

Iodide ions can be measured by extraction at pH 3.5–5.5 into nitrobenzene from aqueous solutions containing excess of tris(1,10-phenanthroline)-Cd(II) [78]. The composition of the extracted complex is $[Cd(1,10\text{-phenanthroline})_3]I_2$. The molar ratio of $Cd(1,10\text{-phenanthroline})_3$ to iodide should be at least 15 in order to achieve quantitative ion-association complex formation. The cadmium content in the nitrobenzene extract is then measured at 228.4 nm in an airacetylene flame. Iodide ions are also indirectly determined by AAS through a reaction with selenium(IV) in acidic media. The equivalent amount of free selenium formed by reduction with iodide is filtered off on a Millipore filter and the remaining selenium(IV) in the filtrate is measured. The decrease in the atomic absorption of selenium is proportional to the iodide concentration [79]. Iodide may also be determined by reaction with chromium(VI) in acidic medium, extraction of the excess of chromium(VI) into methyl isobutyl ketone and AAS measurement of chromium(VI) in either the organic extract or the aqueous phase [79]. Many of the reactions used for the determination of chlorine and bromine may also be used for iodine [80].

REFERENCES

[1] S. S. M. Hassan, Z. Anal. Chem., 257, 345 (1971).
[2] S. S. M. Hassan and M. B. Elsayes, Mikrochim. Acta, 115 (1972).
[3] S. S. M. Hassan, Z. Anal. Chem., 266, 272 (1973).
[4] S. S. M. Hassan, Mikrochim. Acta, 889 (1974).
[5] A. Mitscherlich, Ann. Physik, 116, 499 (1862).
[6] A. Mitscherlich, J. Prakt. Chem., 86, 13 (1862).
[7] A. M. Bond and T. A. O'Donnell, Anal. Chem., 40, 560 (1968).
[8] P. B. Adams and W. O. Passmore, Anal. Chem., 38, 630 (1966).
[9] M. D. Amos and J. B. Willis, Spectrochim. Acta, 22, 1325 (1966).
[10] L. Brewer, G. R. Somoyajulu and E. Brackett, Chem. Rev., 63, 111 (1963).
[11] A. Clearfield, Rev. Pure Appl. Chem., 14, 91 (1964).
[12] I. V. Tananaev, N. S. Nikolaev and Yu. A. Buslaev, Zh. Neorgan. Khim., 1, 274 (1956).
[13] O. Menis and H. P. House, The Encyclopedia of Spectroscopy, G. L. Clark (ed.), p. 343, Reinhold, New York, 1960.
[14] L. Campanella, G. DeAngelis, R. Sbrilli and A. Marabini, Z. Anal. Chem., 307, 411 (1981).
[15] Y. Kidani and E. Ito, Bunseki Kagaku, 25, 57 (1976).
[16] B. Gutsche and R. Herrmann, Z. Anal. Chem., 258, 277 (1972).
[17] B. Gutsche and R. Herrmann, Z. Anal. Chem., 269, 260 (1974).
[18] K. Tsunoda, K. Fujiwara and K. Fuwa, Anal. Chem., 49, 2035 (1977).
[19] K. Tsunoda, K. Chiba, H. Haraguchi and K. Fuwa, Anal. Chem., 51, 2059 (1979).
[20] K. Dittrich, Anal. Chim. Acta, 111, 123 (1979).
[21] G. Merz, J. Prakt. Chem., 80, 487 (1860).
[22] R. Böttger, J. Prakt. Chem., 85, 392 (1862).

[23] G. Salet, *Ann. Chim. Phys.*, **28**, 5 (1873).
[24] V. Castellana, *Gazz. Chim. Ital.*, **36**, 232 (1906).
[25] P. J. Cowan and J. M. Sugihara, *J. Chem. Educ.*, **36**, 246 (1959).
[26] F. K. Beilstein, *Ber.*, **5**, 620 (1872).
[27] H. Jurány, *Mikrochim. Acta*, 134 (1955).
[28] M. Honma, *Anal. Chem.*, **27**, 1656 (1955).
[29] G. E. March, *Appl. Spectrosc.*, **12**, 113 (1958).
[30] M. Maruyama and S. Seno, *Bull. Chem. Soc. Japan*, **32**, 486 (1959).
[31] C. E. Vander Smissen, *U.S. Patent*, 3,025,141 (March 1962).
[32] D. F. Tomkins and C. W. Frank, *Anal. Chem.*, **44**, 1451 (1972); **46**, 1187 (1974).
[33] M. Jureček and F. Mužik, *Chem. Listy*, **44**, 165 (1950).
[34] J. Van Alphen, *Rec. Trav. Chim. Pays-Bas*, **52**, 567 (1933).
[35] H. Milrath, *Chem. Ztg.*, **33**, 1249 (1935).
[36] P. T. Gilbert, *Anal. Chem.*, **38**, 1920 (1966).
[37] P. T. Gilbert, *U.S. Patent*, 3,504,976 (April 1970).
[38] B. Gutsche, R. Herrmann and K. Rüdiger, *Z. Anal. Chem.*, **241**, 54 (1968).
[39] B. Gutsche and R. Herrmann, *Z. Anal. Chem.*, **242**, 13 (1969).
[40] R. Herrmann and B. Gutsche, *Analyst*, **94**, 1033 (1969).
[41] B. Gutsche and R. Herrmann, *Analyst*, **95**, 805 (1970); *Z. Anal. Chem.*, **249**, 168 (1970).
[42] R. M. Dagnall, K. C. Thompson and T. S. West, *Analyst*, **94**, 643 (1969).
[43] E. Yoshimura, Y. Tanaka, K. Tsunoda, S. Toda and K. Keiichiro, *Bunseki Kagaku*, **26**, 647 (1977).
[44] K. Tsunoda, K. Fujiwara and K. Fuwa, *Anal. Chem.*, **50**, 861 (1978).
[45] E. Than, H. Koening and U. Hascher, *Z. Chem. Leipzig*, **15**, 288 (1975).
[46] T. Ozawa, *Tokyo-Toritsu Kogyo Gijutsu Senta Kenkyu Hokuku*, 115 (1980).
[47] H. Bartels, *Atom. Absorp. Newslett.*, **6**, 132 (1967).
[48] J. B. Ezell, Jr., *Atom. Absorp. Newslett.*, **6**, 84 (1967).
[49] J. B. McHugh and J. H. Turner, *U.S. Geol. Surv. Prof. Paper*, No. 1129-A-1, pp. F1–3 (1980).
[50] E. Postorino, R. Franzi, S. Bonanno and A. Di Giacomo, *Essenze Deriv. Agrum.*, **48**, 316 (1978).
[51] W. Reichel and L. Acs, *Anal. Chem.*, **41**, 1886 (1969).
[52] U. Westerlund-Helmerson, *Atom. Absorp. Newslett.*, **5**, 97 (1966).
[53] M. D. Carrido, C. Liaguno and J. M. Carrido, *Am. J. Enol. Viticult.*, **22**, 44 (1971).
[54] L. S. Chuchalina, I. G. Yudelevich and A. A. Chinenkova, *Zh. Analit. Khim.*, **36**, 920 (1981).
[55] R. V. Smith and M. A. Nessen, *Microchem. J.*, **17**, 638 (1972).
[56] Y. Kidani, H. Takemura and H. Koike, *Bunseki Kagaku*, **22**, 187 (1973).
[57] E. D. Truscott, *Anal. Chem.*, **42**, 1557 (1970).
[58] Y. Kidani, H. Takemura and H. Koike, *Bunseki Kagaku*, **22**, 604 (1973).
[59] R. S. Fike and C. W. Frank, *Anal. Chem.*, **50**, 1446 (1978).
[60] M. Maruyama, S. Seno and K. Hasegawa, *Z. Anal. Chem.*, **307**, 21 (1981).
[61] G. Nicholson and A. Syty, *Anal. Chem.*, **48**, 1481 (1976).
[62] K. C. Thompson, *Spectrosc. Lett.*, **3**, 59 (1970).
[63] G. F. Kirkbright, T. S. West and P. J. Wilson, *Atom. Absorp. Newslett.*, **11**, 113 (1972).
[64] G. F. Kirkbright, T. S. West and P. J. Wilson, *Atom. Absorp. Newslett.*, **11**, 53 (1972).
[65] B. V. L'vov and A. D. Khartsyzov, *Zh. Analit. Khim.*, **24**, 799 (1969).
[66] B. V. L'vov and A. D. Khartsyzov, *Zh. Prikl. Spektrosk.*, **11**, 413 (1969).
[67] R. Mavrodineanu and R. C. Hughes, *Proc. Intern. Conf. Spectrosc.*, University of Maryland, 1962.

[68] G. H. Morrison and Y. Talmi, *Anal. Chem.*, **42**, 809 (1970).
[69] Y. Talmi and G. H. Morrison, *Anal. Chem.*, **44**, 1467 (1972).
[70] G. F. Kirkbright, A. F. Ward and T. S. West, *Anal. Chim. Acta*, **64**, 353 (1973).
[71] B. W. Gandrud and R. K. Skogerboe, *Appl. Spectrosc.*, **25**, 243 (1971).
[72] G. F. Kirkbright, T. S. West and P. J. Wilson, *Anal. Chim. Acta*, **68**, 462 (1974).
[73] J. M. Manfield, T. S. West and R. M. Dagnall, *Talanta*, **21**, 787 (1974).
[74] G. F. Kirkbright and P. J. Wilson, *Anal. Chem.*, **46**, 1414 (1974).
[75] M. J. Adams, G. F. Kirkbright and T. S. West, *Talanta*, **21**, 573 (1974).
[76] R. Dagnall, D. Johnson and T. S. West, *Anal. Chim. Acta*, **67**, 79 (1973).
[77] T. Nomura and I. Karasawa, *Anal. Chim. Acta*, **126**, 241 (1981).
[78] T. Kumamaru, *Bull. Chem. Soc. Japan*, **42**, 956 (1969).
[79] G. D. Christian and F. J. Feldman, *Anal. Chim. Acta*, **40**, 173 (1968).
[80] G. F. Kirkbright and M. Sargent, *Atomic Absorption and Fluorescence Spectroscopy*, Academic Press, London, 1974.

10

Organometallic compounds

The growing realization of the important applications and effects of many organometallic compounds in medicine, pharmacy, agriculture and industry has led to an ever increasing demand for accurate determination of their metal contents. Among the various methods used for the analysis of such compounds, AAS has now become one of the most efficient and sensitive approaches.

10.1 DETERMINATION OF METALS IN ORGANOMETALLICS

Determination of organometallics by AAS usually involves direct dissolution of the sample in a suitable solvent or decomposition (mineralization) to yield metal ions, followed by atomization. Most of the decomposition methods are based on acid-digestion, combustion, dry-ashing or fusion. Acid-digestion in micro Kjeldahl flasks or sealed tubes [1] is most frequently employed (Table 10.1). Combustion by the Schöniger oxygen-flask method has also been widely used to oxidize the organic moeity to gases, leaving the metallic element as a

Table 10.1. Acid digestion of organometallic compounds

Acid	Organometallics
HNO_3	Lanthanides, Ag, Cd, Hg
HNO_3 + HCl	Pt, Au
HNO_3 + H_2SO_4	W, Re, Mo, Se, Te, V, Bi, Pb, Rh, Pd, Sb, As, P, Zn, Cu, Co, Ag, Cd, Cr, Sr, Fe
HNO_3 + $HClO_4$	Sn, Sb, Bi, As, P, Pb, Zn, As, Cu, Co, Ag, Cd, Cr, Mo, Sr, Fe
HNO_3 + H_2SO_4 + HCl	Zr, Nb, Hf, Ta
HNO_3 + H_2SO_4 + $HClO_4$	Cr, Mn, Fe, Co, Ni, Cu, Zn, Ti, Al, Se, Ga, Ir, Ti, Li, Na, K, Rb, Cs, B, As, Ag, Cd, Sb, Mo, Sr
H_2SO_4 or HNO_3 + HCl	Mg, Ca, Sr, Ba
H_2SO_4	Ru, Ir, B
H_2SO_4 or HNO_3 + $HClO_4$	Be

solid residue which is brought into solution by using a suitable acid absorbent in the flask. However, some metals, such as arsenic, lead and bismuth, form alloys with the platinum of the ignition basket, and metals such as iron adhere very firmly to the support. For this reason, it has been recommended that a quartz spiral should be used in place of the platinum gauze [2,3]. Nickel and aluminium produce oxides that are difficult to dissolve. Organic compounds containing metalloids are usually decomposed by fusion. Silicon [4] and boron [5] compounds are fused in a metal bomb with an alkali-metal hydroxide or peroxide. There is still much difference of opinion as to which of these decomposition methods is to be preferred. Some workers recommend dissolving the organometallic compound in a suitable solvent and nebulizing or injecting the solution directly into the atomization system.

10.1.1 Organolead Compounds

Organolead compounds have enjoyed a unique position among organometallics since 1937, when Midgley [6] discovered the exceptional effect of these compounds in suppressing knock in the combustion of gasoline in automobile engines. The annual production of alkyl-lead compounds (some millions of tons) far exceeds that of all other organometallics together. The permitted levels of alkyl-lead compounds added to gasoline base stocks to improve the octane rating ($\sim 0.1\%$) are limited by legislation in most countries. Furthermore, the development of catalytic devices for control of pollution from automobile exhaust gases requires the gasoline fuel to contain less than 0.5 g of lead per gallon (~ 0.1 g/l.). These considerations require an efficient method for lead determination, not only in gasoline and the aquatic environment, but also in the atmosphere, owing to the toxicity of lead compounds and their potential impact on the environment.

The usual procedure for the determination of alkyl-lead compounds in petroleum products involves dilution of the sample with an appropriate organic solvent, to yield a solution of sufficiently low viscosity, followed by nebulization into the flame or injection into an electrothermal atomizer. The standards are prepared in the same solvent to match the viscosity and composition of the analyte. Robinson [7] determined tetra-alkyl-lead in gasoline by diluting the sample with iso-octane and aspirating the solution into an oxygen–hydrogen flame. Dagnall and West [8] reported some difficulties experienced if an air-propane flame was used, and demonstrated the effect of both the solvent and the position of the absorption path. These authors recommend methyl isobutyl ketone (MIBK) as the solvent and tetraethyl-lead as a standard. To avoid preparation of different calibration graphs when different tetra-alkyl-lead compounds are to be determined, a mixture of acetone and iso-octane is used as the solvent and the dilution ratio and aspiration rate are carefully controlled [8,9].

Trent [10] found that tetramethyl-lead dissolved in heptane, MIBK or toluene gives a signal 2-3 times that for tetraethyl-lead. He also found that whereas tetraethyl-lead requires aspiration for 30 sec to give a stable signal, tetramethyl-lead gives one immediately. For this reason it has been suggested that the signal should be measured after a compromise aspiration time of 15 sec, to avoid waiting for equilibrium to be reached [11], but it is difficult to see much advantage. Other workers add iodine to the sample [12,13]. Addition of 3 mg of iodine to a 2% gasoline solution in MIBK permits analysis for both tetraethyl- and tetramethyl-lead compounds by flame AAS, with a single calibration graph prepared with either compound [14]. It seems that the iodine exerts a levelling effect on the response by formation of the same species irrespective of the original compound.

$$R_4Pb + 3I_2 \rightarrow 4RI + PbI_2$$

The carbon-rod method has also been used for atomization of some alkyl-lead compounds and gasoline samples in the presence of iodine [15]. The samples are injected, dried at $100°C$ for 20 sec, ashed at $550°C$ for 20 sec and atomized at $2200°C$ for 10 sec in the presence of argon as purge gas. With this technique, levels as low as 0.1 ng/ml can be determined. Lead naphthenate in synthetic lubricating oil [16], and lead cyclohexanebutyrate and tetra-alkyl-lead in a variety of petroleum samples [17] have been determined similarly. These samples are diluted with toluene to reduce the concentration of the carbonaceous matrix of the oil. Samples containing high levels of carbonaceous matrix and possessing high viscosity are treated with 40% nitric acid, and the lead is then extracted with dithizone in toluene, and the extract is atomized. Robbins [18] analysed a series of 18 gasoline samples containing tetraethyl- and tetramethyl-lead or mixtures of alkyl-leads in the range 0.025–2.5 g/gallon. The results obtained in the study compared favourably with those obtained by X-ray fluorescence and indicated that iodine does not have a levelling effect when the carbon rod atomizer is used.

Vickrey *et al.* [19] studied the efficiencies of various commercially available pyrolytic graphite tubes, coated with zirconium, for AAS determination of alkyl-lead compounds. The results showed that the coated tubes gave higher sensitivity than uncoated tubes. Addition of iodine in 400–500-fold amount relative to the alkyl-lead enhances the sensitivity by several orders of magnitude and permits the use of a single lead standard for calibration.

Alkyl-lead compounds in the atmosphere are determined by passing the air sample through a membrane filter to collect lead particulates, and then through a solution of iodine monochloride to convert alkyl-lead compounds into their dialkyl lead halides:

$$R_4Pb + 2ICl \rightarrow R_2PbCl_2 + 2RI$$

The lead in the iodine monochloride solution is then extracted with ammonium pyrrolidine dithiocarbamate in MIBK and nebulized into the flame for AAS [20]. This method permits measurement of lead in air in the concentration range 0.1–4 $\mu g/m^3$. In a similar procedure [21] dialkyl-lead halide is extracted in the presence of EDTA with dithizone in carbon tetrachloride, followed by decomposition of the dithizonate with HNO_3–H_2O_2 mixture and AAS determination of lead in the resulting solution by electrothermal atomization. The limit of detection offered by this technique is 7 ng of lead. Trace levels of alkylated lead are concentrated either by adsorption on 5% Apiezon L on Chromosorb WAW or a similar adsorbent in a stainless-steel U-tube cooled to $-193°C$ [22,23] or by condensation at low temperature in a U-tube filled with glass beads [24].

Tetra-alkyl-lead compounds have also been determined in fish tissue [25]. A homogenate of the sample is extracted with benzene in the presence of EDTA and a portion of the extract is mixed with 50% v/v nitric acid and evaporated under a stream of nitrogen. The fat is removed by extraction with hexane, then the residual aqueous solution is heated at 80-90°C and the lead is measured by electrothermal atomization AAS. Tetra-alkyl-lead in environmental samples may also be extracted at pH 1–7 with hexane in the presence of EDTA, digested and measured similarly [26].

10.1.2 Organomercury Compounds

In recent years, great concern has been expressed about the effect of mercury pollution on the environment. This pollution originates on the one hand from the use of many organomercury compounds in agriculture and horticulture, and on the other from the conversion of inorganic mercury in industrial effluents into methylmercury in the marine environment. The methods used for the determination of organomercury compounds commonly involve decomposition, and reduction of the resulting mercury(II) ions to elemental mercury followed by electrothermal atomization or cold vapour AAS. Mercury compounds are decomposed by wet digestion with: (a) nitric acid [27]; (b) a mixture of sulphuric and nitric acids [28,29]; (c) a mixture of potassium permanganate, nitric acid and sulphuric acid [30]; (d) potassium persulphate [31]; (e) sulphuric acid–hydrogen peroxide mixture [32]; (f) potassium permanganate–potassium persulphate mixture [33]; (g) a mixture of nitric acid and perchloric acid or potassium dichromate [34,35]. Decomposition of organic compounds by ultraviolet irradiation has been proposed by Armstrong et al. [36] and used for destruction of organomercury compounds before cold vapour AAS [37–40]. This method [33] gives a lower blank, lower limit of detection and better reproducibility than chemical methods. Low (15 W) and medium (150 W) pressure ultraviolet lamps can effectively be used for the decomposition (Table 10.2). As long as the analyte solution is efficiently stirred and contains acidified potassium dichromate, ultraviolet irradiation quantitatively destroys a wide range of organomercury compounds in less than 16 min. Breakdown of organomercury compounds by the action of halogens has recently been described [41]. The reactions involved are:

$$R_2Hg + X_2 \rightleftharpoons RHgX + RX$$
$$RHgX + X_2 \rightleftharpoons HgX_2 + RX$$

Table 10.2. AAS determination of some organomercury compounds by decomposition with ultraviolet irradiation [33] (reprinted by permission of the copyright holders, Elsevier Scientific Publishing Co.)

	Hanau TNN 15/32 (low-pressure)			Hanau TQ 150 (high-pressure)		
	Hg added (μg/l.)	Hg found (μg/l.)	Recovery (%)	Hg added (μg/l.)	Hg found (μg/l.)	Recovery (%)
Methylmercury	0.61	0.63	103	0.61	0.57	94
chloride	0.61	0.65	107	0.61	0.61	100
Phenylmercury	0.54	0.55	102	0.54	0.55	102
acetate	0.54	0.58	107	0.54	0.51	95
Phenylmercury	0.50	0.50	100	0.50	0.49	98
borate	0.50	0.46	92	0.50	0.45	90
Phenylmercury	0.50	0.45	90	0.50	0.59	118
nitrate	0.50	0.41	82	0.50	0.59	118
Phenylmercury	0.47	0.43	92	0.47	0.56	119
chloride	0.47	0.43	92	0.47	0.56	119

The samples are treated at room temperature with a mixture of hydrochloric acid and permanganate or bromine monochloride for 5 min and the excess of reagent is then reduced with hydroxylamine hydrochloride. The method has successfully been used in conjunction with AAS for the determination of phenyl-mercury borate ($C_6H_5HgOH-C_6H_5HgBO_2$), thiomersal ($C_2H_5HgSC_2H_4COONa$) and ceresin ($CH_3OC_2H_4HgCl$) in drinking water at mercury levels as low as 60 ng/l.

$$R_2Hg + X_2 \rightleftharpoons RHgX + RX$$

Organomercury compounds in environmental samples such as fish, grass, shellfish, river mud and soil are determined by electrothermal atomization after destruction of the organic matter either by boiling with sulphuric–nitric acid mixture under reflux in a borosilicate glass apparatus (Fig. 10.1) or by heating with nitric acid in a bomb at 140°C for 15 hr, or by combustion in an oxygen-hydrogen flame (Wickbold combustion) [42]. The applicability of these destruction methods to various samples is shown in Table 10.3. Methylmercury in fish tissues [43-45] is determined by homogenizing the tissues with sodium chloride-hydrochloric acid mixture and extracting the mercury compound with toluene. The organic phase is stripped with cysteine acetate solution, which is then digested with potassium permanganate and sulphuric acid at 40°C for 1 hr; the mercury ions released are reduced with tin(II) chloride and measured by the cold vapour technique [43]. The average recovery of the method is 70-80% and limit of detection 2 ng/g. However, the method does not discriminate between other mercury species extractable into cysteine acetate, such as dimethylmercury. For

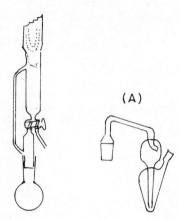

(A)

Fig. 10.1. Apparatus for the wet decomposition of organomercuric compounds; (A) is a trap with ground-glass joint to fit the top of the reflux condenser. [Reprinted from *Anal. Chim. Acta*, **84**, 231 (1976) by permission of the copyright holders, Elsevier Scientific Publishing Co.].

Table 10.3. AAS determination of organomercury compounds in various samples by different decomposition methods [42] (reprinted by permission of the copyright holders, Elsevier Scientific Publishing Co.)

| | | Hg recovery (%) | | |
| | Hg added [a] | Wet | | |
Sample	(mg/kg)	digestion	Bomb	Wickbold
Orchard leaves	0.155 ± 0.015	95	98–107	93–94
Tablets	236 ± 4	95–103	98–101	96–99
as solution in ethanol		94–99	n.r.[b]	99–105
Shellfish, fish	0.30 – 10.0	86–105	94–117	72–120
Fish meal	0.100 – 0.150	95–102	100–105	98–108
Soya bean oil; sunflower oil	2.00	n.a.[c]	n.a.	94–100
Soil	1.00 – 2.00	86	97	n.a.
Grass	0.10 – 0.50	94	n.r.	95–98
Urine	0.20 – 1.00	99	97–113	n.r.
Water	1.00	98–105	98–105	100–105
Methanol, acetone, chlorinated hydrocarbons	0.50 – 36.5	n.a.	n.a.	87–112

[a] Mercury compounds added: methylmercury chloride, phenylmercury chloride, phenylmercury borate or mercury(II) chloride.
[b] n.r., no results.
[c] n.a., method not applicable.

the determination of both methylmercury and inorganic mercury in fish homogenate [46], one aliquot is digested with sulphuric acid–hydrogen peroxide mixture and used for determining the total mercury. The methylmercury is extracted into toluene from a second aliquot of the homogenate and measured. Methylmercury may also be extracted with a toluene solution of dithizone and injected into the graphite furnace [47].

Methylmercury in fish tissues can be determined in the presence of phenylmercury and inorganic mercury compounds. The homogenized samples are injected into an aeration cell followed by treatment with $18M$ sulphuric acid and tin(II) chloride-cadmium nitrate reagent. By this treatment elemental mercury from inorganic and phenylmercury compounds is liberated and swept by nitrogen into a windowless absorption cell for measurement; the residual mixture, containing methylmercury, is then treated with 60% sodium hydroxide and similarly measured [48]. Inorganic mercury, phenylmercury and total mercury in urine have also been determined by similar procedures [49]. Littlejohn *et al.* [50] determined total mercury in urine by the cold vapour technique, using the apparatus shown in Fig. 10.2. The limit of detection (for mercury) obtained with a single-beam AAS instrument without background correction is 0.82 μg/l. A lower limit of detection (~ 20 ng of Hg) and good recovery (~ 98%) have been reported for use of sodium borohydride as reducing agent

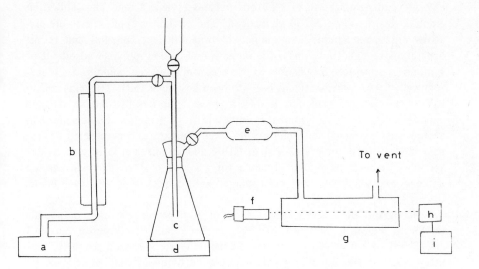

Fig. 10.2. Apparatus for cold vapour determination of mercury: a, air pump; b, air flowmeter; c, reduction flask; d, magnetic stirrer; e, drying reagent (magnesium perchlorate); f, mercury lamp; g, absorption cell; h, detector; i, read-out. [Reprinted from D. Littlejohn, G. S. Fell and J. M. Ottaway, *Clinical Chem.*, 22, 1719 (1976). © 1976 *Clinical Chemistry.*]

and the cold vapour technique [51] and for use of $0.1M$ potassium nitrate-sodium perchlorate as matrix modifier with the Zeeman electrothermal atomization technique [52].

Speciation of several mercury compounds in biological fluids has been achieved by differential AAS [52a].

10.1.3 Organotin Compounds

The industrial applications of organotin compounds as stabilizers against photo-decomposition of poly(vinyl chloride) and other polymers, and later as fungicides, have increased the interest in the analysis of these compounds. Freeland and Hoskinson [53] determined some organotin compounds such as Me_2SnCl_2, $Me_3SnO_2CCH_3$, $Bu_3SnO_2CCH_3$, Bu_3SnCl and Ph_3SnCl by direct atomization of an ethanolic solution of these compounds in an air–acetylene flame. Dibutyltin dilaurate in finished feeds has been determined by extraction with chloroform, concentration of the extract in the presence of methanol, and aspiration of the final methanol extract into an air–acetylene flame [54].

Organotin stabilizers in poly(vinyl chloride) are decomposed and converted into tin acetate before measurement by flame AAS [55]. The method is suitable for samples containing tin at the ~ 300 μg/g level. Bis(tributyltin) oxide, which is used as a wood preservative, is determined by dilution (or extraction from wood) with ethanolic hydrochloric acid, followed by mixing with lithium chloride and nebulization in a nitrous oxide–acetylene flame. The calibration graph covers the range 2–100 μg/ml [56]. The method is applicable to preservatives or timber samples containing chloronaphthalene, phenols and resins. Separation of bis(tributyltin) oxide on activated carbon, polymer adsorbent or synthetic carbonaceous adsorbent, by a continuous-flow column technique, followed by AAS measurement has also been described [57]. Tricyclohexyltin hydroxide pesticide, which is known in most countries by the trivial name cyhexatin, has been determined in apples [58]. A sample is homogenized in acetone and extracted with chloroform, then the extract is evaporated to dryness, and the residue is dissolved in MIBK and atomized in a nitrous oxide–acetylene flame. The results obtained for a number of commercially grown apple samples known to have been sprayed with cyhexatin showed levels below 0.1 mg/kg.

Nanogram and subnanogram levels of 9 different alkyltin compounds ($MeSnCl_3$, Me_2SnCl_2, Me_3SnCl, Et_2SnCl_2, Et_3SnCl, $BuSnCl_3$, Bu_2SnCl_2, Bu_3SnCl and $PhSnCl_3$) have been determined in natural waters, coastal sediments and macro algae by the hydride generation–AAS technique [59]. The samples are reduced with sodium borohydride and the mixture of liberated hydrides is collected in a trap cooled in liquid nitrogen; subsequent warming of the trap volatilizes the hydride, which is swept into a silica-tube atomizer. Since these compounds have boiling points ranging from 1.4°C for $MeSnH_3$ to 250°C for Bu_3SnH, they can be differentially volatilized in order of increasing boiling point, which allows measurement of each in their mixtures (detection limit ranging from 0.4 to 2 ng). Vickrey *et al.* [60] studied the effect of chemical modification of the surface of the graphite cuvette by coating with various metal carbides, in the determination of organotin compounds. Soaking the graphite tube in zirconium, vanadium and molybdenum solutions, followed by drying at 200°C, was investigated. A comparison between the treated and untreated graphite tubes shows an increase by a factor of 2-100 in the sensitivity for various organotin compounds when atomizers treated with zirconium are used.

10.1.4 Organoselenium and Tellurium Compounds
These compounds are generally decomposed either by wet digestion with sulphuric-nitric acid mixture [61] or the Schöniger-flask method [62,63], followed by flame or electrothermal atomization. Electrothermal atomization causes complications in the determination of selenium in complex matrices, however, because of the high volatility of the analyte. Addition of molybdenum, nickel of silver prevents volatilization of organoselenium species from biological fluids at temperatures > 400°C and up to 1300°C [64]. Reasonable recovery of selenium can be obtained if the decomposition conditions are chosen so that most of the oxidation of the compound is completed with perchloric-nitric acid mixture before addition of sulphuric acid [65]. The trimethylselenonium ion in urine has been determined by ion-exchange separation, extraction with dithizone and AAS analysis of the extract [65a].

Organotellurium compounds are determined by decomposition in an oxygen-flask containing hydrogen peroxide-hydrochloric acid mixture as absorbent, followed by flame AAS measurement [66]. This method has been used satisfactorily for the analysis of samples containing 13-40% of tellurium, such as diaryltellurides, diaryltellurium dihalides, aryltellurium trihalides, aryltellurium trihalide complexes with sulphur-ligands, and tellurium tetrahalide complexes. The only class of compounds that is not completely oxidized by the method is that of diaryl ditellurides. An aluminium bomb lined with PTFE can be used for the decomposition of various tellurium compounds. Electrothermal atomization of tellurium compounds requires thermal stabilization. Addition of silver and platinum salts allows the graphite tube to be heated to 1050°C without loss of tellurium [67].

10.1.5 Organo Transition-Metal Compounds
Kidani *et al.* [68] determined some organocopper, nickel, cobalt and zinc compounds by dissolution in 0.1*M* hydrochloric acid or alcohol and direct atomization in an air-acetylene flame. A similar procedure has recently been suggested for the determination of some organoindium compounds by use of a short burner to give a minimal absorption path [69]. The nickel and zinc complexes of nine Schiff's bases have been determined by dissolving 2 mg of the sample in dimethylformamide to give solutions containing ~ 0.5 mg of zinc or

2-3 mg of nickel per l., and aspiration of these solutions into an air–acetylene flame [70]. The coefficient of variation is < 1%. Place [71] studied the effect of dimethylformamide on the AAS determination of organocobalt, copper, nickel, iron, and manganese compounds. He found substantial enhancement of the absorption signals for these metals when the undiluted solvent was used.

10.1.6 Organoantimony Compounds
Of the possible procedures considered by five collaborating laboratories for the determination of organoantimony compounds in the range 0.05–0.5 μg, only AAS with heated tube atomization after hydride generation gave a sensitivity as good as 50 ng/ml [72]. In the recommended procedure, a 2-g sample is oxidized with a mixture of concentrated nitric and sulphuric acids, the cooled solution is treated with 4-5 ml of water and 1 ml of 5% sodium sulphite solution, and heated to fumes of sulphur trioxide to remove nitrogen oxides. The cooled solution is then treated with 30 ml of concentrated hydrochloric acid and diluted, sodium iodide is added, and the antimony is determined at 217.6 nm by hydride-generation AAS. The antimony should be present in the tervalent state before the reduction to form the hydride. This method has been used for the determination of antimony in lime juice, condensed milk and vegetation. Difficulties encountered in the digestion of organoantimony compounds with perchloric acid have been reported [65].

10.2 SPECIATION OF ORGANOMETALLICS
Though satisfactory AAS methods are available for the determination of many organometallic compounds and can easily be extended to cover the analysis of other compounds, determination of individual compounds in mixtures remains a difficult task. Combination of AAS with a preliminary chromatographic separation provides adequate determination of many organometallics in mixtures. This approach has received considerable attention in the analysis of metabolites, man-made pollutants, ambient air, biological substances and industrial products. Biotransformation of metals by biochemical reactions such as alkylation and dealkylation, and volatilization of such compounds across the water–air or lipid-air interface can also be kinetically investigated. In general, a high metal-specificity can be obtained by interfacing the AAS instrument with either gas–liquid (GLC) or liquid chromatographic (LC) equipment. The organometallic mixture is run through the chromatographic column to separate the components, which are then introduced directly into the atomizer of the AAS instrument, one by one. This technique can be used to determine species which can be made to bind to a particular metal.

10.2.1 Speciation by GC–AAS
10.2.1.1 GC-flame AAS. Kolb *et al.* [73] first demonstrated the capabilities of AAS as a detector for GLC. They separated various alkyl-lead compounds in

gasoline by GLC and nebulized the effluent from the column into an air–acetylene flame. The same procedure has been adapted for the determination of some silylated aliphatic alcohols in the presence of non-silylated species, with the oxygen–acetylene or nitrous oxide–acetylene flame [74]. Other workers [75] have used similar methods for the determination of alkyl-lead compounds in ambient air, but the low sensitivity (~ 0.1 μg of Pb) combined with the slow air sampling limits the applicability of these methods in environmental analysis [76]. In all these methods, the interfacing is done by connecting the exit end of the column into the auxiliary gas flow of the burner by a metal tube connector, continuously heated to avoid condensation problems. This type of interfacing has been extensively studied [77–79], and shown to result in appreciable dilution of the sample owing to the mixing process in the burner chamber.

Coker [80] utilized a slightly different method of interfacing, passing the effluent from the chromatographic column into the burner head. He used a short column packed with 10% PEG 20 μm Carbowax on 100–120 mesh Porasil for separation and determination of alkyl-lead compounds in gasoline. The column was heated at 130°C and interfaced by means of a gas union threaded into a hole made at the base of the burner neck. Thus, the gas flow from the column was injected into the burner gases at this port. This approach minimized the peak broadening arising from mixing of the effluent gas with the large volume of the gas in the mixing chamber of the burner. Wolf [81] utilized the same type of interface for the determination of chromium compounds. The samples were digested with sulphuric acid–hydrogen peroxide, then heated with trifluoroacetylacetone. The resulting volatile chromium diketonate complex was extracted into hexane and injected into a PTFE column packed with 10% SE-30 on Chromosorb WHP and heated at 180°C. The effluent was transferred directly into the burner for chromium measurement; the limit of detection was 1 ng.

10.2.1.2 GC-electrothermal AAS. Interfacing of the gas chromatograph with the injection port of electrothermal devices has also been described. Segar [82] interfaced the GC directly to a graphite-tube furnace. A column packed with 4% SE-30/6% OV-210 on Gaschrom G and maintained at 150°C was used for separation of the alkyl-lead. The interface was achieved rather crudely by placing the effluent end of the column as close as possible to the graphite atomizer sample port. A tungsten transfer tube about 10 cm in length was connected to this end of the column and its free end was passed through a somewhat enlarged sample-entry hole in the side of the grooved graphite tube. Alkyl-lead compounds such as Me_4Pb, Me_3EtPb, Me_2Et_2Pb, $MeEt_3Pb$ and Et_4Pb in gasoline are satisfactorily determined, with a sensitivity of about 10 ng. Robinson *et al.* [83] interfaced a home-made hollow-T carbon-tube furnace to a GC by a stainless-steel transfer line. The system was used for the determination of alkyl-lead compounds in gasoline and air. The sensitivity was about

1.5 ng. One interesting application of this technique [83,84] was identification of the source of the gasoline since different brands contain well-defined concentrations of the different alkyl-lead compounds. Radziuk *et al.* [85] interfaced the GC column to a graphite furnace with a tantalum connector machined from a 6.4 mm diameter rod and transferred the effluent to the furnace by Teflon-lined aluminium tubing heated electrically to 80°C (Fig. 10.3). Bye *et al.* [86] compared the flame and graphite-tube atomizer interfaced to a GC for the determination of alkyl-lead compounds. The effluent was pumped either to a three-slot burner (air–acetylene) or a graphite tube with argon purge gas. The detection limit obtained with the graphite tube atomizer was lower by a factor of 100 than that with the flame.

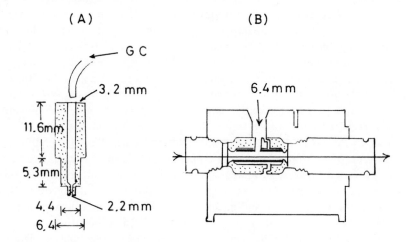

Fig. 10.3. Tantalum connector (A) positioned in the graphite tube furnace (B). [Reprinted from B. Radziuk, Y. Thomassen, J. C. Van Loon and Y. K. Chau, *Anal. Chim. Acta,* **105**, 255 (1979) by permission of the copyright holders, Elsevier Scientific Publishing Co.]

Chau *et al.* [87] used a silica T-tube (4 cm long, 7 mm bore) with open ends for atomization of alkyl-lead compounds separated by GC. The effluent from the column was introduced into the centre of the atomizer (Fig. 10.4) and hydrogen was introduced through a side-arm at a flow-rate of 135 ml/min to improve the sensitivity. Figure 10.5 shows the recorder tracing obtained for a mixture of five tetra-alkyl-lead compounds by use of a column packed with 3% OV-1 on Chromosorb W, 80–100 mesh. Similar results have been reported [88] for use of a column packed with 20% 1,2,3-tris(cyanoethoxy)propane on 60/80-mesh Chromosorb P coated with 1% potassium hydroxide. These methods are also used for the determination of tetra-alkyl-lead compounds in fish tissues [89] and air [90,91].

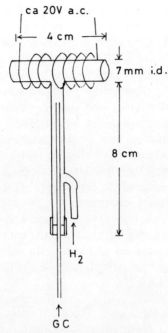

Fig. 10.4. Silica T-furnace. [Reprinted from Y. K. Chau, P. T. S. Wong and P. D. Goulden, *Anal. Chim. Acta,* 85, 421 (1976) by permission of the copyright holders, Elsevier Scientific Publishing Co.]

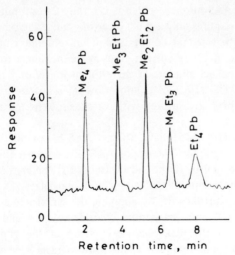

Fig. 10.5. Recorder tracings for a mixture of five tetra-alkyl-lead compounds. Each peak is for an amount of compound containing *ca.* 5 ng of lead. [Reprinted from Y. K. Chau, P. T. S. Wong and P. D. Goulden, *Anal. Chim. Acta,* 85, 421 (1976) by permission of the copyright holders, Elsevier Scientific Publishing Co.]

A quartz T-tube furnace incorporating a volatilization chamber containing a short helix-wound column (~ 120 cm long) has also been interfaced with a GC column [92,93]. Both the volatilization chamber and the quartz tube are electrically heated to about 1000°C. This system is used for the separation and determination of dimethyl selenide, diethyl selenide and dimethyl diselenide at levels of 10–20 ng. Biologically generated volatile selenium, arsenic and tin compounds in the atmosphere have also been determined [94,95]. Mixtures of the highly polar and solvated MeSn, Me_2Sn, Me_3Sn and Sn(IV) are separated and determined by extraction into benzene containing tropolone, followed by butylation with butyl magnesium chloride (Grignard reagent) [96]. The resulting mixture, containing $MeSnBu_3$, Me_2SnBu_2, Me_3SnBu and Bu_4Sn is volatilized in the boiling-point range of the compounds (41–145°C), and separated by GC, and the components are determined. Organoindium and gallium compounds [97], alkylmercury compounds [98], and some other organometallics [99–101] have been determined. High-temperature GC with AAS has been proposed for separation and determination of Na, Cu, Mn and Mg metals [102]. In general the GC–AAS approach is limited to organometallic compounds of sufficient volatility and thermal stability.

10.2.2 Speciation by LC–AAS

10.2.2.1 LC-flame AAS. Mixtures of thermally stable or non-volatile organometallic compounds can be separated and determined by interfacing the AAS instrument with the liquid-chromatography column (LC–AAS). Since the effluent flow from most LC systems can easily be matched with the uptake rate of the nebulizers of AAS burners, which is typically 2–6 ml/min, interfacing of the two instruments requires only a short piece of connecting tube. To overcome flow-rate differences between the column and nebulizer under certain conditions, some workers use an auxiliary solvent flow in the nebulizer for balancing the flow-rates [103] or collect aliquots of the effluent in Teflon cups, followed by nebulization [104]. The effect of column flow-rates slower than 2 ml/min on flame AAS has also been investigated with water as solvent and an air–acetylene flame [105].

Several workers [106–115] have utilized AAS in conjunction with simple ion-exchange chromatography. The use of AAS as a detector for gel chromatography was first suggested by Yoza and Ohashi [116]. Later, Yoza and Koychiyama [117] analysed a mixture of condensed phosphates (e.g. di-, tri-, and tetra-phosphates and Kurrol's salt) by applying the samples to a Sephadex G-25 column pre-equilibrated with magnesium chloride solution and measuring the magnesium content of the magnesium complexes of the phosphates in the effluent. Various zinc compounds in plant tissue extracts have also been determined by gel-permeation chromatography [118]. Thin-layer chromatography has also been used in conjunction with AAS for identification and determination of some organometallic complexes [119]. Tellurium diethyldithiocarbamate was chosen as a model, and its chloroform solution was spotted onto silica-gel plates which were then developed with a hexane–benzene–acetone mixture (65:10:12 v/v). The spots were extracted with methanol and analysed for tellurium at 214.2 nm.

High-pressure liquid chromatography (HPLC) has also been used with AAS (and gradient elution to provide increased speed). Jones and Manahan [120] determined three organochromium complexes in a mixture by direct HPLC–AAS interfacing. Chromium acetylacetonate, tris(2-hydroxyacetophenone)-chromium and chromium hexafluoroacetylacetonate were isolated on a Porasil A column by use of 0.5% pyridine in toluene as mobile phase, and the column effluent was purged into the aspirator input at a rate of 2.5 ml/min. The method gave a detection limit of 40 ng. Polyaminocarboxylic acid chelating agents such as EDTA, NTA, EGTA and DCTA at concentration levels of 0.09–3 μg/ml have similarly been determined after conversion into their copper chelates [121,122]. The chelates are retained by a weak anion-exchange resin (Aminox A-14), eluted with $0.05M$ ammonium sulphate and atomized in an air–acetylene flame, in the order $Cu_2(EGTA)$, $Cu(NTA)^-$, $Cu(EDTA)^{2-}$ and $Cu(DCTA)^{2-}$. Jones and Manahan [123] used known chromatographic values and AAS parameters for calculation of the detection limit.

Organosilicon compounds have been determined by concentration on a porous polymer column, followed by HPLC–AAS [124]. MIBK was used to elute the compounds and the effluent was directly fed to a nitrous oxide-acetylene flame. Trimethylsilanol, triethylsilanol, triphenylsilanol, hexamethyl disiloxane, octamethyl tetrasiloxane, octamethyl cyclotetrasiloxane, decamethyl cyclopentasiloxane and poly(dimethylsiloxane) were determined at levels of 0.5 μg (in terms of total silicon). A recorder tracing for a mixture of four silicon compounds is shown in Fig. 10.6.

Tetramethyl-lead and tetraethyl-lead can be separated by HPLC at 50°C, with water–methanol (3:2 v/v) mixture as mobile phase at a flow-rate of 1 ml/min, followed by introduction of the effluent into the burner [125]. No modification of the burner is made except for removal of the internal flow spoiler to allow a high fraction (*ca.* 80%) of the effluent to reach the flame. This permits determination of 0.25–50 μg of alkyl-lead. The individual tetra-alkyl-lead species (e.g. Me_4Pb, Me_3EtPb, Me_2Et_2Pb, $MeEt_3Pb$, Et_4Pb) in gasoline have been separated and determined by using LC–AAS [126]. A reversed-phase μBondapak C_{18} column and acetonitrile–water mobile phase were used. The effluent from the column was introduced directly into the aspiration uptake capillary of the nebulizer of an air–acetylene flame. The detection limit, based on 20-μl injections, was approximately 10 ng of Pb for each compound. The advantage of using an AAS detector, instead of the conventional ultraviolet detector commonly used with LC, has been demonstrated by direct introduction of the effluent from the ultraviolet detector into the nebulizer of the flame AAS instrument. The chromatogram of commercial leaded 89-octane gasoline diluted fivefold with

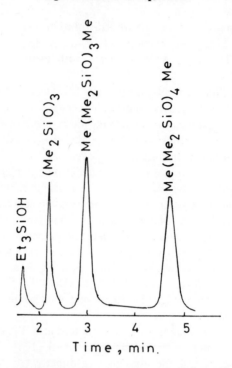

Fig. 10.6. Reversed-phase separation and AAS determination of some organosili-con compounds. [Reprinted from R. M. Cassidy, M. T. Hurteau, J. P. Mislan and R. W. Ashley, *J. Chromatog. Sci.*, **14**, 444 (1976) by permission of the copyright holders, Preston Publishing Co.]

methylene chloride is shown in Fig. 10.7. It is clear that the ultraviolet detector system has little practical value because of the strong absorption at 254 nm displayed by co-eluting unsaturated and aromatic hydrocarbons in the gasoline matrix, which obscures the signals of Me_4Pb and Me_3EtPb. The LC–AAS system, however, shows complete resolution of the five alkyl-lead species. Metal species in petroleum products and amino-acids are similarly determined by HPLC–AAS [127,128].

10.2.2.2 LC–electrothermal AAS. Direct interfacing of electrothermal atom-izers and HPLC is not as easy as interfacing of the flame. This arises from the nature of the two systems. In HPLC, the effluent continuously flows from the column, whereas in the electrothermal atomizer, stepwise processes involving drying, charring and atomization are involved. However, Cantillo and Segar [129] interfaced the column effluent to a furnace atomizer by using multiport sampling and injection valves. With this system, the effluent from the column is stopped at fixed time intervals and collected in sampling cups, and these fractions are injected in sequence into the atomizer.

(A)   (B)

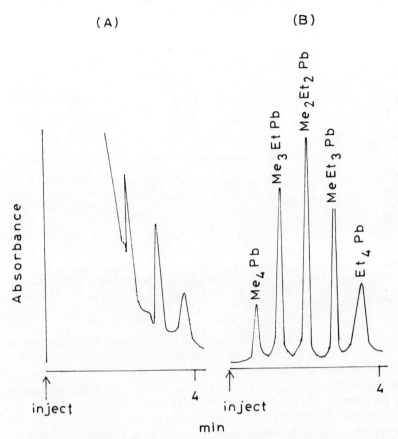

Fig. 10.7. Separation of tetra-alkyl-lead compounds in leaded gasoline by reversed-phase LC and detection by (A) measurement at 254 nm and (B) flame-AAS at 283.3 nm. [Reprinted with permission from J. D. Messman and T. C. Rains, *Anal. Chem.*, **53**, 1632 (1981). Copyright 1981 American Chemical Society.]

Brinckman *et al*. [130] have coupled HPLC with electrothermal–AAS for speciation of trace levels of some organic arsenic, lead, tin and mercury compounds. The effluent from the ultraviolet detector-cell outlet is automatically injected into the graphite-tube atomizer by either of two modes of operation. In the pulsed sampling mode (i.e. with a periodic stream), the column effluent enters at the bottom of a specially constructed Teflon-well sampler through a dead-volume screw fitting where excess of liquid is withdrawn at the top by suction (Fig. 10.8A). The arc of the automatic pipette sampler periodically and reproducibly withdraws a preselected volume of this effluent stream and introduces it into a programmed graphite-tube furnace at intervals. The effluent passes continuously through the well sampler, with the carousel in a fixed

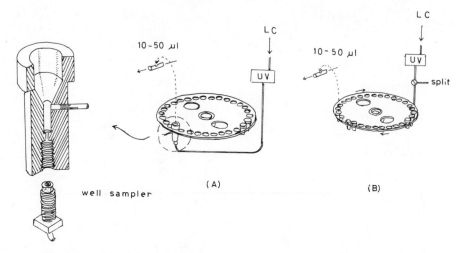

Fig. 10.8. Carousel sample holder, in (A) pulsed (periodic stream sampling) mode, and (B) survey (segmental stream analysis) mode. [Reprinted from F. E. Brinckman, W. R. Blair, K. L. Jewett and W. P. Iverson, *J. Chromatog. Sci.*, **15**, 493 (1977), by permission of the copyright holders, Preston Publishing Co.]

position. Alternatively, a segmental stream can be used (Fig. 10.8B); the carousel revolves normally, at a rate mainly dependent on the column flow, analyte concentration and injected sample volume.

A simple interfacing device has been described, consisting of a sampling valve, dispenser, timing circuit and analyte-addition facility [131]. This system is used for the determination of selenium compounds by use of a Zeeman graphite-tube atomizer. A new graphite furnace (Fig. 10.9) utilizing the Zeeman effect and interfaced with HPLC has been developed and used for the determination of tetra-alkyl-lead in gasoline [132]. It has the ability to suppress interference by other metals. Vickrey *et al.* [133] described a sampling procedure in which the fraction of eluate containing the compound of interest was collected in a capillary tube and stored during the chromatographic run, followed by sequential off-line analysis by graphite-furnace AAS. This procedure was used for the determination of some organic tin and lead compounds. Mixtures containing Me_4Pb, Me_3EtPb, Me_2Et_2Pb, $MeEt_3Pb$ and Et_4Pb were separated by reversed-phase liquid chromatography, digested with methanolic iodine and atomized [134]. Mixtures of tin compounds (e.g. Pr_4Sn, Bu_4Sn, Ph_4Sn and Bu_2SnCl_2) were separated and injected directly.

In a recent study, Burns *et al.* [135] investigated and compared the determination of tetra-alkyltin and alkyltin chlorides by AAS after separation by GLC and HPLC. The GLC effluent was introduced directly into an open-end quartz T-tube furnace heated at 1000°C, whereas the HPLC effluent was directly coupled to a nitrous oxide–acetylene flame. Effluents from both GLC and

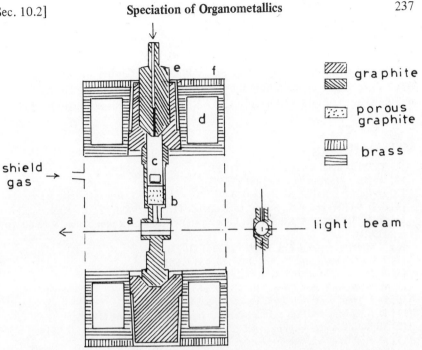

Fig. 10.9. Cross-section of graphite furnace used in conjunction with LC for the determination of organometallic compounds: a, absorption cell; b, porous graphite; c, sample cup (Ta); d, cooling water; e, stopper; f, electrode. [Reprinted with permission from H. Koizumi, R. D. McLaughlin and T. Hadeishi, *Anal. Chem.*, **51**, 387 (1979). Copyright 1979 American Chemical Society.]

Table 10.4. Detection limits of some organotin compounds by use of gas–liquid (GLC) and high-pressure liquid (HPLC) chromatography coupled with AAS [135] (reprinted by permission of the copyright holders, the Royal Society of Chemistry)

Mixture in order of elution	GLC, (μg of Sn)			HPLC, (μg of Sn)		
	Katharometer	AA(P–ETA)	AA(H–ETA)	RI	AA(FA)	AA(H–ETA)
Me_4Sn	8.25	1.0	2.0	50	11	9.6
Me_3SnCl	8.25	1.2	2.0	80	16	9.5
Me_2SnCl_2	8.25	1.0	1.5	90	17	8.6
$MeSnCl_3$	Redistribution	–	–	100	19	8.2
Et_4Sn	8.25	1.4	6.5	50	11	14.0
Et_3SnCl	8.25	1.2	5.3	80	16	13.5
Et_2SnCl_2	12.37	1.1	4.6	90	17	12.0
$EtSnCl_3$	16.50	1.0	2.2	100	19	10.0

AA   = Atomic absorption
FA   = Flame atomization
P–ETA = Pyrolysis–electrothermal atomization
H–ETA = Hydride generation–electrothermal atomization
RI   = Refractive index

HPLC were also mineralized, then reduced with $NaBH_4$, and the SnH_4 released was atomized in the quartz tube. The results (Table 10.4) demonstrated the high sensitivity obtained by using GLC coupled with either the electrothermal atomizer or hydride generation and electrothermal atomization. The advantages and limitations of GLC-AAS and LC-AAS have been discussed [136-137].

REFERENCES

[1] T. T. Gorsuch, *Analyst*, **84**, 135 (1959).

[2] W. Merz, *Mikrochim. Acta*, 456 (1959).

[3] M. Corner, *Analyst*, **84**, 41 (1959).

[4] J. H. Wetters and R. C. Smith, *Anal. Chem.*, **41**, 379 (1969).

[5] H. R. Snyder, J. A. Kuck, J. R. Johnson and V. Campen, *J. Am. Chem. Soc.*, **60**, 105 (1938).

[6] T. Midgley, Jr., *Ind. Eng. Chem.*, **29**, 241 (1937).

[7] J. W. Robinson, *Anal. Chim. Acta*, **24**, 451 (1961).

[8] R. M. Dagnall and T. S. West, *Talanta*, **11**, 1553 (1964).

[9] H. W. Wilson, *Anal. Chem.*, **38**, 920 (1966).

[10] D. J. Trent, *Atom. Absorp. Newslett.*, **4**, 348 (1965).

[11] R. A. Mostyn and A. F. Cunningham, *J. Inst. Petrol.*, **53**, 101 (1967).

[12] M. Kashiki and S. Oshima, *Anal. Chim. Acta*, **55**, 436 (1971).

[13] M. Kashiki and S. Oshima, *Bunseki Kagaku*, **20**, 1398 (1971).

[14] M. Kashiki, S. Yamazoe and S. Oshima, *Anal. Chim. Acta*, **53**, 95 (1971).

[15] M. Kashiki, S. Yamazoe, N. Ikeda and S. Oshima, *Anal. Lett.*, **7**, 53 (1974).

[16] K. G. Brodie and J. P. Matousek, *Anal. Chem.*, **43**, 1557 (1971).

[17] M. P. Bratzel, Jr. and C. L. Chakrabarti, *Anal. Chim. Acta*, **61**, 25 (1972).

[18] W. K. Robbins, *Anal. Chim. Acta*, **65**, 285 (1973).

[19] T. M. Vickrey, G. V. Harrison and G. J. Ramelow, *Atom. Spectrosc.*, **1**, 116 (1980).

[20] L. J. Purdue, R. E. Enrione, R. J. Thompson and B. A. Bonfield, *Anal. Chem.*, **45**, 527 (1973).

[21] S. Hancock and A. Slater, *Analyst*, **100**, 422 (1975).

[22] E. Rohbock and J. Müller, *Mikrochim. Acta*, 423 (1979I).

[23] R. M. Harrison, R. Perry and D. G. Slater, *Atmos. Environ.*, **8**, 1187 (1974).

[24] S. G. Jiang, D. Chakrabarti, W. De Jonghe and F. Adams, *Z. Anal. Chem.*, **305**, 177 (1981).

[25] G. R. Sirota and J. F. Uthe, *Anal. Chem.*, **49**, 823 (1977).

[26] B.-W. Zeng, *Huanching K'o Hsueh*, **1**, 55 (1980).

[27] L. T. Hallet, *Ind. Eng. Chem., Anal. Ed.*, **14**, 956 (1942).

[28] D. L. Tabern and E. F. Shelberg, *Ind. Eng. Chem., Anal. Ed.*, **4**, 401 (1932).

[29] A. K. Klein, *J. Assoc. Off. Agr. Chem.*, **35**, 537 (1952).

[30] C. S. Yeh, *Microchem. J.*, **14**, 279 (1969).

[31] H. A. Sloviter, W. M. McNabb and E. C. Wagner, *Ind. Eng. Chem., Anal. Ed.*, **13**, 890 (1941).

[32] P. Neske, A. Hellwing, B. Thriene and G. Schumann, *Z. Gesamte Hyg. Ihre Grenzgeb.*, **26**, 796 (1980).

[33] *Anal. Chim. Acta*, **109**, 209 (1979).

[34] Y. Kimura and V. L. Miller, *Anal. Chim. Acta*, **27**, 325 (1962).

[35] T. C. Rains and O. Menis, *J. Assoc. Off. Anal. Chem.*, **55**, 1339 (1972).

[36] F. A. J. Armstrong, P. M. Williams and J. D. H. Strickland, *Nature*, **211**, 481 (1966).

[37] F. Frimmel and H. A. Winkler, *Z. Wasser Abwasser Forsch.*, **8**, 67 (1975).

[38] M. Dujmovic and H. A. Winkler, *Chem. Ztg.*, **98**, 233 (1974).

[39] F. Frimmel and H. A. Winkler, *Z. Wasser Abwasser Forsch.*, **9**, 126 (1976).

[40] A. M. Kiemeneij and J. G. Kloosterboer, *Anal. Chem.*, **48**, 575 (1976).

[41] O. Szakács, A. Lásztity and Zs. Horváth, *Anal. Chim. Acta*, **121**, 219 (1980).

[42] *Anal. Chim. Acta*, **84**, 231 (1976).

[43] I. M. Davies, *Anal. Chim. Acta*, **102**, 189 (1978).

[44] J. L. Kacprzak and R. Chvoyka, *J. Assoc. Off. Anal. Chem.*, **59**, 153 (1976).

[45] K. Tanaka, K. Fukaya, S. Fukui and S. Kanno, *Bunseki Kagaku*, **20**, 349 (1974).

[46] *Analyst*, **102**, 769 (1977).

[47] G. T. C. Shum, H. C. Freeman and J. F. Uthe, *Anal. Chem.*, **51**, 414 (1979).

[48] N. Velghe, A. Campe and A. Claeys, *Atom. Absorp. Newslett.*, **17**, 139 (1978).

[49] A. Campe, N. Velghe and A. Claeys, *Atom. Absorp. Newslett.*, **17**, 100 (1978).

[50] D. Littlejohn, G. S. Fell and J. M. Ottaway, *Clin. Chem.*, **22**, 1719 (1976).

[51] S. Terada, T. Inui and H. Tamura, *Eisei Kagaku*, **27**, 179 (1981).

[52] B. Magyar and H. Vonmont, *Spectrochim. Acta*, **35B**, 177 (1980).

[52a] J. W. Robinson and E. M. Skelly, *J. Environ. Sci. Health*, **17A**, 391 (1982).

[53] G. N. Freeland and R. M. Hoskinson, *Analyst*, **95**, 579 (1970).

[54] G. M. George, M. A. Albrecht, L. J. Frahm and J. P. McDonnell, *J. Assoc. Off. Anal. Chem.*, **56**, 1480 (1973).

[55] H. Fassy and P. Lalet, *Chim. Anal. (Paris)*, **52**, 1281 (1970).

[56] *British Standards Institution*, BS 5666, Part 7 (1980).

[57] D. Bhattacharyya, C. Adema and D. Jackson, *Sepn. Sci. Technol.*, **16**, 495 (1981).

[58] J. L. Love and J. E. Patterson, *J. Assoc. Off. Anal. Chem.*, **61**, 627 (1978).

[59] V. F. Hodge, S. L. Seidel and E. D. Goldberg, *Anal. Chem.*, **51**, 1256 (1979).

[60] T. M. Vickrey, G. V. Harrison, G. J. Ramelow and J. C. Carver, *Anal. Lett.*, **13**, 781 (1980).

[61] T. S. Ma and W. G. Zoellner, *Mikrochim. Acta*, 329 (1971).

[62] E. Meier and N. Shaltiel, *Mikrochim. Acta*, 580 (1960).

[63] E. Debal, G. Madelmont and S. Peynot, *Talanta*, **23**, 675 (1976).

[64] J. Alexander, K. Saeed and Y. Thomassen, *Anal. Chim. Acta*, **120**, 377 (1980).

[65] C. A. Watson, *Anal. Proc.*, **18**, 482 (1981).

[65a] N. Oyamada and M. Ishizaki, *Sangyo Igaku*, **24**, 320 (1982).

[66] E. R. Clark and M. A. Al-Turaihi, *J. Organometall. Chem.*, **118**, 55 (1976).

[67] G. Weibust, F. J. Langmyhr and Y. Thomassen, *Anal. Chim. Acta*, **129**, 23 (1981).

[68] Y. Kidani, M. Noji and H. Koike, *Bunseki Kagaku*, **21**, 1652 (1972).

[69] L. G. Mashireva, N. G. Ryabtsev, M. M. Legasova, N. V. Kuznetsova, A. S. Orlov and T. Z. Bir, *Zavodsk. Lab.*, **47**, 39 (1981).

[70] S. L. Davydova, O. N. Domanina, Z. D. Voronina and L. M. Rapoport, *Zh. Analit. Khim.*, **32**, 1171 (1977).

[71] A. R. Place, *Anal. Chem.*, **49**, 2122 (1977).

[72] *Analyst*, **105**, 66 (1980).

[73] B. Kolb, G. Kemmner, F. H. Schleser and E. Wiedeking, *Z. Anal. Chem.*, **221**, 166 (1966).

[74] R. W. Morrow, J. A. Dean, W. D. Shults and M. R. Guerin, *J. Chromatog. Sci.*, **7**, 572 (1969).

[75] T. Katou and R. Nakagawa, *Bull. Inst. Environ. Sci. Technol.*, **1**, 19 (1974).

[76] R. M. Harrison and R. Perry, *Atmos. Environ.*, **11**, 847 (1977).

[77] J. G. Gonzalez and R. T. Ross, *Anal. Lett.*, **5**, 683 (1972).

[78] J. E. Longbottom, *Anal. Chem.*, **44**, 111 (1972).

[79] N. K. Rudneveskii, D. A. Vyakhirev, V. T. Demarin, M. V. Zueva and A. I. Lukyanova, *Dokl. Akad. Nauk. SSSR*, **223**, 887 (1975).

[80] D. T. Coker, *Anal. Chem.*, **47**, 386 (1975).

[81] W. R. Wolf, *Anal. Chem.*, **48**, 1717 (1976).
[82] D. A. Seger, *Anal. Lett.*, **7**, 89 (1974).
[83] J. W. Robinson, L. E. Vidarreta, D. K. Wolcott, J. P. Goodbread and E. Kiesel, *Spectrosc. Lett.*, **8**, 491 (1975).
[84] W. R. A. De Jonghe, D. Chakrabarti and F. Adams, *Anal. Chim. Acta*, **115**, 89 (1980).
[85] B. Radziuk, Y. Thomassen, J. C. Van Loon and Y. K. Chau, *Anal. Chim. Acta*, **105**, 255 (1979).
[86] R. Bye, P. E. Paus, R. Solberg and Y. Thomassen, *Atom. Absorp. Newslett.*, **17**, 131 (1978).
[87] Y. K. Chau, P. T. S. Wong and P. D. Goulden, *Anal. Chim. Acta*, **85**, 421 (1976).
[88] P. R. Ballinger and I. M. Whittermore, *Proc. Am. Chem. Soc. Division of Petroleum Chem.*, Atlantic City Meeting, September 1968.
[89] Y. K. Chau, P. T. S. Wong, G. A. Bengert and O. Kramar, *Anal. Chem.*, **51**, 186 (1979).
[90] W. R. A. De Jonghe, D. Chakrabarti and F. C. Adams, *Anal. Chem.*, **52**, 1974 (1980).
[91] J. W. Robinson and E. L. Kiesel, *J. Environ. Sci. Health*, **12A**, 411 (1977).
[92] J. C. Van Loon and B. Radziuk, *Can. J. Spectrosc.*, **21**, 46 (1976).
[93] B. Radziuk and J. C. Van Loon, *Sci. Total Environ.*, **6**, 251 (1976).
[94] Y. K. Chau, P. T. S. Wong and P. D. Goulden, *Anal. Chem.*, **47**, 2279 (1975).
[95] G. E. Parris, W. R. Blair and F. E. Brinckman, *Anal. Chem.*, **49**, 378 (1977).
[96] Y. K. Chau, P. T. S. Wong and G. A. Bengert, *Anal. Chem.*, **54**, 246 (1982).
[97] A. F. Shushunova, V. T. Demarin, G. I. Makin, L. V. Sklemina, N. K. Rudenskii and Yu. A. Aleksandrov, *Zh. Analit. Khim.*, **35**, 349 (1980).
[98] R. Bye and P. E. Paus, *Anal. Chim. Acta*, **107**, 169 (1979).
[99] P. T. S. Wong, Y. K. Chau and P. L. Luxon, *Nature*, **253**, 263 (1975).
[100] Y. K. Chau, P. T. S. Wong, B. A. Silverberg, P. L. Luxon and G. A. Bengert, *Science*, **192**, 1130 (1976).
[101] D. A. Siemer, P. Koteel and V. Jariwala, *Anal. Chem.*, **48**, 836 (1976).
[102] K. Ohta, B. W. Smith and J. D. Winefordner, *Anal. Chem.*, **54**, 320 (1982).
[103] N. Yoza and S. Ohashi, *Anal. Lett.*, **6**, 595 (1975).
[104] J. C. Atwood, G. J. Schmidt and W. Slavin, Pittsburgh Conference on Analytical Chemistry and Applied Spectroscopy, Cleveland, 1979.
[105] D. R. Jones IV, H. C. Tung and S. E. Manahan, *Anal. Chem.*, **48**, 7 (1976).
[106] S. E. Manahan and D. R. Jones IV, *Anal. Lett.*, **6**, 745 (1973).
[107] J. F. Pankow and G. E. Janouer, *Anal. Chim. Acta*, **69**, 97 (1974).
[108] J. C. Van Loon and B. Radziuk, Federation of Analytical Chemistry and Spectroscopy Societies, Philadelphia Meeting, November 1976.
[109] D. K. Galle, *Appl. Spectrosc.*, **25**, 664 (1971).
[110] R. A. Baetz and C. T. Kenner, *J. Agr. Food Chem.*, **21**, 436 (1973).
[111] R. A. Baetz and C. T. Kenner, *J. Assoc. Off. Anal. Chem.*, **57**, 14 (1974).
[112] D. G. Biechler, *Anal. Chem.*, **37**, 1055 (1965).
[113] J. S. Fritz and E. M. Moyers, *Talanta*, **23**, 590 (1976).
[114] J. P. Riley and D. Taylor, *Anal. Chim. Acta*, **40**, 479 (1968).
[115] S. Sukiman, *Anal. Chim. Acta*, **84**, 419 (1976).
[116] N. Yoza and S. Ohashi, *Anal. Lett.*, **6**, 595 (1973).
[117] N. Yoza and K. Kouchiyama, *Anal. Lett.*, **8**, 641 (1975).
[118] M. Umebayashi and K. Kitagishi, 5th International Conference on Atomic Spectroscopy, Melbourne, Australia, August 1975.
[119] H. J. Issaq and E. W. Barr, *Anal. Chem.*, **49**, 189 (1977).
[120] D. R. Jones IV and S. E. Manahan, *Anal. Lett.*, **8**, 569 (1975).
[121] D. R. Jones IV and S. E. Manahan, *Anal. Chem.*, **48**, 502 (1976).
[122] D. R. Jones IV and S. E. Manahan, *Anal. Chem.*, **49**, 10 (1977).
[123] D. R. Jones IV and S. E. Manahan, *Anal. Chem.*, **48**, 1897 (1976).

[124] R. M. Cassidy, M. R. Hurteau, J. P. Mislan and R. W. Ashley, *J. Chromatog. Sci.*, **14**, 444 (1976).

[125] C. Botre, F. Cacace and R. Cozzani, *Anal. Lett.*, **9**, 825 (1976).

[126] J. D. Messman and T. C. Rains, *Anal. Chem.*, **53**, 1632 (1981).

[127] J. C. Van Loon, B. Radziuk, N. Kahn, F. J. Fernandez and J. D. Kerber, Pittsburgh Conference on Analytical Chemistry and Applied Spectroscopy, Cleveland, 1977.

[128] J. C. Van Loon, B. Radziuk, N. Kahn, J. Lichwa, F. J. Fernandez and J. D. Kerber, *Atom. Absorp. Newslett.*, **16**, 79 (1977).

[129] A. Y. Cantillo and D. A. Segar, *Proc. Intern. Conf. Heavy Metals in the Environment*, Toronto, 1975.

[130] F. E. Brinckman, W. R. Blair, K. L. Jewett and W. P. Iverson, *J. Chromatogr. Sci.*, **15**, 493 (1977).

[131] T. M. Vickrey, M. S. Buren and H. E. Howell, *Anal. Lett.*, **11**, 1075 (1979).

[132] H. Koizumi, R. D. McLaughlin and T. Hadeishi, *Anal. Chem.*, **51**, 387 (1979).

[133] T. M. Vickrey, H. E. Howell and M. T. Paradise, *Anal. Chem.*, **51**, 1880 (1979).

[134] T. M. Vickrey, H. E. Howell, G. V. Harrison and G. J. Ramelow, *Anal. Chem.*, **52**, 1743 (1980).

[135] D. T. Burns, F. Glockling and M. Harriott, *Analyst*, **106**, 921 (1981).

[136] J. C. Van Loon, *Anal. Chem.*, **51**, 1139A (1979).

[137] F. J. Fernandez, *Atom. Absorp. Newslett.*, **16**, 33 (1977).

11

Pharmaceutical compounds

The high selectivity and sensitivity offered by AAS has stimulated a widespread interest in this technique for direct and indirect determination of various pharmaceutical compounds and preparations. Some of these methods have recently been reviewed [1-4]. In general, organic pharmaceutical compounds are determined by their reaction with metal species to form either a precipitate or an extractable ion-association compound or other complex, the metal content of which is then measured by AAS. Metals present in pharmaceutical preparations as one of the main ingredients, impurities or preservatives can be determined by direct methods. These metal species are commonly present as simple salts or organometallic compounds and require only a simple pretreatment to dissolve or extract the metal component. Although some AAS and atomic-emission methods are already given in international pharmacopoeias, the replacement of many of the pharmacopoeial titrimetric, spectrophotometric and gravimetric procedures by AAS methods has been advocated [5-9].

11.1 ANTIBACTERIAL AND ANTIFUNGAL DRUGS

11.1.1 Penicillins

Benzylpenicillin forms an ion-association complex with tris(1,10-phenanthroline)cadmium. The complex is extractable into nitrobenzene and has the composition $[Cd(phen)_3]$ $[Penicillin]_2$. AAS measurement of cadmium in the nitrobenzene phase at 228.8 nm in an air–acetylene flame [10] gives a linear relationship between cadmium absorbance and penicillin concentration in the range 19–112 $\mu g/ml$ (relative standard deviation 2.1%). For stoichiometric reaction the pH should be adjusted to 5-6 and the phenanthroline cadmium chelate added in at least 16-fold molar ratio to penicillin.

Hassan *et al.* [11] described a method for the determination of penicillins, based on desulphurization with alkali-metal plumbite followed by AAS measurement of the excess of lead. The penicillin is heated with solid potassium hydroxide at 250-280°C followed by addition of alkali-metal plumbite. Under these conditions the thiazolidine-ring of the penicillin is quantitatively decomposed, with the formation of one mole of lead sulphide per mole of penicillin. The use of lead is necessary not only to remove the sulphur but also to ensure complete degradation. The lead sulphide is isolated and the excess of lead is measured in an air–acetylene flame at 217 nm after acidification and dilution. The method is applied to the determination of as little as 20 μmole of penicillin. The results obtained compare favourably with those obtained by the iodometric procedure of the British Pharmacopoeia (Table 11.1).

Table 11.1. AAS determination of some penicillins by desulphurization with alkali plumbite [11] (reprinted by permission of the copyright holders, Pergamon Press)

Sample	Labelled active amount	Found (mg)	
		BP	AAS
Ampicillin	250 mg per capsule	258	256
		255	252
Dexacillin,	250 mg per capsule	–	240
'Epicillin'		–	237
'Ospen',	600 mg per tablet	598	592
(phenoxymethyl penicillin)		594	593
Penicillin G, sodium	600 mg per vial	583	578
		580	580
'Prostaphlin',	250 mg per capsule	253	256
(oxacillin sodium)		252	254

B.P. British Pharmacopoeia method.

11.1.2 Nystatin (Mycostatin)

Nystatin is a polyene antibiotic used in the treatment of yeast fungal infections, including intestinal moniliasis, but principally for *Candida albicans* infections of skin and mucous membranes. It is known to inhibit the growth of fungi and yeasts by a mechanism involving binding with the sterol component of the cytoplastic cell to form pores within the cell membrane [12,13]. The formation of these pores results in a change in the permeability of the membrane, which induces a loss of cellular components, including potassium ions [14]. The rate of potassium release as a function of nystatin concentration is monitored by AAS [15]. Yeast cells cultivated from *Saccharomyces Cerevisiae* are used for this reaction. Yeast cells grown at 30–32°C are the most suitable; those grown at below 20°C or above 37°C give a non-linear response to nystatin. The yeast cells are washed, centrifuged and suspended in 0.03M 'tris'–hydrochloric acid buffer (pH 7.4) before use. A working yeast-cell solution (10^8 viable cells/ml), is prepared by diluting 3 volumes of the concentrated suspension of the cells with 7 volumes of the buffer.

Nystatin is determined by transferring an aliquot of the cell suspension to a 10-ml test-tube placed in a water-bath at 30°C, leaving it for ~ 30 min to come to the bath-temperature, and then adding the nystatin solution. After gentle shaking for 60 min, the reaction mixture is centrifuged and the potassium content of the supernatant liquid is measured in an air–acetylene flame. The amount of potassium released under controlled conditions of reaction time and temperature is linearly related to nystatin concentration in the range 0.5–1 μg/ml [15]. The precision of the method is very close to that obtained by microbiological assay. Other polyene antibiotics such as amphotericin B and candicidin, which also induce the leakage of potassium ions from yeast cells, may also be determined by this method.

11.1.3 Chloramphenicol

Most of the available methods for the determination of chloramphenicol are based on its structure, since the NO_2 group is a rather unusual structural feature in biological systems. Hassan and Eldesouki [16] determined chloramphenicol compounds in pure powders, suspensions, injections, eye drops, capsules and oral suspensions by an AAS procedure involving reduction by cadmium metal. In the presence of $0.05M$ hydrochloric acid, cadmium metal quantitatively reduces chloramphenicol within 15 min, with the release of three moles of cadmium ions per mole of chloramphenicol:

$$O_2N\!-\!\!\left\langle\!\!\!\bigcirc\!\!\!\right\rangle\!\!-\!\underset{\underset{OH}{|}}{CH}\!-\!\underset{\underset{NHCOCHCl_2}{|}}{CH}\!-\!CH_2OH + 3Cd + 6HCl \longrightarrow$$

$$H_2N\!-\!\!\left\langle\!\!\!\bigcirc\!\!\!\right\rangle\!\!-\!\underset{\underset{OH}{|}}{CH}\!-\!\underset{\underset{NHCOCHCl_2}{|}}{CH}\!-\!CH_2\!-\!OH + 3CdCl_2 + 2H_2O$$

The solution is then aspirated into an air–acetylene flame. Chloramphenicol stearate, palmitate and succinate esters require prior hydrolysis with $1M$ alcoholic potassium hydroxide at room temperature.

The results obtained with pure powders show an average recovery of 99.5% (mean standard deviation 1.1%). Pharmaceutical additives and diluents commonly used in drug formulations, such as acacia, sucrose, Tween-80, ethylene glycol, carboxymethylcellulose, glycerol, vanillin, cocoa butter and lactose have no effect. The results (Table 11.2) compare favourably with those obtained by measuring the cadmium ions with a cadmium ion-selective electrode, or the amine by the spectrophotometric method based on diazotization and coupling with N-(1-naphthyl)ethylenediamine, and are much better than those obtained by the U.S. Pharmacopeia procedure, which fluctuate within ± 3.5% (probably on account of the extraction step used).

Table 11.2. Determination of chloramphenicol and its esters in some pharmaceutical preparations by atomic-absorption spectrometry (AAS), spectrophotometry and potentiometry with the cadmium electrode [16] (reprinted by permission of the copyright holders, Pergamon Press)

Trade name	Nominal amount, as chloramphenicol	Recovery (%)			
		U.S.P. methods	AAS	Cd-electrode	Spectrophotometry
Cidocetine (eye drops)	0.5% aqueous solution	104.3 / 103.5	104.3 / 104.8	105.1 / 105.8	104.8 / 105.3
Cidocetine[b] (suppositories)	125 mg each	98.3 / 101.7	94.1 / 95.3	95.3 / 95.0	96.1 / 95.5
Cidocetine succinate[a] (injections)	1 g/vial	96.3 / 95.0	97.2 / 96.8	97.5 / 96.9	97.3 / 96.6
Chloramphenicol[b] (eye drops)	0.5% aqueous solution	96.3 / 94.1	95.1 / 94.8	96.2 / 95.6	95.8 / 96.4
Cloramidina[b] (capsule)	250 mg/capsule	98.0 / 97.2	95.8 / 96.7	96.4 / 87.0	96.3 / 95.7
Cloramidina palmitate[c] (suspensions)	25 mg/ml	99.8 / 104.2	100.5 / 102.0	100.0 / 99.6	100.9 / 99.0
Globenicol (suppositories)[b]	125 mg each	100.8	99.8	101.0	98.5
Levocol[b] (capsule)	250 mg/capsule	100.2 / 101.6	99.1 / 98.5	99.5 / 98.8	100.8 / 99.8
Miphenicol palmitate[c]	30 mg/ml	100.4 / 105.5	97.0 / 98.2	98.0 / 97.5	98.8 / 97.8
Synthomycetine[a] (injections)	1 g/vial	97.3 / 96.6	96.0 / 96.8	96.4 / 96.8	96.1 / 96.9

The active ingredients are (a) chloramphenicol succinate; (b) chloramphenicol; (c) chloramphenicol palmitate

11.1.4 Sulphonamides

Determination of sulphonamides in pharmaceutical preparations by direct titration with sodium nitrite, in acidic media at low temperature, is given in the U.S. Pharmacopeia and the British Pharmacopoeia and by the American Pharmaceutical Association Foundation. The method suffers from lack of selectivity, and the stability of the titrant is affected by acidity, temperature, dissolved oxygen and atmospheric carbon dioxide. Reaction of sulphonamides with silver or copper ions at pH 8, followed by AAS measurement of the excess of metal ion in the supernatant liquid, at 328.1 and 324.8 nm respectively, has been described by Hassan and Eldesouki [17]. The method is used for various sulphonamides, even those which give slightly dissociated silver and copper salts.

The selectivity and convenience of this method in comparison with the official ones have been demonstrated by analysing sulphapyridine powders contaminated with the precursor aminopyridine. The U.S. Pharmacopeia procedure gave results with a 2% positive bias, whereas elemental sulphur determination and AAS both gave results with a consistent 6% negative bias, in good agreement with the results obtained by potentiometric titration with copper or silver nitrate (94.4% recovery). The presence of aminopyridine contamination was shown by thin-layer chromatography. The AAS procedure overcomes the problems associated with the determination of sulphonamides in the presence of amines, for which nitrosation, acid–base titration and bromometry are not useful. The average recovery in determination of sulphonamide in various pharmaceutical tablets, syrups, injection, drops and ointments compared favourably with that obtained by direct potentiometric titration of sulphonamides with silver or copper (Table 11.3).

11.1.5 Isoniazid

Isonicotinylhydrazide (isoniazid) is used as an antituberculous drug. Concentrations of this compound as low as 0.1mM can be determined by an AAS method based on a reaction with copper(II) and measurement of the copper content of the chelate [18]. Excess of copper(II) chloride is added to the sample in acetate–acetic acid buffer of pH 4-4.8, followed by extraction of the chelate (isoniazid)$_2$Cu into methyl isobutyl ketone and aspiration of the organic layer into an air–acetylene flame. Good linearity is obtainable between the concentration of isoniazid (0.2-1.7 $\times$ 10^{-4}M) and the copper absorbance. The recovery of copper by back-extraction with 1M hydrochloric acid from the organic layer is 98.5%.

11.1.6 Ethambutol

Ethambutol is another antituberculous drug often used in conjunction with isoniazid. Determination of ethambutol by AAS involves reaction with copper(II) in sodium hydroxide medium at pH 8-11.5 to form a copper chelate extractable into methyl isobutyl ketone. The absorbance of the copper in the extract is linearly related to ethambutol concentrations up to 400 μg/ml [19].

Table 11.3. AAS and potentiometric determinations of sulphonamides in some pharmaceutical preparations [17] (reprinted by permission of the copyright holders, the Association of Official Analytical Chemists)

Trade name	Nominal amount	Average recovery (%)				
		USP XIX	Reaction with silver		Reaction with copper	
			Ag–ISE	AAS	Cu–ISE	AAS
Bactrim, tablet	0.4 g per tablet	97.4	97.7	96.9	97.9	97.5
Bayrena, ampoule (sulphamethoxydiazine)	0.5 g per ampoule	96.9	97.3	96.2	97.2	97.4
Diazovit, suspension (sulphadiazine)	5.5 g per 100 ml	101.0	100.4	102.6	100.3	101.3
Longactin, tablet (sulphamethoxypyridazine)	0.5 g per tablet	100.1	99.9	102.1	99.9	99.2
Mecozine, tablet (sulphadimethoxine)	0.5 g per tablet	95.3	94.9	96.4	95.3	95.9
Sulphadimidine, tablet	0.5 g per tablet	96.9	97.0	97.8	96.8	96.5
Sulphamerazine, tablet	0.5 g per tablet	99.9	99.7	100.4	99.6	98.9
Sulphapyridine, tablet	0.5 g per tablet	99.3	99.3	99.7	99.4	100.1

11.1.7 Chinoform (clioquinol)

Chinoform (5-chloro-8-hydroxy-7-iodoquinoline, clioquinol) forms metal chelates with various metal ions. It reacts with zinc to form a yellow 2:1 (Chinoform: Zn) chelate, which is extracted into methyl isobutyl ketone, from a borax buffer at pH 7.9-9; the zinc content of the extract is determined by AAS [20]. The zinc absorbance is linearly related to chinoform concentrations in the range 6-30 μg/ml provided that the zinc concentration in the aqueous reaction mixture is 5-20 times that of the chinoform, and the volume ratio of aqueous to organic phase is about 2. Analysis of some antiamoebic drugs containing 18.2% of chinoform gives a mean recovery of 98.5%. The sensitivity of the method (6 μg/ml) compares favourably with that obtained by colorimetry (2-8 μg/ml), spectrophotometry (20 μg/ml) and infrared spectrometry (2 mg/ml).

11.2 SEDATIVES AND HYPNOTIC DRUGS

11.2.1 Barbiturates

The quantitative reaction of barbituric acid and its derivatives with copper-pyridine reagent to give insoluble complexes has been adapted for AAS measurement. This involves reaction of the barbiturate ($\sim$ 0.5-16 mg) with 1-3 ml of copper solution (3.4 mg/ml) and 1 ml of pyridine in a total volume of 10 ml in the presence of sodium carbonate (to eliminate some interferences) [21]. The reaction mixture is allowed to stand for about 30 min to ensure complete precipitation, and a 1-ml aliquot of the filtrate is mixed with 10 ml of concentrated nitric acid and diluted to 50 ml with water before aspiration into the flame. Alternatively, the precipitate is washed thoroughly with 5% pyridine solution mixture, then dissolved in concentrated nitric acid, and the copper is measured by AAS. Both procedures give acceptable results without interference from compounds such as sulphonal, bromovalerylurea and bromodiethylacetylurea.

The method was later modified [21a].

11.2.2 Benzodiazepines

These drugs are used to treat neurosis, anxiety, insomnia, muscular spasm and countless functional disorders ranging from headache to dyspareunia. They are also used as tranquillisers and hypnotics in the daily practice of medicine. Tens of thousands of patients in hospital for treatment of alcohol withdrawal, tetanus and strychnine poisoning receive benzodiazepines as premedication for endoscopic or surgical procedures or as induction agent prior to general anaesthesia.

The protonated forms of these compounds behave as cations able to react with anionic reagents. This behaviour has been utilized for AAS determination of micromolar levels of these compounds.

A number of compounds in the diazepine series form 1:1 adducts with sodium bis(2-ethylhexyl)sulphosuccinate (NaEHSS) at pH 3-4.5. The excess of

NaEHSS is reacted with copper-1,10-phenanthroline chelate (Cu–Phen) to form a complex extractable into methyl isobutyl ketone. Azepines that form NaEHSS complexes more stable than the NaEHSS complex with Cu–Phen are determined by treatment of the sample (1–4 μmole) with 20 ml of 0.25mM aqueous NaEHSS solution containing 15 g of sodium chloride, followed by addition of 25mM Cu–Phen, extraction with MIBK and AAS measurement of the copper content of the organic layer [22]. Benzodiazepines with an oxo-group in the 2-position are similarly determined after hydrolysis with 50% hydrochloric acid for 30 min. However, azepines that form NaEHSS complexes less stable than the NaEHSS complex with Cu–Phen are determined by the same procedure after prior extraction of the NaEHSS–azipene complex with carbon tetrachloride before addition of the Cu–Phen.

11.3 HORMONES

11.3.1 Ethinyloestradiol
Ethinyloestradiol, a sex hormone used in contraceptive tablets, forms a cobalt complex by reaction with $Na_3Co(NO_2)_6$. This reagent acts as a nitrosating and complexing agent by which the cobalt 3-hydroxy-4-nitroso derivative of the hormone is formed. Extraction of the metal chelate into methyl isobutyl ketone followed by aspiration into an air–acetylene flame for cobalt measurement at 240.7 nm provides a sensitive method for AAS determination of the hormone in tablets [23]. The accuracy of this method is similar to that obtained by the U.S. Pharmacopeia method.

11.3.2 Insulin
Various zinc insulin injections are extensively used in therapeutic medicine. The official methods for the analysis of these pharmaceuticals are usually concerned with the zinc content as an important criterion for both the nature and concentration of the hormone, and usually performed by spectrophotometric and gravimetric procedures. AAS determination of the zinc is simpler, however. The samples are dissolved in water, diluted and aspirated into an air–acetylene flame [24]. Results as good as those obtained by the pharmacopoeia methods are obtainable. An evaluation of the spectrophotometric, polarographic and AAS methods for determination of zinc in insulin injections has shown that the AAS procedure gives more accurate and precise results with less time consumption. The coefficient of variation is 0.9% compared to 1.7% and 4.2% for the polarographic and spectrophotometric methods, respectively [25].

11.4 OTHER DRUGS

11.4.1 Methylamphetamine and Ephedrine
Methylamphetamine hydrochloride is determined in mixtures with ephedrine hydrochloride by an AAS method based on its selective reaction with bismuth

[26]. The acidified test solution is treated with bismuth chloride–potassium iodide reagent (7% potassium iodide solution saturated with bismuth chloride) and allowed to stand for 30 min. The reaction mixture is then filtered, the filtrate is diluted with hydrochloric acid, and its bismuth content is measured by AAS. The limit of detection is 16 μg/ml and the recoveries of methylamphetamine are in the range 97.2–102.2%. Ephedrine hydrochloride does not respond to this reaction.

Methylamphetamine and ephedrine hydrochlorides have been determined by reaction with carbon disulphide, followed by complexation with copper and extraction of the chelates into methyl isobutyl ketone. Aspiration of the extract into the flame for copper measurement gives linear calibration graphs in the concentration range 1.2–8 μg/ml for ephedrine hydrochloride and 1.9–8 μg/ml for methylamphetamine hydrochloride [27]. The results show recoveries in the range 93.3–103.2% without any significant interference due to the presence of a 3-5-fold w/w amount of methylamine and diphenylamine hydrochlorides.

Reinecke's salt has also been used for determination of ephedrine, strychnine, quinine, emetine, procaine and tetracaine [27a] (see also Section 11.4.3).

11.4.2 Strychnine and Brucine

Heteropoly acids are extensively used for precipitation and determination of many alkaloids and organic bases. AAS determination of strychnine and brucine by reaction with phosphomolybdic acid followed by measurement of the molybdenum associated with the alkaloids has been described [28]. In solutions with a final acidity of $0.3N$ sulphuric acid, strychnine and brucine are quantitatively precipitated as the salts $(C_{21}H_{22}N_2O_2H)_3PMo_{12}O_{40}$ and $(C_{23}H_{36}N_2O_4H)_3PMo_{12}O_{40}$, respectively. Addition of 10% citric acid solution to the reaction mixture destroys the excess of phosphomolybdic acid and allows quantitative extraction of the salts into methyl isobutyl ketone. The organic layer should be isolated immediately, since a long contact time between the aqueous and organic phases (> 20 min) affects the stability of the extracted salt. Stripping of the molybdate content of the salt by treatment with a basic buffer and aspiration of the aqueous phase into the flame may also be used. The method is useful for the determination of up to 40 ppm of either alkaloid, with a relative standard deviation of 2-3%.

11.4.3 Noscapine and Chlorprothixene

Precipitation of alkaloids with Reinecke's salt [ammonium tetrathiocyanatodiamminochromate(III)] is a well known reaction commonly used for gravimetric, colorimetric and titrimetric determination of many alkaloids. The general procedure involves treatment of the analyte with 2% ammonium reineckate solution and isolation of the precipitate. The reaction has been used by Minamikowa and Matsumusa [29] for AAS determination of noscapine, a cough suppressant drug. The reaction is done in the presence of tartaric acid at pH 1.7 and the complex formed, $C_{22}H_{24}NO_7[Cr(NH_3)_2(SCN)_4] \cdot x H_2O$, is extracted into chloroform, followed by measurement of the chromium absorbance at 357.9 nm. This method has been used for the determination of noscapine in pharmaceutical preparations without interference from at least 64 different basic ingredients.

Chlorprothixene has been determined by precipitation of its reineckate followed by AAS measurement of the chromium in the excess of reagent [30]. The results show recoveries in the range 98.6-100.9%. Although noscapine and chlorprothixene are the only two alkaloids, so far, determined by the reineckate reaction in conjunction with AAS, many other alkaloids can be similarly determined. Optimum conditions for isolation of the reineckates of atropine, antistin, cocaine, codeine, cinchonidine, cinchonine, chlorathen, doxylamine, diphenhydramine, dibucaine, emetine, hydrastine, hyoscine, methapyrilene, narcotine, pilocarpine, procaine, panthesin, papaverine, pyranisumine, prophenpyridamine, phenindamine, quinidine, strychnine, sparteine, tetracaine and xylocaine have been reported [31-35]. Sympathomimetic amines (e.g. ephedrine, amphetamine, desoxyephedrine), norcolchicine and morphine, however, do not precipitate quantitatively under these conditions.

11.4.4 Chlorpheniramine maleate (MCP)

MCP reacts with zincon and copper to form a ternary complex extractable into chloroform at pH 4.3. Since zincon is known to form a 1:1 blue complex with copper, MCP can be determined indirectly by measuring the copper content in the organic phase by AAS. The MCP solution is added to $1.3 \times 10^{-3}M$ zincon and $3 \times 10^{-3}M$ copper chloride solution in Britton–Robinson buffer at pH 4.3. The reaction product is extracted into chloroform, diluted with methanol and aspirated into an air–acetylene flame [36]. The relationship between the copper absorbance and MCP concentration is linear in the range 1.9-27.4 μg/ml and the average recovery is 100.2%. No interferences are caused by salicylamide, acetoaminophene, caffeine, β-hydroxybutyl-p-phenetidine, lactose and starch.

11.4.5 Metoclopramide

Metoclopramide forms a stable water-insoluble ion-association complex with ammonium thiocyanatocobaltate. The complex is soluble in most organic solvents, and is quantitatively extracted by 1,2-dichloroethane. The method is used for AAS determination of the drug, by the electrothermal atomization technique [37]. A linear relationship is obtained between cobalt absorbance and drug concentration in the range 0.01-0.1mM. The coefficient of variation is 1.8% and many excipients used in pharmaceutical preparations do not interfere.

11.4.6 Theobromine

Theobromine sodium salicylate has been determined by reaction with excess of silver and measurement of the unreacted metal by AAS [38]. The reaction involves displacement of the imino hydrogen atom of theobromine by silver; there is no interference from salicylate. The reproducibility of the method is quite acceptable but the uncertainty in the exact composition of the sample tested makes it difficult to evaluate the accuracy.

11.5 VITAMINS

11.5.1 Vitamin B_1

Hassan *et al.* [39] studied the reaction of vitamin B_1 with alkali-metal plumbite under various conditions, followed by AAS measurement of the unreacted lead ions. The results revealed that the sulphur in the vitamin is quantitatively precipitated as lead sulphide by heating the sample with solid potassium hydroxide followed by addition of alkali-metal plumbite. Since the unreacted lead is not precipitated as hydroxide or decomposed in the strongly alkaline medium, its measurement can be used for determination of the vitamin B_1 concentration. After precipitation of lead sulphide is complete, EDTA is added, the mixture is made up to standard volume and filtered and an aliquot of filtrate is diluted with nitric acid to standard volume and aspirated into an air–acetylene flame for lead measurement at 217 nm. Results with an average recovery of 99.1% and a mean standard deviation of 0.8% are obtainable. Vitamins B_2, B_6 and B_{12}, nicotinamide, and many of the excipients commonly present in dry formulations, do not interfere. The accuracy of the method compares favourably with that of the official fluorometric method of the U.S. Pharmacopeia.

11.5.2 Vitamin B_{12}

The pharmacopoeial methods for the determination of cyanocobalamin (B_{12}) are based on measurement of the absorbance of an aqueous solution of the vitamin at 361 nm. Spectrographic methods using arc excitation between carbon [40] or copper [41] electrodes have also been reported. AAS determination of the cobalt content at 240.7 nm with an oxygen–acetylene turbulent flame and the standard addition technique has also been attempted [42]. Tablets containing the vitamin are pulverized, then extracted with water, and the filtered solution is aspirated into the flame. Vitamin injections are directly nebulized into the flame after dilution. The method offers the advantages of a lower detection limit and applicability to turbid sample solutions. However, the AAS method measures the total cobalt present and does not differentiate between vitamin-cobalt and any other cobalt salts that might be present.

Kidani *et al.* [43] reported that the accuracy of AAS measurement of the vitamin-cobalt depends on the nature of the solvent used for sample dissolution. They recommended 0.1M hydrochloric acid even for samples readily soluble in water. Vitamin B_{12} in samples containing organic solvents or materials that affect the viscosity of an aqueous solution are first mineralized with sulphuric–nitric acid mixture [44]. Cobalt salts, if present, can be separated first by extraction with 8-hydroxyquinoline in chloroform. Vitamin B_{12} in dry feeds is determined [45] after extraction with water, filtration, pH-adjustment to 7 with ammonia, and addition of charcoal to adsorb the vitamin. The charcoal is filtered off on ashless paper, then ashed at 600°C, and cobalt oxide in the ash is dissolved with $5M$ nitric acid and the solution is aspirated into an air–acetylene flame for cobalt measurement at 240.72 nm.

One of the limitations of flame AAS lies in the size of sample required for analysis. Electrothermal atomization, however, permits measurement of lower vitamin concentrations [46]. The sample is dissolved in water, acidified with hydrochloric acid and diluted to give a cobalt concentration of 15–19 ng/ml (or vitamin concentration of 345–437 ng/ml). A 20-μl portion is then injected into the graphite furnace, where it is charred at 750–800°C and atomized at 2700°C. Multivitamin drops, tablets, syrups and injections are satisfactorily analysed by this technique. The microbiological and AAS results agree within ± 5% for samples which have not exceeded their expiry date. Polarized Zeeman-AAS coupled with high-pressure liquid chromatography has been utilized for the analysis of mixtures containing vitamin B_{12} and cobalt nitrate [47]. The samples are automatically injected into an anion-exchange column and eluted with $2M$ acetic acid. At a flow-rate of 1.0 ml/min and a pressure of 20 kg/cm^2, the retention times for the vitamin and cobalt salt are 3 and 9 min, respectively. After separation, aliquots from the collected fractions of eluate are sequentially introduced into the furnace.

11.5.3 Vitamin C

L-Ascorbic acid (vitamin C) reduces the copper(II)–neocuproine chelate to the Cu(I)-neocuproine chelate. This forms an ion-association complex with nitrate which is extractable into chloroform. The concentration of copper in the organic phase is measured by AAS, as a function of the vitamin level [48]. A linear calibration graph for the range 3.75×10^{-7}–$6.25 \times 10^{-6}M$ vitamin C, and an average recovery of 99.3% have been reported. The reaction is done at pH 4.5–7, and the nitrate must be present in a concentration at least equal to that of the copper. Vitamins B_1 and B_6, nicotinamide, calcium pantothenate, biotin and glucose in concentrations as high as 100 times that of the vitamin C do not interfere. The method may also be used for measuring the activity of ascorbic acid oxidase.

11.5.4 Folic Acid

One of the products of oxidation of folic acid with potassium permanganate, namely 2-amino-4-hydroxypteridine-6-carboxylic acid, reacts with nickel to form a stable chelate extractable into methyl isobutyl ketone (MIBK) in the presence of bathophenanthroline. This permits quantitative AAS determination of folic acid [49]. The analyte is dissolved in 0.01M sodium hydroxide and treated with 4% potassium permanganate solution, then an aliquot of the reaction mixture is treated with nickel sulphate in borate–sulphuric acid buffer of pH 8.5. The mixture is shaken with MIBK in the presence of bathophenanthroline and the nickel content of the organic layer is measured at 232 nm. A linear relationship in the folic acid range 1.2–20 $\mu g/ml$ and a relative standard deviation of 1.8% are obtainable, without interference from many compounds normally present or used in mixtures with folic acid.

11.6 METALS IN PHARMACEUTICAL PRODUCTS

11.6.1 Aluminium

Aluminium compounds such as the hydroxide, phosphate, lactate, salicylate, acetate, silicate, allantoinate and alums are frequently used in a number of pharmaceutical products. However, no official method is available for the determination of low concentrations of this element, especially in dermatological products. The early AAS procedure using an air-acetylene flame is not useful, owing to the formation of insufficiently dissociated refractory products [50]. The use of a nitrous oxide-acetylene flame permits the measurement of trace levels of aluminium [51]. Creams, lotions and powders are digested with hydrochloric acid before aspiration into the flame [52]. A recovery of 100.8% is obtained with tablets (Al_2O_3 135 mg/tablet), powders [51.8% $Al_2(SO_4)_3$] and creams (70 ppm Al) without interference from diluents or excipients normally used in these formulations. Determination of aluminium in antacids containing calcium, magnesium, bismuth, titanium and silicon has also been reported [53].

11.6.2 Barium

Barium sulphate is used for X-ray examination of the gastrointestinal tract. The U.S. Pharmacopeia recommends a gravimetric method for the determination of barium, based on fusion with sodium carbonate, followed by precipitation of barium chromate. AAS, however, offers an attractive approach for sensitive and rapid assay. One of the procedures available is based on the same principle, but entails AAS measurement of the barium [54]. A simpler method involves direct dissolution of barium sulphate by heating with alkaline EDTA at 80°C for 30 min, followed by aspiration of the clear solution into an air-acetylene flame [55]. The recovery is 99.2% (standard deviation 0.6%) and calcium and strontium do not interfere.

11.6.3 Calcium

Calcium compounds (e.g. glycerophosphate, glycoheptonate, ascorbate, gluconate, lactate, carbonate, phosphate and chloride) are used in pharmaceutical preparations as therapeutic agents for calcium deficiency, and as anti-anaphylactic and antacid agents. The most commonly used pharmacopoeial method for the determination of calcium in these preparations is titration with EDTA, Hydroxynaphthol Blue being used as indicator. This method suffers, however, from interference by phosphate and/or the organic compounds present in some drugs. These substances have no effect on the AAS procedure [56-61]. Syrups, suspensions, injections, elixirs and other drugs containing water-insoluble calcium compounds are analysed by dissolution in hydrochloric acid followed by addition of lanthanum(III) as releasing agent to offset the effect of phosphate [62]. The significant enhancement effect of sodium on the atomic absorption of calcium is compensated for by addition of sodium to the standard [63,64].

Determination of calcium in some pharmaceutical preparations by both the direct and standard-addition techniques shows some discrepancies. The direct technique gives results approximately 2% higher than the real values, owing to interference by large organic molecules such as gelatin or large amounts of dissolved solids which may scatter or absorb light. Preparations of this nature require prior ashing. It has been reported that pharmaceutical preparations packed in soft red gelatin capsules usually give low calcium recovery by the AAS method, probably because of calcium migration or absorption on the capsule material [62]. Analysis of the entire sample, including the capsule, gives results close to the real values. Calcium in a barium sulphate diagnostic meal is determined by dissolving the sample in perchloric acid, followed by treatment with EDTA and aspiration into the flame [64].

11.6.4 Copper and Manganese

Copper and manganese in ointments and multivitamin-mineral tablets containing cobalt and iron are determined by AAS with good precision and without significant interference from other pharmaceutical ingredients [65]. The effect of fuel mixtures and mineral acids on the AAS of copper and manganese has been carefully studied [66], and optimized for the determination of these metals in multivitamin-mineral tablets [67]. A portion of the powdered tablets is dissolved in 10% hydrochloric acid and the solution is aspirated into an air–acetylene flame for copper and manganese measurements at 324.7 and 279.5 nm, respectively. Measurement of manganese at the resonance line of lead at 280.3 nm may be used to check and compensate for possible interference by molecular species. Copper and manganese in multivitamin-mineral preparations containing cobalt, zinc, iron and molybdenum have been determined either directly by using electrothermal atomization [68] or after prior separation by anion-exchange resin [69].

11.6.5 Germanium

Germanium in medicinal plants is determined by electrothermal atomization AAS [70]. The plant sample is digested with nitric acid and germanium is extracted with carbon tetrachloride. The organic phase is back-extracted with water and the aqueous solution is made alkaline and aspirated into the flame for germanium measurement at the 265.12-265.16 nm doublet. The calibration graph for germanium is linear from 10-80 ng/ml; the limit of detection is 1 ng/ml.

11.6.6 Iron

Iron preparations generally fall into three categories: iron carbohydrate complexes, iron chelates and iron salts. AAS is one of the methods of choice for accurate determination of iron in many of these preparations. This is due to the relatively few interferences that can be observed [71-73] and the fact that iron displays at least fifteen lines useful for the determination of a wide range of iron concentrations with various sensitivities [74]. Preparations containing fumarate and gluconate are ashed, dissolved in $6M$ hydrochloric or nitric acid and aspirated into an air-acetylene flame for iron measurement at 248.3 nm [75,76]. The accuracy and precision compare favourably with those obtained by the official titrimetric methods [75]. Iron can also be determined in multivitamin preparations containing copper and manganese [77].

11.6.7 Mercury

Pharmaceutical compounds containing mercury, such as thiomersal, mersalyl, merbromin, nitromersal, sodium mercaptomerin and phenylmercury borate, nitrate, benzoate and acetate are commonly used as diuretics, and anti-infective and bacteriostatic agents. Conventional flame AAS gives poor sensitivity for mercury [78]. Mercurial formulations containing a sufficient level of the element (6-8%) can be directly determined with little difficulty [79]. In contrast, determination of mercury in ophthalmic solutions containing 10-20 ppm of phenylmercuric nitrate is inherently more difficult, requiring solvent extraction with ammonium pyrrolidine dithiocarbamate [80].

　　Mercury in pharmaceutical powders, tablets, gels, injections, tinctures, suspensions and ointments has been determined [81,82]. This involves demercuration by heating with 65% nitric acid, concentrated hydrochloric acid or hydrochloric–nitric acid mixture for between 25 min and 1 hr. The mercury ions released are reduced with tin(II) chloride to elemental mercury which is then allowed to pass through a flow-through cell for mercury measurement at 253.7 nm. Solutions and tinctures are first evaporated to dryness under a current of air, whereas ointment is dissolved in diethyl ether and extracted with 10% hydrochloric acid [81].

11.6.8 Tin

Butynorate (dibutyltin dilaurate) is an active drug used to remove large roundworms, caecal worms and several species of tapeworms from chickens and turkeys. It is used at the dosage level of 0.2-1.4 mg/ml as an aid in the protection of turkeys from coccidiosis and hexamitiasis. Determination of this drug in finished feed involves extraction with chloroform at 55-60°C, concentration of the extract in the presence of methanol until all traces of chloroform are removed, followed by aspiration of an aliquot of the extract into an air-acetylene flame [83,84]. The absorbance of tin is then measured at 286.3 nm and compared with a graph prepared with standards made from dibutyltin bis(2-ethylhexanoate).

Feed additives and antibiotics do not interfere. The recoveries of 0.02 and 0.04% levels of the drug are 101.4 and 100.5%, respectively.

11.6.9 Titanium

Titanium at levels < 120 ppm in sun-screen formulations containing iron oxide and complex organic bases has been determined. The samples are digested with hot concentrated sulphuric acid and ammonium sulphate, followed by addition of 30% hydrogen peroxide until all the organic materials are completely decomposed [85]. The resulting titanium solutions are then measured at 364.3 nm in a nitrous oxide–acetylene flame. A linear calibration graph is obtained for up to 120 ppm titanium. It should be noted that iron is usually present in these formulations. Its concentration in the aspiration solutions ranges between 11 and 22 ppm, depending on the hue of the screen. This level does not interfere with the titanium determination since up to 200 ppm of iron can be tolerated.

11.6.10 Zinc

Zinc stearate, naphthalenate, undecenoate, oxide and salts are often used in ointments and pastes for external application. Methods for the AAS determination of zinc in many of these formulations yield results with accuracy and precision comparable to or better than those obtained by the gravimetric procedure of the British Pharmacopoeia. Furthermore, impurities such as iron, aluminium and copper, as well as some common excipients such as talc, starch and emollient bases, do not interfere. Zinc in creams, ointments and pastes (e.g. calamine powder, calamine lotion and dusting powders) is determined by extracting the organic ingredients with diethyl ether and digesting the remaining solids (principally zinc oxide) with hot hydrochloric acid and analysing by AAS at 213.9 nm, with an air–acetylene flame [86]. Dusting powder and lotion are directly dissolved in acid, whereas mouth washes and eye drops containing zinc chloride and/or zinc sulphate are diluted and analysed.

   Zinc undecenoate in ointments is determined by a method based on the formation of an oil-in-water emulsion [87]. The sample solution in benzene is shaken with water and Tween-80, emulsified by ultrasound, and aspirated into the flame. The sensitivity for zinc in the emulsion depends on both the nature and percentage of the emulsifier. This method minimizes most of the known problems associated with the introduction of organic solvents or acidic solutions into the flame.

11.7 NON-METALS IN PHARMACEUTICAL PRODUCTS

11.7.1 Arsenic and Halogens

Pharmaceutical preparations containing arsenamide and lead arsenate have been analysed by AAS [75]. Detailed procedures for the determination of the various organoarsenicals are described in Chapter 6. Smith and Nessen have determined the chloride content of various pharmaceutical bases by AAS [88]. Silver nitrate is added to aqueous solutions of amine hydrochlorides to precipitate silver chloride, and the excess of silver is measured. The results compare favourably with those obtained by mercurimetric and argentimetric titrations.

11.7.2 Silicon

Fluids containing polydimethylsiloxane (dimethicone) as a major component are used in some commercial beauty aids such as hand lotions for skin protection. Silicon in the form of magnesium trisilicate is used in the formulation of some antacids [89,90]. Determination of dimethicone in these preparations requires complete removal of water, extraction with benzene and aspiration of the organic extract into the flame [90]. Since extraction of dimethicone from suspensions is difficult, it is often necessary to prepare and treat standard solutions in the same manner as the samples. On the other hand, silicon has a tendency to form refractory oxides. This renders an air-acetylene flame useless. The nitrous oxide-acetylene flame is satisfactory for silicon measurement. Preparations containing 50-100 mg of polydimethylsiloxane per tablet can be analysed after extraction with carbon tetrachloride [91]. This approach gives a reasonable degree of accuracy and is less time-consuming than the previously reported gravimetric [92] and infrared spectrometric [93] methods.

11.8 INFUSION FLUIDS AND DIALYSIS SOLUTIONS

Aqueous electrolyte solutions containing sodium, potassium, calcium and magnesium chlorides are used for extra-renal dialysis and in Ringer's solutions and Ringer's injections. Analysis of these solutions requires accurate and precise measurement of the constituent elements, since the electrolyte balance is critical. Atomic-absorption and flame-emission spectrometry provide potentially useful techniques for the determination of these cations. However, it is known that one problem in measuring mixtures of elements by AAS arises from mutual interferences. Enhancement or depression of the absorption of sodium, potassium and calcium in their mixtures is well described and known to be dependent on their relative concentrations [60,94-97]. Fortunately, these effects tend to become constant over a wide range of metal concentrations, which allows compensation by appropriate dilution or addition of the interfering element to the reference standard.

Since potassium and calcium exert no influence on the absorbance of sodium, the reference standard for sodium can be prepared from solutions containing sodium only. It is known that the absorbance of potassium is affected by relatively high concentrations of sodium but not affected by calcium. Thus sodium should be added to the potassium reference standard. Sodium is also known to enhance the absorption of calcium, so the calcium standard should contain sodium. For measurement, Ringer's solution is diluted 100 times for sodium and potassium measurement, and calcium is measured in a second aliquot diluted 10 times [98]. Similarly, sodium, potassium, calcium and magnesium ions have been determined in peritoneal dialysis solutions [99]. Infusion and dialysis solutions containing dextrose, citrate and lactate are similarly analysed, the standard-addition technique being used to overcome possible matrix interferences [100].

11.9 METALLIC PRESERVATIVES IN PHARMACEUTICAL PRODUCTS

Organomercurial compounds such as phenylmercuric acetate or nitrate and thiomersal are commonly used at a concentration level of 20 μg/ml as preservatives in eye drops, multidose injections, vaccines and antisera to prevent the growth of micro-organisms during the prescribed period of use. Determination of mercury in these products by AAS has been widely used since the introduction of the cold vapour technique. The main disadvantage of the procedure is the preliminary decomposition step which involves wet digestion [101-103] or extraction with an organic reagent [104]. Reviews of the copious literature on mercury determination have been published [105-106]. Evidence has been presented by Meakin and Khammas [107] that cold digestion of pharmaceutical mercury compounds gives low recovery. May et al. [108] satisfactorily used concentrated sulphuric acid–potassium permanganate mixture for decomposition of organomercurial preservatives in vaccines and globulins. Similarly, trace levels of thiomersal, phenylmercuric acetate and phenylmercuric nitrate can be measured [109-110].

Loss of organomercurial preservatives by adsorption on the rubber teats of eye-drop bottles [111-113] and on plastic containers [114-116] has been examined by AAS. Calder and Miller [117] verified this loss by determining the content of phenylmercuric nitrate (PMN) in a number of commercial and hospital preparations. The samples were digested for 5-10 min with a mixture of sulphuric acid and perchloric acid (5:2 v/v) followed by cold vapour measurement of mercury. The results showed that PMN was below its theoretical bactericidal level in many of the preparations contained in conventional glass eye-drop bottles with rubber teats, and in a plastic squeeze bottle. This apparently high deficiency constitutes a hazard to patients making prolonged use of eye drops. Thus accurate AAS determination of mercury in such preparations is necessary. Substances present in such formulations are unlikely to interfere in the determination of mercury by AAS.

11.10 IMPURITIES IN PHARMACEUTICAL PRODUCTS

A number of recent papers demonstrate the use of AAS for the determination of trace metal impurities and contaminants in medical compounds and pharmaceutical products. Contamination of phenylbutazone, sodium aminosalicylate, carbromal, methyl phenobarbitone, chlorpromazine hydrochloride and phenytoin with metals has been examined by AAS. Zinc, iron, nickel and cobalt have been found at levels of 0.87-87, 8.1-87.5, 1.9-50 and 10-15 $\mu g/g$, respectively [118]. Lead, copper and zinc at levels of 10-20 $\mu g/g$ in barbitone, caffeine and amidopyrine have similarly been determined by direct flame AAS [119]. Trace zinc, iron and lead in linseed, sunflower or castor oil used for pharmaceutical preparations are determined by shaking 10% nitric acid with the oil and chloroform, and aspirating the resulting emulsion directly into an air–propane–butane flame [120]. Dyes and antioxidants authorized for use in drugs are analysed for lead, zinc, copper and cadmium after mineralization with sulphuric acid and dissolution of the reisdue in nitric acid [121]. The level of lead contamination in bismuth subcarbonate [122], zinc oxide and carbonate [123], and aspirin, salicylamide, caffeine, phenobarbitone and codeine [124] has been measured by flame AAS.

Mercury in bottled Austrian medicinal and table water is determined down to 1.5 ng/ml by electro-deposition from the solution onto a gold cathode maintained at -3.0 V $vs.$ a platinum anode [125]. The cathode is then heated at $500°C$ in a stream of nitrogen, and the mercury vapour is passed into a flow-through cell for its AAS determination at 253.7 nm. Palladium, which is used as a catalyst during the preparation of semi-synthetic penicillins, is measured in the final products by AAS, to be sure that its level is less than 20 $\mu g/ml$ [125]. Ethylenediaminetetra-acetic acid (EDTA), which is added in small quantities to some pharmaceutical products to remove the alkaline-earth metals and improve the efficiency of the production process, has been determined in streptomycin [126]. The method involves addition of nickel to the sample solution at pH 6-6.5, to form the nickel-EDTA complex, and the excess of nickel is removed by precipitation with dimethylglyoxime. The nickel complexed by the EDTA is then released by pH adjustment and measured in an air-acetylene flame at 232.4 nm. For EDTA levels of 10-24 $\mu g/g$, the average reproducibility is $\pm$ 1.3 $\mu g/g$. The origin and degree of purity of heroin has been determined by measuring the levels of the associated metals. The samples are diluted with water or acid and copper, manganese, iron, lead, calcium, magnesium, strontium, potassium, rubidium and caesium are then determined. The AAS data serve for comparison of different batches of heroin and identification of illicit products from a common source [127].

11.11 DRUG METABOLITES IN BODY FLUIDS

Analysis of biological material for metals and other drug metabolites is a distinctly uncommon procedure in pharmaceutical analysis. However, close monitoring of therapy with some drugs by measuring the level of the drug or its metabolites in biological fluids or tissues is essential, because individual patients vary in their tolerance and sensitivity of response to the drug.

11.11.1 Antimony-Containing Drugs

Antimony compounds (e.g. sodium antimonyl gluconate and sodium stiboglucochnate) have been incorporated in the free or liposome-bound form into a variety of drugs for treatment of parasitic diseases. The antimony associated with these drugs has been measured in the blood, liver and spleen of mice at various time intervals after injection *in vivo* [128]. The antimony-containing drugs are incubated at 37°C for 2 hr with human whole blood and samples of the blood and of lysed erythrocytes are then analysed for antimony by the hydride-generation and electrothermal atomization technique. The calibration graph covers the range 0.8–5.8μM antimony.

11.11.2 Disulphiram

Disulphiram or antabuse (tetraethylthiuram disulphide) is used for the treatment of alcoholism. It metabolizes in man into various metabolites, amongst which is diethyldithiocarbamate. This metabolite is measured by AAS of the patient's urine after acute intoxication or overdosage by the drug [129,130]. The assay procedure involves reaction with copper(II) solution at pH 4, extraction of the copper chelate into carbon tetrachloride, and measurement of copper by electrothermal atomization AAS. The method is not applicable for measurement of the metabolite of this drug in protein-containing fluids such as plasma, because of the formation of unbreakable emulsions during the extraction step. Prior deproteinization decomposes the metabolites and renders the AAS method useful for the determination.

11.11.3 Gold-Containing Drugs

Gold compounds such as sodium aurothiomalate (Myocrisin®) and sodium aurothioglucose (Solganol®) have been used in the treatment of rheumatoid arthritis over the last five decades and their therapeutic effect was proved in the 1960s. It is always necessary to control the effect of gold and to establish the optimal dosage regimen to avoid toxicity. The therapy with these drugs is usually discontinued when the gold level in the blood reaches 3–6 μg/ml. The current editions of the British Pharmacopoeia and U.S. Pharmacopeia describe an assay method for sodium aurothiomalate based on decomposition with concentrated sulphuric–nitric acid mixture to form elemental gold, followed by ignition of the precipitate to constant weight. The use of AAS has greatly simplified the method of analysis. Sodium aurothiomalate injections, which contain about 12 mg of gold per ml, are simply diluted between 500 and 5000-fold with demineralized water and aspirated directly into an air–acetylene flame for gold measurement at 242.8 nm [131]. Recoveries in the range 97.6–100% are obtainable by this method [131,132].

Serum and urinary gold in rheumatoid patients receiving gold therapy has been measured by flame and electrothermal atomization AAS. Whole blood and serum samples are diluted 1:1 with demineralized water and aspirated into an air–acetylene flame [133] or a graphite furnace [134]. The coefficient of variation at the 5-μg/ml level is 1.8%. Urine samples are treated with tellurium chloride and tin(II) chloride and the precipitate is dissolved in *aqua regia*. Gold(III) is then extracted into methyl isobutyl ketone (MIBK) and injected into the graphite furnace [133]. Serum samples may also be diluted fourfold with 0.1% Triton X-100 solution, and urine with 0.01M hydrochloric acid and methanol, to prevent foaming before flame AAS determination of gold [135]. Extraction of gold from human serum and urine with dimorpholinethiuram disulphide in MIBK and injection of the organic phase into a graphite-furnace atomizer can also be used [136]. Determination of gold in the different protein fractions of blood serum has been described. The protein fractions are separated into a globulin (γ, β, α_2 and α_1) and an albumin fraction by combined gel filtration and electrophoresis, and then gold is measured in each fraction by AAS [137]. Samples of tissue and faeces from rats that had received sodium aurothiomalate have been analysed similarly [138]. The results obtained by AAS agree with those obtained by neutron-activation analysis.

11.11.4 Lithium-Containing Drugs

Therapy with lithium is now generally accepted as one of the major regimens for treatment of acute mania as well as for the prevention of relapse in recurrent bipolar mood disorders. Because of possible side-effects, including polyurea, ataxia, hypothyroidism, and weight gain, the concentration of lithium in the serum must be controlled within narrow limits. The normal concentration of lithium in serum is $\sim 3\mu M$ and the suggested therapeutic range is 0.3–1.3mM. Serum values exceeding 3.5mM are regarded as potentially lethal. AAS forms the basis of most of the published methods for the determination of therapeutic concentrations of lithium in serum [139,140].

Rocks *et al.* [141] suggested a method for direct determination of therapeutic concentrations of lithium in serum by flow-injection analysis and AAS detection. The samples are manually injected into a continuously flowing stream of demineralized water, which is pumped through a dispersion tube to the nebulizer. Hisayasu and Cohen [142] showed that the erythrocyte lithium concentration correlates better than serum lithium concentration with the clinical response in manic-repressive patients and with lithium-induced electro-encephalographic changes. However, direct measurement of lithium in erythrocytes is too complex. The relatively small difference in intra- and extra-erythrocyte lithium concentration suggested the use of the haematocrit (the volume % of erythrocytes) in flame AAS analysis. The use of a microhaematocrit at high applied centrifugal force significantly decreases the errors ascribable to trapped plasma. Erythrocyte lithium concentrations are indirectly calculated from the haematocrit. The detection limit is 10μM lithium and the calibration graph is linear for the 0.25–2mM range [142].

11.11.5 Platinum-Containing Drugs

The potential of platinum compounds as a new class of antitumour agents and the remarkable anti-neoplastic activity of some of these compounds have been demonstrated in many animal tumour-screening systems [143]. *cis*-Dichloro-diamminoplatinum(II) (CDDP) is the agent that has been most extensively studied and clinically tested in patients. This drug has been determined in plasma and urine by AAS by direct injection of the sample solutions into the graphite furnace and measurement of platinum at 265.9 nm [144]. Total platinum derived from CDDP in tissues [145] or albumin [146] is determined by complete combustion, dissolution of the residue in *aqua regia*, and electrothermal atomization AAS. Sera and ultrafiltrates containing the drug are analysed by digestion with concentrated nitric acid, extraction of the platinum with ammonium pyrrolidine dithiocarbamate into methyl isobutyl ketone and atomization in a graphite furnace [147]. Both wet and dry digestion procedures followed by electrothermal atomization AAS have been reported to eliminate matrix interferences in the determination of platinum in biological samples [148–150].

Separation of the protein-bound platinum species from freely circulating species by centrifugal ultrafiltration, followed by AAS measurement, has been described [151]. This process removes 98% of the plasma protein and other macromolecules without loss of CDDP in the filtrate. The freely circulating platinum species in the ultrafiltrate is then converted into a stable cation by reaction with ethylenediamine:

$$\underset{H_3N}{\overset{H_3N}{>}}Pt\underset{Cl}{\overset{Cl}{<}} + \begin{matrix} CH_2-NH_2 \\ | \\ CH_2-NH_2 \end{matrix} \longrightarrow \left[\underset{H_3N}{\overset{H_3N}{>}}Pt\underset{NH_2-CH_2}{\overset{NH_2-CH_2}{<}}\right]^{2+}$$

The product is collected on a paper disc loaded with cation-exchange resin and eluted with $5M$ hydrochloric acid; the eluate is atomized. The detection limit for platinum is 35 ng/ml in plasma, which is about 10 times the detection limit obtained by X-ray fluorescence. Although this method is useful for monitoring platinum in the blood of patients receiving CDDP chemotherapy, it is still not sufficient to support rigorous pharmacokinetic studies, because it fails to discriminate between the various platinum compounds which may be present. To differentiate between the drug and other metabolites or breakdown platinum species, the ultrafiltrate of the plasma is fractionated by HPLC on a strong-base anion-exchange column filled with Partisil-10 SAX and operated at a flow-rate of 1 ml/min with 1:1 v/v methanol–acetone mixture buffered to pH 3.8 as mobile phase [152]. The effluent is pumped through a dual-channel detector; the fractions containing CDDP are isolated and evaporated to dryness, then the residue is treated with $0.01M$ sodium cyanide to convert the platinum compound into the corresponding negatively charged and water-soluble tetracyano complex $[Pt(CN)_4]^{2-}$, followed by electrothermal atomization AAS.

11.12 PHARMACEUTICAL CONTAINERS

It has been reported that drug contamination by heavy metals mainly originates from the containers used for preparation or packing of the products. Leaching of some substances from the containers may cause significant contamination. Lead and tin compounds are often added to plastic materials and ferric, copper and zinc compounds are used in the manufacture of rubber closures. Contamination of intravenous fluids, packed in glass bottles with rubber closures, with iron, copper and zinc has been tested by AAS [153]. Miller *et al.* [154] evaluated the extent of contamination of parenteral solutions dispensed from disposable plastic syringes treated with silicon lubricants. The results of the AAS measurements show that 0.1–0.3 and 2.5–9.4 mg of silicon are obtained from 0.5- and 20-ml syringes, respectively. Traces of barium leached from rubber stoppers or glass vessels used for parenteral solutions have been measured semi-automatically by electrothermal atomization AAS [155]. Results for drugs stored at various temperatures for 7–28 months in glass cylinders with rubber closures are discussed.

The extent to which pharmaceutical products are liable to contamination with lead from plastic containers is difficult to define. Determination of lead in plastics by AAS generally requires a preliminary ashing procedure. The Sub-Committee of the Analytical Methods Committee of the Royal Society of Chemistry recommended wet oxidation with 50% hydrogen peroxide in the presence of hot concentrated sulphuric acid [156,157]. Direct aspiration of an aqueous solution obtained after ashing does not usually provide sufficient sensitivity [158]. Extraction of lead from these solutions into an organic solvent and aspiration of the organic extract into the flame may be used [158,159]. Dissolution of poly(vinyl chloride) in dimethylacetamide followed by direct aspiration of the organic solution into the flame has been reported by Musha *et al.* [160]. Lead in a variety of plastic pharmaceutical containers has been determined by direct atomization of 1–4 mg of the solid sample on a carbon-rod atomizer [161]. Ashing and atomization temperatures of 800°C for 60 sec and 1600°C for 25 sec, respectively are generally sufficient, but inadequate ashing is liable to cause apparent increase in the recovery of lead. Typical results are shown in Table 11.4. Other methods for testing plastic containers have been described [162,163].

REFERENCES

[1] R. V. Smith, *Am. Lab.*, **5**, No. 3, 27 (1973).
[2] J. R. Leaton, *J. Assoc. Off. Anal. Chem.*, **53**, 237 (1970).
[3] J. H. M. Miller, *Am. Lab.*, **10**, No. 8, 41 (1978).
[4] F. Rousselet and F. Thuillier, *Prog. Anal. Atom. Spectrosc.*, **1**, 353 (1979).
[5] E. Glatzel and A. Yersin, *Zentralbl. Pharm. Pharmakother. Laboratoriumsdiagn.*, **119**, 1282 (1980).

Table 11.4. AAS determination of lead in various plastics by direct atomization of solid samples with a carbon rod atomizer [161] (reprinted by permission of the copyright holders, the Royal Society of Chemistry)

Plastic	Sample weight (mg)	Temperature, °C and (time, sec)		Lead found (ppm)	Std. devn. (ppm)
		Ashing	Atomization		
Amber polystyrene vials	4.0	800 (60)	1600 (2)	0.026–0.052	0.009
Brown polyethylene bottles	3.0	650–800 (30)	1600 (2)	0.65–1.05	0.07–0.11
Poly(vinylchloride) tubing	2.0	380–800 (60)	1800 (2)	0.067–0.092	0.008
White polyethylene bottle	3.0	650–800 (30)	1600 (2)	0.017–0.107	0.03–0.04
White polypropylene bottle	1.3	600 (120) or 800 (60)	1800 (2)	0.067–0.107	0.012–0.015

[6] D. Meissner, H. Dawczynski, E. Glatzel, E. Preu, K. Winnefeld and A. Yersin, *Zentralbl. Pharm. Pharmakother. Laboratoriumsdiagn.*, 119, 1285 (1980).

[7] D. Meissner, H. Dawczynski, E. Glatzel, I. Kulick, E. Preu, L. H. Schmidt, K. Winnefeld and A. Yersin, *Zentralbl. Pharm. Pharmakother. Laboratoriumsdiagn.*, 119, 1267 (1980).

[8] W. Rehpenning, H. Dawczynski, E. Glatzel, D. Meissner, E. Preu, K. Winnefeld and A. Yersin, *Zentralbl. Pharm. Pharmakother. Laboratoriumsdiagn.*, 119, 1279 (1980).

[9] M. Roschig, W. Bach, K. Graesser and S. Rackow, *Zentralbl. Pharm. Pharmakother. Laboratoriumsdiagn.*, 119, 1375 (1980).

[10] K. Kidani, K. Nakamura, K. Inagraki and H. Koike, *Bunseki Kagaku*, 24, 742 (1975).

[11] S. S. M. Hassan, M. T. Zaki and M. H. Eldesouki, *Talanta*, 26, 91 (1979).

[12] A. Finkelstein and R. Holz, in G. Eisenman (ed.), *Membranes*, Vol. 2, *Lipid Bilayers and Antibiotics*, Dekker, New York, 1973, p. 377ff.

[13] B. Dekruijff and R. A. Demel, *Biochem. Biophys. Acta*, 339, 57 (1974).

[14] W. Zygmunt, *Appl. Microbiol.*, 14, 953 (1966).

[15] I. Ndzinge, S. D. Peters and A. H. Thomas, *Analyst*, 102, 328 (1977).

[16] S. S. M. Hassan and M. H. Eldesouki, *Talanta*, 26, 531 (1979).

[17] S. S. M. Hassan and M. H. Eldesouki, *J. Assoc. Off. Anal. Chem.*, 64, 1158 (1981).

[18] Y. Kidani, K. Inagaki, T. Saotome and H. Koike, *Bunseki Kagaku*, 22, 896 (1973).

[19] A. V. Kovatsis and M. A. Tsougas, *Arzneim. Forsch.*, 28, 248 (1978).

[20] Y. Kidani, K. Inagaki, N. Osugi and H. Koike, *Bunseki Kagaku*, 22, 892 (1973).

[21] T. Mitsui and Y. Fujimura, *Bunseki Kagaku*, 24, 575 (1975).

[21a] Y. Minami, T. Mitsui and Y. Fujimura, *Bunseki Kagaku*, 31, 604 (1982).

[22] J. Alary, A. Villet and A. Coeur, *Ann. Pharm. Franc.*, 34, 419 (1976).

[23] M. M. Amer, M. I. Walash, I. A. Haroun and F. M. Ashour, *J. Pharm. Belg.*, 33, 297 (1978).

[24] G. I. Spielholtz and G. C. Toralballa, *Analyst*, 94, 1072 (1969).

[25] K. Szivós, L. Pólos, L. Bezur and E. Pungor, *Acta Pharm. Hung.*, 43, 90 (1973).

[26] T. Mitsui, Y. Fujimura and T. Suzuki, *Bunseki Kagaku*, 24, 244 (1975).

[27] T. Mitsui and Y. Fujimura, *Kogyo Kagaku Zasshi*, 1908 (1974).

[27a] Y. Minami, T. Mitsui and Y. Fujimura, *Bunseki Kagaku*, 30, 811 (1981).

[28] S. J. Simon and D. F. Boltz, *Microchem. J.*, 20, 468 (1975).

[29] T. Minamikawa and K. Matsumura, *Yakugaku Zasshi*, 96, 440 (1976).

[30] S. Tammilehto, *Acta Pharm. Fenn.*, 88, 25 (1979).

[31] F. J. Bandelin, *J. Am. Pharm. Assoc., Sci. Ed.*, 39, 493 (1950).

[32] H. Böhme and H. Lampe, *Arch. Pharm.*, 284, 227 (1951).

[33] H. Böhme and R. Strohecker, *Arch. Pharm.*, 285, 422 (1952).

[34] P. Duquenois and M. Faller, *Bull. Soc. Chim. France*, 6, 998 (1939).

[35] H. Schmidt-Hebbel and P. A. Benavides, *Pharm. Zentralhalle*, 79, 526 (1938).

[36] Y. Kidani, T. Saotome, M. Kato and H. Koike, *Bunseki Kagaku*, 23, 265 (1974).

[37] M. K. Park, B. R. Lim, K. S. Yu and K. H. Yong, *Yakhak Hoeji*, 22, 27 (1978).

[38] H. K. L. Gupta and D. F. Boltz, *Mikrochem. J.*, 16, 571 (1971).

[39] S. S. M. Hassan, M. T. Zaki and M. H. Eldesouki, *J. Assoc. Off. Anal. Chem.*, 62, 315 (1979).

[40] F. M. Farhan and M. Makhani, *J. Pharm. Sci.*, 59, 1200 (1970).

[41] J. Ramírez-Muñoz, *Atomic Absorption Spectroscopy*, Elsevier, Amsterdam, 1968, p. 328.

[42] F. J. Diaz, *Anal. Chim. Acta*, 58, 455 (1972).

[43] Y. Kidani, K. Takeda and H. Koike, *Bunseki Kagaku*, 22, 719 (1973).

[44] B. Mandrou and J. Bres, *J. Pharm. Belg.*, 25, 3 (1970).

[45] L. L. Whitlock, J. R. Melton and T. J. Billings, *J. Assoc. Off. Anal. Chem.*, 59, 580 (1976).

[46] E. Peck, *Anal. Lett.*, **11**, 103 (1978).

[47] H. Koizumi, T. Hadeishi and R. McLaughlin, *Anal. Chem.*, **50**, 1701 (1978).

[48] Y. Kidani, R. Umemoto and K. Inagaki, *Mikrochim. Acta*, 329 (1981 **II**).

[49] Y. Kidani, K. Nakamura and K. Inagaki, *Bunseki Kagaku*, **25**, 509 (1976).

[50] H. L. Kahn, *J. Chem. Educ.*, **43**, A103 (1966).

[51] T. V. Ramakrishna, P. W. West and J. W. Robinson, *Anal. Chim. Acta*, **39**, 81 (1967).

[52] P. P. Karkhanis and J. R. Anfinsen, *J. Assoc. Off. Anal. Chem.*, **56**, 358 (1973).

[53] A. Stahlavska, K. Propokova and M. Tuzar, *Pharmazie*, **29**, 140 (1974).

[54] E. M. Cochran, E. B. Inskip, P. King and H. W. Ziegler, *J. Pharm. Sci.*, **57**, 1215 (1968).

[55] R. A. Sharp and A. M. Knevel, *J. Pharm. Sci.*, **60**, 458 (1971).

[56] D. T. David, *Analyst*, **84**, 536 (1959); **85**, 495 (1960).

[57] J. B. Willis, *Anal. Chem.*, **33**, 556 (1961).

[58] E. G. Gimblet, A. F. Marney and R. W. Bonsens, *Clin. Chem.*, **13**, 204 (1967).

[59] D. T. Trudeau and E. F. Freier, *Clin. Chem.*, **13**, 101 (1967).

[60] J. B. Willis, *Spectrochim. Acta*, **16**, 259 (1960).

[61] G. K. Billings and J. A. S. Adams, *Atom. Absorp. Newslett.*, **3**, 65 (1964).

[62] B. A. Dalrymple and C. T. Kenner, *J. Pharm. Sci.*, **58**, 604 (1969).

[63] R. V. Smith and M. A. Nessen, *J. Pharm. Sci.*, **60**, 907 (1971).

[64] T. A. Bluemer and J. A. Patel, *Am. J. Hosp. Pharm.*, **29**, 80 (1972).

[65] J. R. Leaton, *J. Assoc. Off. Anal. Chem.*, **53**, 237 (1970).

[66] W. B. Barnett, *Anal. Chem.*, **44**, 695 (1972).

[67] Y. S. Chae, J. P. Vacik and W. H. Shelver, *J. Pharm. Sci.*, **62**, 1838 (1973).

[68] P. O. Kosonen, A. M. Salonen and A.-L. Nieminen, *Finn. Chem. Lett.*, 136 (1978).

[69] J. Korkisch and H. Hübner, *Mikrochim. Acta*, 311 (1976 **II**).

[70] Y. Mino, N. Ota, S. Sakao and S. Shimomura, *Chem. Pharm. Bull.*, **28**, 2687 (1980).

[71] J. E. Allan, *Spectrochim. Acta*, **15**, 800 (1959).

[72] D. J. David, *Atom. Absorp. Newslett.*, **1**, 9 (1962).

[73] K. E. Curtis, *Analyst*, **94**, 1068 (1969).

[74] W. Slavin, *Atomic Absorption Spectroscopy*, Wiley, New York, 1968, pp. 78–189.

[75] J. R. Leaton, *J. Assoc. Off. Anal. Chem.*, **53**, 237 (1970).

[76] H. I. Tarlin and M. Batchelder, *J. Pharm. Sci.*, **59**, 1328 (1970).

[77] E. Van Den Eeckhout and P. de Moerloose, *Pharm. Weekblad*, **106**, 749 (1971).

[78] W. R. Hatch and W. L. Ott, *Anal. Chem.*, **40**, 2085 (1968).

[79] J. R. Leaton, *J. Assoc. Off. Anal. Chem.*, **53**, 237 (1970).

[80] W. H. Harper, *Proc. Soc. Anal. Chem.*, **7**, 104 (1970).

[81] R. D. Thompson and T. J. Hoffmann, *J. Pharm. Sci.*, **64**, 1863 (1975).

[82] W. Luethi, W. Kuenzle and M. Sahli, *Pharm. Acta Helv.*, **54**, 60 (1979).

[83] G. M. George, L. J. Frahm and J. P. McDonnell, *J. Assoc. Off. Anal. Chem.*, **60**, 1054 (1977).

[84] G. M. George, M. A. Albrecht, L. J. Frahm and J. P. McDonnell, *J. Assoc. Off. Anal. Chem.*, **56**, 1480 (1973).

[85] J. T. Mason, Jr., *J. Pharm. Sci.*, **69**, 101 (1980).

[86] R. R. Moody and R. B. Taylor, *J. Pharm. Pharmacol.*, **24**, 848 (1972).

[87] L. P. Díez, J. H. Méndez and J. A. R. González, *Analyst*, **106**, 737 (1981).

[88] R. V. Smith and M. A. Nessen, *Microchem. J.*, **17**, 638 (1972).

[89] A. Stahlavska, K. Propoková and M. Tuzar, *Pharmazie*, **29**, 140 (1974).

[90] E. Mario and R. E. Gerner, *J. Pharm. Sci.*, **57**, 1243 (1968).

[91] T. Rihs, *Pharm. Acta Helv.*, **46**, 550 (1971).

[92] D. Ridge and M. Todd, *J. Chem. Soc. Ind.*, **69**, 49 (1950).

[93] A. Pozefsky and M. E. Grenoble, *Drug Cosmetic Ind.*, **80**, 752 (1957).

[94] J. B. Willis, *Spectrochim. Acta*, **16**, 551 (1960).

[95] H. Sanui and N. Pace, *Appl. Spectrosc.*, **20**, 135 (1966).

[96] D. T. David, *Analyst*, **85**, 495 (1960).

[97] J. Ramírez-Muñoz, *Anal. Chem.*, **42**, 517 (1970).

[98] R. V. Smith and M. A. Nessen, *J. Pharm. Sci.*, **60**, 907 (1971).

[99] T. A. Bluemer and J. A. Patel, *Am. J. Hospit. Pharm.*, **29**, 80 (1972).

[100] A. Stahlevska, *Pharmazie*, **28**, 238 (1973).

[101] L. Magos and A. A. Cernik, *Br. J. Ind. Med.*, **26**, 144 (1969).

[102] L. Magos, *Analyst*, **96**, 847 (1971).

[103] Analytical Methods Committee, *Analyst*, **90**, 515 (1965).

[104] P. G. Takla and V. Valijanian, *Analyst*, **107**, 378 (1982).

[105] D. C. Manning, *Atom. Absorp. Newslett.*, **9**, 97 (1970).

[106] S. Chilov, *Talanta*, **22**, 205 (1975).

[106a] A. M. Ure, *Anal. Chim. Acta*, **76**, 1 (1975).

[107] B. J. Meakin and Z. M. Khammas, *J. Pharm. Pharmacol.*, **31**, 653 (1979).

[108] J. C. May, J. T. C. Sih and A. J. Mustafa, *J. Biol. Stand.*, **6**, 339 (1978).

[109] P. W. Woodward and J. R. Pemberton, *Appl. Microbiol.*, **27**, 1094 (1974).

[110] B. Aaro and B. Salvesen, *Medd. Nor. Farm. Selsk.*, **35**, 83 (1973).

[111] S. J. Weiner, *J. Pharm. Pharmacol.*, **7**, 118 (1955).

[112] G. Sykes, *J. Pharm. Pharmacol.*, **10**, 40 T (1958).

[113] J. Ingversen and V. Sten-Andersen, *Dansk. Tidsskr. Farm.*, **42**, 264 (1968).

[114] L. Lachman, *J. Bull. Parenteral Drug Assoc.*, **22**, 127 (1968).

[115] T. J. McCarthy, *Pharm. Weekblad*, **105**, 1139 (1970).

[116] K. Eriksson, *Acta Pharm. Suecica*, **4**, 261 (1967).

[117] I. T. Calder and J. M. McB. Miller, *J. Pharm. Pharmacol.*, **28**, 25 (1976).

[118] J. Pawlaczyk and M. Makowska, *Acta Pol. Pharm.*, **36**, 59 (1979).

[119] J. Mohay, M. Veress and G. Szasz, *Magy. Kem. Foly*, **85**, 465 (1979).

[120] J. Mohay, M. Veress and G. Szasz, *Magy. Kem. Foly*, **84**, 492 (1978).

[121] F. Pellerin and J. P. Goulle, *Am. Pharm. Franc.*, **35**, 189 (1977).

[122] O. Joens and L. Toft, *Acta Pharm. Chem. Sci. Ed.*, **1**, 14 (1973).

[123] A. Smith and K. Bache-Hansen, *Acta Pharm. Suecica*, **10**, 254 (1973).

[124] K. A. Kovar, W. Lautenschlaeger and R. Seidel, *Dt. Apothz.*, **115**, 1855 (1975).

[125] F. Rousselet, V. Courtois and M. L. Girard, *Analusis*, **3**, 132 (1975).

[126] R. J. Hurtubise, *J. Pharm. Sci.*, **63**, 1131 (1974).

[127] M. J. Pro and R. L. Brunelle, *J. Assoc. Off. Anal. Chem.*, **53**, 1137 (1970).

[128] R. J. Ward, C. D. V. Black and G. J. Watson, *Clin. Chim. Acta*, **99**, 143 (1979).

[129] F. K. Martens and A. M. Heyndrickx, *J. Anal. Toxicol.*, **2**, 269 (1978).

[130] F. K. Martens and A. M. Heyndrickx, *Meded. Fac. Landbowwet. Rijksuniv. Gent*, **41**, 1393 (1976).

[131] H. T. Smart and D. J. Campbell, *Can. J. Pharm. Sci.*, **4**, 73 (1969).

[132] F. Rousselet, V. Courtois and M. L. Girard, *Analusis*, **3**, 132 (1975).

[133] J. V. Dunckley, D. M. Grennan and D. G. Palmer, *J. Anal. Toxicol.*, **3**, 242 (1979).

[134] M. J. Barrett, R. DeFries and W. M. Henderson, *J. Pharm. Sci.*, **67**, 1332 (1978).

[135] D. M. Schattenkirchner and Z. Grobenski, *Atom. Absorp. Newslett.*, **16**, 84 (1977).

[136] O. Wawschinek and F. Rainer, *Atom. Absorp. Newslett.*, **18**, 50 (1979).

[137] H. Kamel, D. M. Brown, J. M. Ottaway and W. E. Smith, *Analyst*, **102**, 645 (1977).

[138] R. M. Turkall and J. R. Bianchine, *Analyst*, **106**, 1096 (1981).

[139] V. Lehmann, *Clin. Chim. Acta*, **20**, 523 (1968).

[140] T. B. Cooper, G. M. Simpson and D. Allen, *Atom. Absorp. Newslett.*, **13**, 119 (1974).

[141] B. F. Rocks, R. A. Sherwood and C. Riley, *Clin. Chem.*, **28**, 440 (1982).

[142] G. H. Hisayasu, J. L. Cohen and R. W. Nelson, *Clin. Chem.*, **23**, 41 (1977).

[143] B. Rosenberg, L. Van Camp and T. Krigas, *Nature*, **205**, 698 (1965).

[144] A. F. LeRoy, M. L. Wehling, H. L. Sponseller, W. S. Friauf, R. E. Solomon, R. L. Dedrick, C. L. Litterst, T. E. Gram, A. M. Guarino and D. A. Becker, *Biochem. Med.,* **18**, 184 (1977).

[145] M. L. Denniston, L. A. Sternson and A. J. Repta, *Anal. Lett.,* **14**, 451 (1981).

[146] N. K. Bel'skii, L. I. Ochertyanova and L. K. Shubochkin, *Zh. Analit. Khim.,* **34**, 814 (1979).

[147] D. A. Hull, N. Muhammad, J. G. Lanese, S. D. Reich, T. T. Finkelstein and S. Fandrich, *J. Pharm. Sci.,* **70**, 500 (1981).

[148] R. G. Miller and J. U. Doerger, *Atom. Absorp. Newslett.,* **14**, 66 (1975).

[149] N. F. Pera and H. C. Harder, *Clin. Chem.,* **23**, 1245 (1977).

[150] D. Priesner, L. A. Sternson and A. J. Repta, *Anal. Lett.,* **14**, 1255 (1981).

[151] S. J. Bannister, Y. Chang, L. A. Sternson and A. J. Repta, *Clin. Chem.,* **24**, 877 (1978).

[152] Y. Chang, L. A. Sternson and A. J. Repta, *Anal. Lett.,* **11**, 449 (1978).

[153] W. H. Thomas and Y. K. Lee, *Acta Pharm. Suecica,* **11**, 495 (1974).

[154] J. R. Miller, J. J. Helprin and J. S. Finlayson, *J. Pharm. Sci.,* **58**, 455 (1969).

[155] F. J. Szydlowski and F. R. Vianzon, *Anal. Lett.,* **11**, 161 (1978).

[156] Analytical Methods Committee, *Analyst,* **92**, 403 (1967).

[157] Analytical Methods Committee, *Analyst,* **101**, 62 (1976).

[158] R. K. Roschnik, *Analyst,* **98**, 596 (1973).

[159] Analytical Methods Committee, *Analyst,* **100**, 899 (1975).

[160] S. Musha, M. Munemori and Y. Nakanishi, *Bunseki Kagaku,* **13**, 330 (1964).

[161] P. Girgis-Takla and I. Chroneos, *Analyst,* **103**, 122 (1978).

[162] D. C. Hunt, *Lab. Pract.,* **24**, 411 (1975).

[163] H. Endriss, *Dt. Lebensmit. Rdsch.,* **70**, 243 (1974).

12

Biological materials

Application of AAS in biological analysis has resulted in an enormous number of important publications covering a wide range of materials. Methods are now available for the determination of trace elements, metalloproteins, protein-thiol and enzyme activity. Immunology, histochemistry and physiology have also benefited from this technique.

12.1 TRACE ELEMENTS IN BIOLOGICAL MATERIALS

There is a copious literature on the determination of various elements in biological materials by AAS. Lithium [1,2], sodium [3-6], potassium [6-8], magnesium [5,6,9-32], calcium [3,6,32-67], barium [68], strontium [69-75], copper [50,76-119], cadmium [30,102,116,119-146], cobalt [78,147-151], nickel [114,120,152-160], chromium [118,120,161-178], iron [89,101,117,118, 179-195], lead [100,102,119,121-123,146,156,196-149], manganese [74, 250-261], zinc [10,30,84,89,94,101-104,117,119,262-296], aluminium [297-300], silver [301-303], vanadium [304], arsenic [305-328], thallium [146, 329], selenium [330-338], tellurium [338-342], boron [120,343], beryllium [344,345], bismuth [148,342,346,347], platinum [348-350], gold [351,352] and mercury [353-379] have been determined in various biological substances by the flame and electrothermal atomization techniques.

Determination of many of these elements in human biological materials obtained either by biopsy or post-mortem is becoming an increasingly important function of clinical and forensic laboratories. It is well established that accumulation of some metals in biological fluids and soft and hard tissues interferes with the body function and is linked with respiratory ailments, hypertension and damage to bone and liver. Metal toxicity can also be confirmed by the level of the toxic element in certain biological organs, and especially in hair and nails. Metal levels are also implicated in the pathogenesis of a number of clinical disorders. Thus assessment of the status of some of these elements in the body provides valuable information for the diagnosis of a wide range of nutritional and inherited or acquired metabolic disorders and in the investigation of cause in suspicious death. For example, in Wilson's disease the copper level in liver increases and contributes to the dysfunction of liver and brain. Measurement of copper by AAS in a small biopsy specimen of the liver helps in diagnosis of the disease in its early stages [112,380]. Relationships have been established, by use of AAS, between hypofertility and high lead and cadmium levels in semen, between poisoning and high level of arsenic in nails and hair, and between a high level of aluminium and Alzheimer's disease, a progressive encephalopathic syndrome in haemodialysis patients [381–383].

Accurate determination of various elements in biological material depends on a number of factors. These are: (a) the type of sample to be analysed; (b) the nature and level of the element of interest; (c) the method used in the analysis; (d) the possible contamination or loss caused by the method used for sample collection, storage and pretreatment. No general method can be used for pretreatment of all biological materials before AAS measurement. However, the following approaches are commonly utilized.

(1) Dilution of the biological fluid with water [384–388], alcohol [389], buffer [390], surfactant [391–394], acid [395], or masking and enhancement reagents such as EDTA [382,396], KCl [32,397], $LaCl_3$ [398,399], $(NH_4)_2HPO_4$ [125,381,400] and $SrCl_2$ [401]. Prior ultrafiltration [401, 402] or precipitation of the protein content with trichloroacetic acid [80, 271,403–405] and dilution of the filtrate with one of the above-mentioned diluents may also be used.

(2) Extraction of the metal of interest from the fluid or tissue into an appropriate solvent containing a suitable chelating agent. This method can be used with biological samples either directly [247,330,406–410] or after their decomposition [411–416]. Such methods offer the advantage of eliminating the interference of some elements and yielding the desired element in concentrations high enough for accurate measurement. Table 12.1 summarizes some of the extraction systems used in the AAS analysis of biological materials.

(3) Wet digestion at 60–250°C in open [425,433–435] or pressure vessels [420,436–441] with nitric, sulphuric and perchloric acids or their mixtures. A mixture of these three acids in 3:1:1 ratio (in the order given) is widely used [166,417,422,443]. Hydrogen peroxide, permanganate, molybdate and vanadate [444–447] may be added to the acid digest to catalyse the decomposition and to keep the metal in the digest in the highest oxidation state.

(4) Dry-ashing by heating the sample to a relatively high temperature (300–700°C) in an oxygen atmosphere in an open vessel [321,448,449]. Some ashing aids such as magnesium salts or oxides may be added to the samples to prevent metal volatilization. The residue remaining after ashing is dissolved in a suitable mineral acid.

Table 12.1. AAS determination of some trace elements in biological materials after extraction

Element	Biological material	Extraction system			References
		pH	Reagent	Solvent	
Arsenic	Tissues		Diethylammonium diethyldithiocarbamate	$CHCl_3$	306
Bismuth	Urine	2.8	APDC	MIBK	148
Cadmium	Blood, urine	1–5	APDC	MIBK	130,131
	Tissues		APDC	CCl_4	406,416
	Blood, urine	5–11	Diethyldithiocarbamate (DDC)	MIBK	129
	Blood, urine	9	Dithizone (HDz)	$CHCl_3$	417
Chromium	Urine	4.5	APDC	MIBK	409
	Tissues		Methyltrioctylammonium chloride	MIBK	418
Cobalt	Tissues	2–4	APDC	MIBK	419
	Tissues		APDC	Toluene	420
Copper	Blood	6	Tri-n-octylamine (TOA)	CCl_4	411
	Haemodialysate	4.3	APDC	MIBK	408
Lead	Urine		APDC	MIBK	148,410
	Blood	2–4	APDC	MIBK	230,247, 407,421
	Urine		APDC	MIBK	148,407
	Tissues		APDC	MIBK	422
Manganese	Blood, urine	8–11	Dithizone	MIBK,$CHCl_3$	417
Molybdenum	Blood	7.1	Cupferron	MIBK	423
	Blood, plasma		Zinc-dithiol	MIBK	424
Nickel	Urine	2–8	APDC	MIBK	148,414
	Tissues	9	APDC	MIBK	425
	Urine	9	Dimethylglyoxime	MIBK	414
Silver	Serum, urine	9	Furildioxime	MIBK	426
	Tissues	8–8.5	Benzylxanthate	MIBK	427

Table 12.1. *continued*

Element	Biological material	Extraction system			References
		Reagent	pH	Solvent	
Selenium	Tissues	4-Chloro-*o*-phenylenediamine	1–2	Toluene	428
	Blood	Diethyldithiocarbamate	4–6	CCl_4, cyclohexane	429
	Blood, tissues	Dithizone		CCl_4	430
Vanadium	Blood, urine	*N*-Cinnamoyl-*N*-(2,3-xylyl)-hydroxylamine (*N*-2,3-xylylcinnamohydroxamic acid)		CCl_4	431
	Blood	*N*-*o*-Tolylbenzohydroxamic acid		CCl_4	432
Zinc	Haemodialysate	APDC	4.3	MIBK	408
	Blood, urine	Dithizone	9	$CHCl_3$	417
	Blood, plasma, serum	1-(2-Pyridylazo)-2-naphthol		MIBK	405

APDC = Ammonium pyrrolidine dithiocarbamate

(5) Ashing in an oxygen atmosphere in a closed vessel, either at a high temperature in an oxygen-filled flask [335,430] or a Parr bomb, or at a low temperature with an oxygen plasma asher operated at 300–500 W radiofrequency power [450–455]. The radiofrequency discharge of this asher produces radicals reactive enough to attack the organic matter at a low temperature ($< 100°C$).

(6) Solubilization of the sample at temperatures up to $60°C$ with tetramethylammonium hydroxide [456,457] or other alkaline digestants such as 'Soluene 100' or 'Soluene 350' (mixture of aromatic hydrocarbons and strong organic base) [458,459]. The sample solution is then diluted with water or methyl isobutyl ketone and atomized.

(7) Homogenization of the sample either ultrasonically or in an agate mortar with buffer, acid or surfactant. The last of these reagents [460,461] has been advantageously used with the electrothermal atomization technique to prevent residue build-up in the graphite furnace. Slurried or freeze-dried samples may also be atomized directly.

(8) Enzymatic digestion of soft tissue samples by incubation at $55°C$ with crystalline subtilisin A [462]. The digest is then centrifuged and a portion of the supernatant liquid is used for analysis.

Volatility losses during the digestion step depend on the nature of the element, method of decomposition, temperature used and the material of the containers in which the decomposition is conducted. In the dry-ashing methods, temperature control is important since some metals such as mercury, selenium, silver and zinc are more likely to volatilize at high temperature. A radiochemical study of the loss of such elements during ashing of blood and liver tissue shows that no loss takes place at temperatures up to $700°C$ [463]. Various ashing procedures for AAS determination of lead and cadmium in animal tissue have been compared by statistical techniques [464]. Digestion with acids followed by calcination at $650°C$ [464] and the effect of some acids, peroxide and surfactants on digestion of biological materials have also been examined [465].

During the ashing step some metals such as cerium, copper, iron, silver, and manganese may be retained by containers made of silica. Elements such as silicon, aluminium, calcium, copper, tin, bismuth and iron may form acid-insoluble compounds after ashing. This may be circumvented by performing the digestion in closed vessels made of platinum or coated with polypropylene or PTFE and addition of ashing aids such as magnesium oxide, magnesium nitrate, nitric acid and sulphuric acid to the sample before or during heating. Acid-digestion procedures are less troublesome. High-purity grades of acid, free from metal impurities, are required, and charring during the digestion should be avoided, especially when hydride-forming elements such as selenium, arsenic and antimony are the elements sought. Precautions should also be taken when perchloric acid is used. It is customary to pretreat the sample with other acids before addition of perchloric acid and to avoid boiling to dryness.

Despite the large number of procedures involving prior digestion of biological substances, direct atomization of biological fluids and tissue slurries in the flame [466], hollow T-shaped atomizer [467], Delves cup [125,140,392,468], carbon rod [111,123,285,436], metal ribbon [130,230,272,469] and graphite tube [299,382,383,390,471] have been suggested. Electrothermal atomization is usually preferred owing to the low limits of detection obtainable. A further improvement can be obtained by using the L'vov platform [472] and Zeeman background correction [139,389,473,474]. The reliability, sensitivity, matrix effect, sample preparation and range of detection of some of these methods [475-479] and interferences by major elements have been reviewed [480].

Sources of gain and loss of analyte in trace analysis have been the subject of a recent 'paper symposium' [480a].

12.1.1 Biological Fluids

Trace metals in whole blood, serum, plasma, saliva, gastric juice, cerebrospinal fluids, synovial fluids, semen and urine are satisfactorily determined by AAS (Table 12.2). Blood samples are collected by a syringe with a stainless-steel needle and stored under refrigeration. When whole blood and plasma are to be analysed, a chemical anticoagulant such as heparin or EDTA is added. For serum analysis, the blood is allowed to stand until it has clotted, and the serum is isolated. Haemolysis must be avoided because of the difference in the concentrations of some elements in the red cell fluid and serum or plasma. Hinks *et al.* [480b] have described a method for AAS determination of zinc and copper in leucocytes. The white cells are isolated by using a dextran sedimentation technique to eliminate the red cells, followed by digestion with nitric acid at 155°C for 30 min, and electrothermal atomization of the digest.

Urine samples are collected either at random or after a specific time. Although it is most common to collect the urine during a 24-hr period, analysis performed on timed specimens yields excretion rates that are of value in the interpretation of metabolic process, renal function and renal clearance of intoxicants. In both cases, the pH of the urine sample is kept at 4.5 to prevent precipitation of calcium phosphate and loss of some metal ions. Gastric secretion is collected either under basal physiological conditions or following pharmacological stimulation. It should be noted that for AAS analysis of most biological fluids, only simple pretreatment is required, since these samples are obtainable with good homogeneity. Thus dilution, centrifugation and extraction of the metal of interest may be sufficient and 10-200 μl of the sample is commonly used for analysis.

Table 12.2. AAS determination of some trace elements in biological fluids

Element	Biological fluid	Sample preparation	Atomization system	References
Aluminium	Blood	Digestion with HNO_3 at 90°C	GF	481
	Blood serum or plasma	Dilution with water	GF	386,482
	Cerebrospinal fluid	Dilution with ammonia–Teepol-70	GF	299
		Digestion with HNO_3–H_2SO_4–$HClO_4$ at 160°C	F	298
	Serum	None	GF	297,383,470, 483
		Dilution with Tris–NaCl buffer	GF (L'vov platform)	390
		Dilution with EDTA–NH_3–Triton X-100	GF	382
		Dilution with Triton X-100	GF	484
	Urine	None	GF	383,470
Antimony	Blood	Digestion with H_2SO_4	HG	485
		Digestion with HNO_3–$HClO_4$, extraction	GF	486
	Urine	Digestion with H_2SO_4	HG	485
		Digestion with HNO_3–H_2SO_4–$HClO_4$	HG	486
Arsenic	Blood	Ashing with $Mg(NO_3)_2$ at 550°C, extraction	GF	308
	Urine	Ashing with $Mg(NO_3)_2$–NH_4NO_3 at 500°C, extraction	F	488
		Digestion with HNO_3–$HClO_4$–H_2SO_4	HG	489
		Addition of anti-foaming agent	HG	322
		Digestion with HNO_3–$HClO_4$	HG	326
		Extraction as arsenomolybdic acid	F	490
		Digestion	HG	324,491
Beryllium	Blood	Digestion with HNO_3	GF	345
		Centrifugation	GF	402
		Digestion with HNO_3	GF	345
		Dilution with HNO_3–H_2SO_4	GF	492
Bismuth	Blood	Digestion with HNO_3 at 180°C	HG	493
		Digestion with $HClO_4$–HNO_3	HG	494
		Digestion with H_2SO_4–HNO_3	HG	495
	Cerebrospinal fluid, haemolysed blood	Extraction	GF	496
	Urine	Digestion with $HClO_4$–HNO_3	HG	494
		Digestion with H_2SO_4–HNO_3	HG	495

Element	Material	Extraction	F	F
Boron	Blood	Ashing at 600°C		148
Cadmium	Blood	Dilution with Triton X-100	DC	343
			GF	392
			ZAAS	126
		Dilution with Triton X-100, ashing at 500°C	GF	139
		Dilution with water	GF	121
				102,108,132, 497
		Dilution with EDTA	F	385
		Dilution with EDTA, extraction	CB	396
		Dilution with HNO$_3$-(NH$_4$)$_2$HPO$_4$	GF	129
		Dilution with (NH$_4$)$_2$HPO$_4$	GF	400
		Dilution with NH$_4$NO$_3$-HF	DC	125
		Digestion with HNO$_3$-H$_2$SO$_4$-H$_2$O$_2$, extraction	CR	122
		Digestion with HNO$_3$ at 85–95°C	GF	415
		Digestion with HNO$_3$-H$_2$O$_2$, extraction	CR	123
		Digestion with quaternary ammonium hydroxide, extraction	GF	417
		Ashing at 300°C	ZAAS	379
		Plasma ashing	GF	498
		Extraction	DC	137,140
			GF	129,323,406, 407,416
	Gastric contents		TR	130
	Saliva	Digestion with HNO$_3$-H$_2$O$_2$, extraction	DC	131
	Semen	Dilution with water	GF	417
	Serum	Digestion with HNO$_3$ at 80°C	GF	499
	Urine	None	GF	381
		Dilution with Triton X-100	GF	124
		Dilution with HNO$_3$	ZAAS	139
		Dilution with water	GF	395
		Digestion with HNO$_3$ and dry ashing at 430°C	CB	143
		Digestion with HNO$_3$ at 80°C	CF	132
		Digestion with HNO$_3$-H$_2$O$_2$, extraction	GR	142
		Ashing, at 300–400°C	GF	380
		Extraction	GF	417
			GF	498,500
			GF	129,323,406, 407,501

Table 12.2. *continued*

Element	Biological fluid	Sample preparation	Atomization system	References
Calcium	Cerebrospinal fluid	Dilution with water	F	47
	Plasma	Dilution with SrCl$_2$	F	502
		Dilution with EDTA	F	503
	Saliva	None	F	46
		Dilution with LaCl$_3$	F	399
		Dilution with EDTA	F	504,41,42
	Serum	Dilution with water	F	505
		Dilution with KCl	F	397
		Dilution with LaCl$_3$ or SrCl$_2$	F	398,40,507
		Dilution with NaCl–KCl–LaCl$_3$	F	508
		Dilution with LaCl$_3$(butanol–octanol)	F	44
		Treatment with trichloroacetic acid	F	509
	Urine	Dilution with water	F	505
		Dilution with KCl	F	32
		Dilution with LaCl$_3$ or SrCl$_2$	F	40,43
		Dilution with NaCl–KCl–LaCl$_3$	F	508
		Treatment with trichloroacetic acid	F	509
Chromium	Blood	Digestion with HNO$_3$, extraction	F	510
	Plasma and serum	None	GF	511
		Digestion with HNO$_3$–H$_2$O$_2$ at 80°C	GF	166
		Digestion with HNO$_3$–H$_2$SO$_4$–HClO$_4$ or dry ashing at 55°C	F	168
	Urine	Digestion with HNO$_3$, extraction	F	510
		Digestion with HNO$_3$ at 80°C	GF	164
		Digestion with HNO$_3$–H$_2$O$_2$ at 80°C	GF	166,442,443
		Digestion with HClO$_4$–H$_2$O$_2$	GF	162
		Digestion with HNO$_3$	CR	161
		Digestion with HNO$_3$–H$_2$SO$_4$–HClO$_4$ or dry ashing at 55°C	F	168
		Ashing at low temperature (O$_2$-plasma asher)	GF	165,178
		Dilution with HNO$_3$	GF	512
		Extraction	GF	409
Cobalt	Blood	Digestion with H$_2$SO$_4$–HNO$_3$	GF	150
	Urine	Ashing at 550°C, extraction	GF	411
		Ashing at 550°C, extraction	GF	412
		Digestion with HNO$_3$	GF	150

Element	Material	Treatment	Technique	References
Copper	Aqueous humour	None	GF	512
	Blood	Dilution with water	GF	102
		Dilution with Triton X-100	CR	393
		Extraction	F	408
		Heparinized	GF	108
	Plasma	Dilution with NH_3–EDTA, HPLC	GF	514
		Ultrafiltration	GF	111
		Dilution with water	F	388,515
	Saliva	Dilution with water	GF	499
	Serum	Dilution with water or alcohol	GF	384
		Dilution with acids	F	89,505, 516–518
		Dilution with $LaCl_3$	F	519
		Dilution with H_3PO_4–NH_4NO_3–Triton X-100	GF	112,520
		None	F	96,101
		Deproteination with trichloroacetic acid	F	85,86,404, 521
	Synovial fluid	Incubation with hyaluronidase	F	522
	Urine	Dilution with acid	F	182,505,517
		Dilution with H_3PO_4–NH_4NO_3–Triton X-100	GF	112
		Extraction	GF	410
		Extraction	F	85,523
Gold	Blood	Dilution with water or extraction	CR,GF	111
	Serum	Dilution with H_3PO_4–NH_4NO_3–Triton X-100	GF	520
		Dilution with water	GF	524
		Dilution with water	F	525
		Digestion with $KMnO_4$–HCl, extraction	GF	526
	Synovial fluid	Dilution with water	F	525
		Digestion with $KMnO_4$–HCl, extraction	GF	526
		Dilution with water	F	525
		Precipitation as telluride, dissolution and extraction	GF	471
Iron	Aqueous humour	None	GF	513
	Haemodialysate	Freeze-drying, dilution with water	F	183
		Dilution with water–Sterox SE	F	190
	Haemoglobin	None	GR	179,323
	Serum	Dilution with water	F	101,527
		Dilution with $LaCl_3$	F	519

Table 12.2. *continued*

Element	Biological fluid	Sample preparation	Atomization system	References
	Urine	Dilution with water	GF	195
		Dilution, reduction and deproteination	GF	528
		Deproteination	GF	180
		Digestion with H_2SO_4, extraction	F	529
		Dilution with HCl	GF	530
		None	GF	323
Lead	Blood	Dilution with water	GF	102,108,202, 231,497,530a
		Dilution with acids	CR	218
			ZAAS	474,531
		Dilution with NH_3,-isopropyl alcohol or NH_4NO_3	GF	197
		Dilution with Triton X-100	GF	100
			CR	229
			TR	469
			DC	219
		Dilution with Triton X-100, ashing at 500°C	GF	394
			GF	121,498
			GF	532
		Dilution with HNO_3 or $HClO_4$, KCN–citric buffer	HG	533
		Digestion with HNO_3 at 85–95°C	CR	123
		Digestion with HNO_3–HCl	DC	214
		Digestion with HNO_3–HF	CR	122
		Digestion with HNO_3–H_2O_2 at 250°C, extraction	GF	417
		Digestion with H_2O_2 at 140°C	DC	201
		Digestion with quaternary ammonium hydroxide	GF	456,534
		Ashing at low temperature	TR	137
		Extraction	GF	247,407,535
			TR	230
	Gastric contents	Addition of antifoaming agent	CR	421
		None	GF	536
	Urine	Digestion with HNO_3–H_2O_2 at 250°C, extraction	GF	537,538
		Dilution with H_2O, acids	GF	417
		Dilution with NH_4NO_3	CR	218
		Dilution with acids	GF	121
			ZAAS	531

Element	Material	Treatment	Technique	References
		Addition of phosphoric acid	GF	539
		Digestion with HNO_3 or dry ashing at 430°C	GF	142
		Digestion with HNO_3-H_2O_2 at 250°C, extraction	GF	417
		Extraction	F	148
			GF	407
		Treatment with Triton X-100, ashing at 500°C	GF	498
Lithium	Erythrocytes, plasma, serum	Dilution with water	F	174,505, 540–542
Magnesium	Cerebrospinal fluid	Dilution with water	F	28,47,543
	Erythrocytes	Deproteination with trichloroacetic acid and/or dilution	F	18,25,60, 508,544,545
	Plasma	Ultrafiltration, addition of $SrCl_2$	F	401
	Saliva	None	F	546
	Serum	Dilution with water	F	11,15,16,19, 27,29,59,61, 505,547–550
	Urine	Dilution with $LaCl_3$	F	398
		Dilution with KCl	F	397
		Dilution with KCl	F	32
		Dilution with water	F	43,505
Manganese	Blood	Digestion with $HClO_4$, extraction	GF	423
		None	GF	551
	Cerebrospinal fluid	None	GF	254
	Serum	None	GF	254
		Dilution with water	GF	387
	Urine	Dilution with water	GF	387
		Dilution with Triton X-100, extraction	F	552
		Digestion with $HClO_4$, extraction	GF	423
		Digestion with HNO_3, and Cummins' reagent	F	256
Mercury	Blood	Dilution with water, reduction	CV	553
		Dilution with acids, reduction	CV	554–557
		Concentration on copper powder, decomposition	GF	558
		Ashing with CaO at 600–900°C	GF	559
		Decomposition in a stream of oxygen	GF	560
	Urine	Deproteination, reduction	CV	561
		Decomposition, reduction	CV	562

Table 12.2. *continued*

Element	Biological fluid	Sample preparation	Atomization system	References
Molybdenum	Plasma	Ashing with CaO at 600–900°C	GF	559
		Dilution with acid, reduction	CV	554,556,557
		Deproteination, reduction	CV	561
	Plasma	Digestion with HNO_3-$HClO_4$, extraction	F	424
Nickel	Serum	Deproteination with trichloroacetic acid	F	403
		Deproteination, extraction	GF	414
		Deproteination or digestion with HNO_3-H_2SO_4, extraction	F	155,563
		Dilution with Triton X-100	GF	391
		Digestion with HNO_3-H_2SO_4-$HClO_4$, extraction	GF	413,426
		Ashing at 560°C, extraction	GF	159
	Urine	Deproteination, extraction	GF	414
		Digestion with acids, extraction	F,GF	155,413,426, 564
		Extraction	F	148
Palladium	Blood, urine	Digestion with HNO_3-$HClO_4$	GF	565
Platinum	Blood, urine	Digestion with HNO_3-$HClO_4$	GF	350,565
Potassium	Serum, urine	Dilution with water	F	505
Selenium	Blood	Digestion with HNO_3-$HClO_4$, reduction	HG	334
		Dilution with Triton X-100	GF	566
		Addition of rhodium	GF	567
		Extraction	GF	330
	Urine	Digestion with HNO_3-$HClO_4$-H_2SO_4, reduction	HG	337,487
Silicon	Blood, serum, urine	Dilution with water	GF	568
Silver	Blood	None	DC	468
Sodium	Serum, urine	Dilution with water	F	505,569
Strontium	Serum, urine	Co-precipitation of calcium oxalate, addition of $LaCl_3$	F	570

Element	Material	Treatment	Technique	References
Thallium	Blood, urine	Digestion with HNO$_3$–H$_2$SO$_4$, extraction, dilution	F	339,340
	Urine	Dilution	F	571
Tin	Blood, urine	Digestion with H$_2$SO$_4$–HNO$_3$–H$_2$O$_2$ at 250°C, extraction	GF	417
Vanadium	Blood, urine	Digestion with HNO$_3$–HClO$_4$, extraction	GF	431,432
	Urine	Digestion with HNO$_3$, extraction	GF	572
Zinc	Blood	Dilution with water	F	262,405,573,574
		Dilution with water	GF	102,108,497
		Dilution with Triton X-100	F	264
		Digestion with HNO$_3$–H$_2$O$_2$ at 250°C	F	417
	Cerebrospinal fluid	Digestion with HClO$_4$, ion-exchange	CR	285
	Gastric contents	Digestion with HNO$_3$–H$_2$O$_2$ at 250°C	F	417
	Haemodialysate	Dilution with citrate buffer, extraction	F	408
	Plasma	Dilution with water	F	262,264,388,575,576
		Dilution with water	CR	270
		Dilution with water-glycerol	F	265
		Deproteination with trichloroacetic acid	F	405
	Saliva	Dilution with water	GF	499
	Serum	Dilution with water	F	86,101,276,385,516,578-580
		Deproteination	GF	269
			F	80,271,384,581-583
	Synovial fluid	Addition of hyaluronidase	F	522
	Urine	Dilution with water	F	95,148,262,516,584,585
		Dilution with water	GF	269
		Digestion with HNO$_3$–H$_2$O$_2$ at 250°C	GF	417

GF = Graphite tube furnace
F = Flame
CR = Carbon rod
HG = Hydride generation
CV = Cold vapour

ZAAS = Zeeman flameless
DC = Delves cup
CB = Carbon bed
TR = Tantalum ribbon

12.1.2 Soft Tissues

Trace metals in brain, liver, kidney, cardiac, placental, pulmonary, spleenic, pancreatic and intestinal tissues are determined by AAS, in samples collected by dissection with stainless-steel or chrome-plated knives or scissors. The blood clots are removed and excess of fluids is blotted gently from the tissue. The fresh or freeze-dried tissue samples are dried at ~ 100°C and a 1–500 mg portion is used for subsequent pretreatment and measurement. The dried tissue is ground, homogenized and decomposed. The use of graphite-furnace atomizers permits decomposition of the organic matrix in the charring stage and thus avoids the pretreatment step. However, prior decomposition of some tissues has been recommended to ensure complete homogeneity of the sample. Table 12.3 summarizes some methods used for AAS determination of trace elements in mammalian soft tissues. Similar methods for plant tissues have been described [621-641].

12.1.3 Hair and Fingernails

Analysis of human hair offers a unique approach to the monitoring of trace metal levels in the body. The ease with which hair samples are collected and stored as necropsy and biopsy specimens and the accumulation of metals in hair at concentrations higher than those in blood, serum and urine, render this type of analysis useful in biomedical and forensic studies such as metal metabolism, pollution control, nutritional status, clinical diagnosis and suspicion of poisoning. However, sample preparation has been reported as one of the major problems in hair analysis. Because grease and dust on the hair increase the surface contamination by other metals (exogenous metals), a pretreatment to remove these metals without affecting the metals present within the hair structure (endogenous metals) is an important step.

Washing the hair with water, ionic and non-ionic surfactants, organic solvents, combinations of aqueous surfactants, organic solvents such as diethyl ether, and chelating agents has been proposed and all are known to differ in their effect. A mild wash with a 1-5% solution of ionic surfactant (e.g. sodium dodecyl sulphate) [642] or non-ionic surfactant (e.g. Triton X-100, 7X O-Matic) [643, 644] at or near the physiological pH is an acceptable treatment. The effect of different washing procedures and successive washing on the concentration of some trace metals in human scalp hair indicates that for every element there is a level below which the concentration cannot be reduced by further washing [643]. In some cases, however, the washing step is completely omitted, provided that interference or contamination risk from exogenous metals is minimal in the determination of the endogenous metals [645].

The basic quantitative scheme for trace-metal analysis of hair involves collection of 0.1-0.5 g of the hair sample, grinding in a micro ball-mill or segmentation into 0.1-1 cm lengths, washing, drying, followed by wet digestion with concentrated nitric acid alone [645-647] or mixed with perchloric acid in various proportions [648-650]. The digestion is done in a closed pressurized PTFE vessel to reduce both volatilization of some metals, and the blank values. Dry decomposition at 450-500°C followed by dissolution of the ash in a suitable acid may also be used [643,651]. It has been reported, however, that the best method for hair analysis involves direct introduction of 1-cm long samples of hair into the graphite-tube furnace and ashing at 1100°C before atomization. This technique minimizes sample-handling and contamination problems from the digestion acids, and allows the study of anatomical and longitudinal variation of trace elements within the same hair [652]. The technique has been adapted by Renshow *et al.* [653,654] for the determination of copper and lead and by Alder *et al.* [652,655,656] for the simultaneous determination of up to 13 elements. The carbon rod atomizer has similarly been used for the determination of cadmium [657].

Various metals have been determined in hair by AAS either individually or simultaneously, by the flame [644,658-667], electrothermal [365,646,650,651, 653,654,668-671] and cold-vapour [647,649,672] techniques (Table 12.4). However, the major difficulties in the assessment of many of these methods are: (a) the absence of a reference standard of a similar nature and composition (an attempt to use silk was not successful [673]); (b) the change of metal concentration from hair to hair and along the length of the same hair; (c) the amount of metal leached, owing to the nature of the washing reagent and washing time. Some of these problems can be overcome by sampling from a large bulk of hair and by using at least two different methods of analysis for comparison.

On the other hand, some interesting conclusions have been drawn from the AAS determination of various metals in hair.

1. The level of certain metals can reflect community exposure trends within a single metropolitan area [665].
2. Systematic differences in the average trace-metal concentrations in hair from different anatomical sites are difficult to substantiate [652,674].
3. Trace-element correlation exists only within one individual's hair and similar longitudinal trace-element profiles can be found within an individual [652].
4. Interpretation of data obtained from multielement analysis appears to have potential in the field of discrimination between individuals [656].
5. Sex differentiation by means of the metal levels of the scalp, and differences between vegetarians and non-vegetarians may be observed [675,676].

Trace metals in fingernails have been determined by AAS. The occurrence of white spots in the fingernails of many children, teenagers and a few adults, and their possible relationship to zinc deficiency has been investigated by use of electrothermal atomization AAS [678]. The nail samples, after washing with acetone and drying, are either decomposed by treatment with concentrated

Table 12.3. AAS determination of some trace elements in mammalian tissues

Element	Tissue	Sample preparation	Atomization system	References
Aluminium	Brain, liver	Dry ashing at 500–700°C	GF	438,586
	Lung	Digestion with HNO_3–HCl–HF at 150°C	GF	587
	Miscellaneous	Digestion with HNO_3–$HClO_4$ at 110°C	GF	438
Antimony	Liver	Digestion with HNO_3–H_2SO_4–$HClO_4$	GF	588
		Digestion with HNO_3–H_2SO_4–H_2O_2	GF	589
Arsenic	Liver, kidney	Ashing at 550°C	GF	321
	Miscellaneous	Digestion with HNO_3–H_2SO_4	GF	441
		Digestion with HNO_3–$HClO_3$–$HClO_4$ at 140°C	GF	590
		Digestion with HNO_3–$HClO_3$–H_2SO_4, extraction	GF	306
Barium	Pancreatic	Direct injection of freeze-dried samples	GF	591
Bismuth	Brain, liver	Ashing at low temperature with plasma asher	GF	361
Cadmium	Brain, liver, lungs	Digestion with HNO_3 at 135°C	F	440
	Liver	None	ZAAS	592
	Liver, kidney	Ashing at < 600°C	F	593
	Kidney cortex	Digestion with HNO_3	ZAAS	260
	Liver, kidney	Digestion with crystalline subtilisin A	F	462
	Lung	Digestion with HNO_3 at 140°C	CR	436
		Ashing at low temperature (plasma asher)	GF	450
	Miscellaneous	None	GF	594
		Ashing at 450°C, extraction	F	448,449
		Digestion with HNO_3–H_2SO_4	GF	595
		Homogenization with tetramethylammonium hydroxide	F	119
Calcium	Brain	Solubilization with Soluene-350	F	459
	Cardiac and skeletal	Digestion with HNO_3 at 68°C	F	596
	Liver	Digestion with HNO_3–$HClO_4$ at 90–120°C	F	437
		Digestion with H_2SO_4–HNO_3–H_2O_2, or dry ashing	F	451
	Kidney	Homogenization with butanol–HCl–$LaCl_3$ or dry ashing	F	65
	Pancreas	Washing with buffer	GF	597
	Placenta and muscles	Digestion with H_2SO_4–H_2O_2	GF	598
	Miscellaneous	Ashing at 300–400°C	F	449

Element	Tissue	Procedure	Technique	Reference
Chromium	Brain, lung, liver, intestine	Digestion with HNO_3 at 135°C	F	440
	Liver	Digestion with HNO_3–$HClO_4$–H_2SO_4, extraction	GF	418
	Liver	Digestion with HNO_3–$HClO_4$	GF	599
	Lung	Ashing with plasma asher	GF	450
	Miscellaneous	Digestion with HNO_3 at 150°C	GF	594,600
Cobalt	Kidney	Digestion with HNO_3, extraction	GF	601
	Liver	Dry ashing, extraction	F	419
	Lung	Ashing with plasma asher	GF	450
Copper	Brain	Digestion with H_2SO_4–H_2O_2	GF	435
	Brain	Digestion with HNO_3–$HClO_4$	F	17,94,602
		Ashing at 450–600°C	F	79,603
		Homogenization with Soluene 350	CR	459
	Kidney	Digestion with HNO_3	ZAAS	260
	Kidney, liver	Digestion with crystalline subtilisin A	F	462
	Liver	Direct atomization of the homogenate	F	466
	Lung	Ashing with plasma asher	GF	450
	Lung, intestine	Digestion with HNO_3 at 135°C	F	440
	Miscellaneous	Digestion with HNO_3 at 150°C or HNO_3 + H_2O_2	F, GF	594,600, 604,605
		Digestion with tetramethylammonium hydroxide	F	119
Gallium	Liver	Digestion with HNO_3	GF	606
Gold	Kidney, liver, spleen	Digestion with H_2SO_4–HNO_3 at 70°C	GF	607
	Miscellaneous	Digestion with Soluene-100	CR	352
Iron	Aorta	Digestion with HNO_3 at 60°C	GF	439
	Brain	Digestion with Soluene-350	F	459
	Kidney, liver	Dry ashing at < 450°C	F	593
	Liver	Digestion with HNO_3–$HClO_4$ at 225°C	F	608
	Liver	Digestion with H_2SO_4–HNO_3–H_2O_2 or ashing at 500°C	CF	451
	Lung	Ashing with plasma asher	GF	450
	Spleen	Extraction	F	609
	Miscellaneous	Ashing at 300°C	F	449
	Miscellaneous	Digestion with HNO_3 at 150°C	F	594

Table 12.3. *continued*

Element	Tissue	Sample preparation	Atomization system	References
Lead	Aorta, brain, heart, kidney, liver	Digestion with HNO_3, extraction	F	422
	Brain, intestine, liver, lung	Digestion with HNO_3 at 135°C	F	440
	Kidney cortex	Dissolution in HNO_3	ZAAS	260
	Kidney, liver	Digestion with subtilisin A	F	462
	Kidney, liver, lung	Homogenization with water	F	610
	Liver	Digestion with H_2SO_4, then calcination at 650°C	F	464
		Direct atomization	ZAAS	592
	Lung	Digestion with HNO_3 at 140°C	GF	436
		Ashing with plasma asher	GF	450
		Freeze-drying and atomization	GF	611
	Miscellaneous	Digestion with HNO_3 then ashing at 300–400°C	F	449
		Digestion with HNO_3 at 150°C	GF	594
		Digestion with tetramethylammonium hydroxide	F	119,612
Magnesium	Brain	Digestion with HNO_3–$HClO_4$	F	10,17,94
	Cardiac, skeletal	Digestion with HNO_3 at 68°C	F	596
	Liver	Ashing with plasma asher	F	451
	Myocardial	Digestion with Soluene-100, extraction	F	458
	Miscellaneous	Digestion with HNO_3 at 400°C	F	449
Manganese	Brain	Digestion with HNO_3 at 60°C	GF	433
		Digestion with Soluene-350	CR	459
	Kidney cortex	Digestion with HNO_3	ZAAS	260
	Liver	Digestion with Triton X-100 at 60°C	GF	461
		Homogenization	F	466
		Digestion with H_2SO_4–H_2O_2	GF	435
	Lung	Ashing with plasma asher	GF	450
	Miscellaneous	Digestion with HNO_3–$HClO_4$, extraction	F	434
		Solubilization	GF	600
Mercury	Liver	Digestion with HNO_3–V_2O_5	CV	447
		Reduction	GF	613

Element	Tissue		Technique	
	Liver, lung, intestine	Digestion with HNO_3	F	440
	Miscellaneous	Reduction	CV	614
		Pyrolysis in stream of N_2	GF	615
Nickel	Brain, intestine, liver, lung	Digestion with HNO_3 at 135°C	F	440
	Lung	Ashing with plasma asher	GF	450
	Palatine tonsils	Digestion with HNO_3–$HClO_4$–H_2SO_4 at 250°C	GF	425
	Miscellaneous	Digestion with HNO_3 or HNO_3–H_2SO_4–$HClO_4$	GF, F	594,616
Platinum	Kidney, liver	Digestion with HNO_3	GF	617
Praseodymium	Kidney, liver	Digestion with tetramethylammonium hydroxide at 70°C	GF	460
Rubidium	Liver	Digestion with acids	F	618
Selenium	Kidney, liver	Digestion with acids	F	618
	Liver	Digestion with HNO_3–H_2SO_4	HG	519
Silver	Kidney, liver, lung, pancreas	Extraction	F	427
Thallium	Liver	Digestion with crystalline subtilisin A	F	462
Tin	Brain, kidney, liver	Digestion with Soluene-350 at 65°C	F	457
	Miscellaneous	Digestion	GF	620
Zinc	Brain	Digestion with HNO_3–$HClO_4$	F	17,94
	Brain, intestine, liver, lung	Digestion with HNO_3 at 135°C	F	440
	Liver	Digestion with H_2SO_4–HNO_3–H_2O_2	F	451
		Homogenization	F	466
			ZAAS	592
	Miscellaneous	Ashing at 450–600°C	F	79,603
		Digestion with tetramethylammonium hydroxide	F	119
		Digestion with acids	GF	600

GF = Graphite tube furnace   HG = Hydride generation
CR = Carbon rod   ZAAS = Zeeman flameless technique
F = Flame   CV = Cold vapour

Table 12.4. AAS determination of some trace elements in hair

Element	Decomposition method	Atomization method	References
Aluminium	Wet (HNO_3)	GF	655
	None	GF	655,669
Beryllium	Dry (500°C)	GF	651
	Wet ($HNO_3 + HClO_4$)	GF	650,671
	Wet (HNO_3)	GF	647
Bismuth	Wet (HNO_3)	GF	655,669
	None	GF	655,669
Cadmium	Wet ($HNO_3 + HClO_4$)	GF	649
	None	CB	657
	Dry (450°C)	F and GF	643
Chromium	Wet (HNO_3)	GF	655,669
	None	GF	655,669
	Wet	GF	649
Cobalt	Wet (HNO_3)	GF	655,669
	None	GF	655,669
Copper	Wet ($HNO_3 + HClO_4$)	GF	642
	Dry (450°C)	F and GF	643
Iron	Wet (HNO_3)	GF	646
	Wet ($HNO_3 + HClO_4$)	F	644
	None	GF	655,669
	Wet (HNO_3)	F and GF	655,669
	Dry (450°C)	F and GF	643
Lead	Wet ($HNO_3 + HClO_4$)	GF	648,649
	None	GF	655
Magnesium	Wet ($HNO_3 + HClO_4$)	F	644
Manganese	None	GF	655
	Wet (HNO_3)	GF	646,655,669
	Dry (450°C)	F and GF	643
Mercury	Wet (HNO_3)	CV	649,672
	Wet ($HNO_3 + HClO_4$)	CV	649,667
Nickel	Wet (HNO_3)	GF	655,669
	None	GF	655,669
Silicon	Wet (HNO_3)	GF	655
	None	GF	669
Silver	Wet (HNO_3)	GF	655,669
	None	GF	655,669
Thallium	Wet (HNO_3)	GF	645
Zinc	Wet ($HNO_3 + HClO_4$)	F	642,644,649
	None	GF	655
	Dry (450°C)	F and GF	643

GF = graphite furnace, F = Flame, CB = Carbon bed, CV = Cold vapour nitric acid at 65°C for 1 hr or placed in a tantalum boat and directly introduced into the graphite furnace for drying at 125°C (30 sec), ashing at 500°C (30 sec) and atomization at 1900°C (10 sec). It has been observed that atomization of solid samples shows a double peak for zinc, probably owing to the presence of two discrete forms of zinc which volatilize and dissociate at different rates. This is not observed for the same sample after wet digestion. Beryllium in nails is determined by digesting the samples with nitric–perchloric acid mixture and treatment with lanthanum as a releasing and enhancement reagent [650,671]. The furnace programme used is 100°C (50 sec) for drying, 1000°C (60 sec) for ashing and 2600°C (10 sec) for atomization. Contamination of dental personnel with mercury can also be monitored from their fingernails. The nail samples are decomposed by combustion in an oxygen-filled flask containing 50% nitric acid as absorbent and the mercury ions are reduced to the metal, which is passed through a gas-cell and measured by the cold-vapour technique [670].

12.1.4 Teeth and Bones

Correlation of the concentration and distribution of certain trace elements in dental enamel and dentine with the occurrence of dental caries has been extensively investigated by various instrumental techniques, including AAS. As a general procedure, the enamel is isolated by mechanical means, washed, dried and pulverized and a portion of the powder is dissolved in a suitable acid and atomized. Owing to the complex matrix of the tooth, some interferences may be expected. Determination of strontium in human tooth enamel by flame AAS is influenced by the calcium and phosphate matrix of the enamel. If the air-acetylene flame is used, the interference can be reduced by dissolving the sample in perchloric acid and adding lanthanum [679]. With electrothermal atomization, however, hydrochloric acid is used for sample dissolution since perchloric acid significantly reduces the absorption of strontium and results in non-reproducible absorption [680].

Calcium, magnesium and carbonate have been determined in dental enamel and dentine by acidifying the sample powder with hydrochloric acid and absorbing the liberated carbon dioxide in alkaline barium chloride. The barium carbonate precipitated is isolated, and the excess of barium (in the filtrate) is determined by atomic-emission spectrometry; the acidified solution is analysed for calcium by atomic-emission and for magnesium by AAS [681]. Mercury in roots of teeth and in jaw bones has also been determined by dissolving the sample in nitric-hydrofluoric acid mixture at 180°C in a PTFE bomb. The solution is then treated with boric acid and the mercury measured by the cold-vapour technique [682]. Electrothermal atomization has been used for the determination of zinc, copper, aluminium, iron, lead and manganese in the surface enamel [683].

Langmyhr and Sundli [684] determined cadmium and lead in whole erupted or unerupted human teeth by direct atomization of the solid sample. The pulverized sample (0.5-3 mg) is applied to a graphite furnace with a programme of 150°C (60 sec) for drying, 350°C (60 sec) for ashing and 1400-1600°C (30 sec) for atomization [684]. Direct atomization from powdered solid dental samples mixed with graphite for the determination of manganese, silver and zinc has also been reported [685]. The possibility of using the carbon-cup atomization technique for the determination of aluminium in dental enamel has been investigated. Increasing the rate of heating of the carbon cup in the atomization step and the presence of calcium phosphate in the matrix resulted in an enhancement of the aluminium absorption [686]. The sensitivity of the method (3.5 × 10^{-11} g of Al) permits *in vitro* and *in vivo* measurements.

Determination of trace elements in bone is commonly approached by washing the sample with diethyl ether to remove lipids, and drying at 110°C, followed by decomposition. Wet digestion with either nitric acid [20] or nitric-perchloric acid mixture [687,688], dry-ashing at 500-600°C [603,689] and digestion with nitric acid-hydrogen peroxide followed by ashing at 350-400°C [690] have all been used for the decomposition. This is followed by either extraction of the element of interest with a suitable organic reagent or direct atomization of the bone digest. Since the presence of calcium and phosphorus in the bone matrix affects measurement of some metals, a separation step involving extraction, ion-exchange or precipitation may be used.

Ishino *et al.* [690] described a method for the determination of copper, cadmium and zinc by extraction with ammonium pyrrolidine dithiocarbamate in methyl isobutyl ketone (MIBK) and aspiration or the organic phase into an air-acetylene flame. Beryllium is determined by extraction into acetylacetone-MIBK mixture in the presence of EDTA, and back-extraction with hydrochloric acid, followed by electrothermal atomization [689]. Okano *et al.* [688] determined trace quantities of cadmium in bone by prior separation on Dowex 50W-X8 resin and flame-AAS measurement. Separation of calcium as calcium sulphate and phosphorus as molybdophosphate before extraction of cadmium with diethyldithiocarbamate has also been attempted. On the other hand, direct atomization of the solid samples has been advocated by some workers. A tungsten crucible made from thoriated tungsten rod is used for electrothermal atomization of copper in calcined bone samples [691]. Cadmium, lead and manganese have also been determined [692]. The interference effect of chloride on lead and manganese is minimized by heating the sample twice in the graphite furnace with nitric acid at ~ 100°C for 1 min before ashing and atomization.

12.2 METALLOPROTEINS

The distribution of metals bound to the various fractions of proteins in biological materials has been evaluated by AAS after a suitable separation step. Fractionation of metalloproteins by gel permeation chromatography and electrophoresis is successfully used in conjunction with AAS.

12.2.1 Chromatographic AAS Methods

Iron proteins in needle biopsy specimens of human liver are separated on a carboxymethylcellulose column and the iron-containing protein fractions are eluted with citrate buffer. The transferrin, ferritin, haemprotein and haemosiderin fractions are analysed for iron by AAS [693]. Gradient chromatography on a DEAE-cellulose column is combined with AAS for the study of the distribution of iron, zinc, manganese, nickel, calcium, magnesium and strontium in dialysed human serum [694]. Heparinized plasma from normal adults is chromatographically fractionated on DEAE-Sepharose CL-6B, with elution by 'tris'–hydrochloric acid buffer [695]. The solution is fed to the top of the column by a variable speed peristaltic pump and the atomic absorption of sodium, potassium, magnesium, copper and zinc in the eluate is measured. This study indicates that over 90% of the copper in human plasma is firmly bound to α_2-globulin and caeruloplasmin, and the zinc is mainly bound to globulins, although small amounts of both metals are bound to albumin.

Protein-bound zinc and copper are separated on Sephadex G-100 and measured by electrothermal atomization AAS [287]. The zinc is associated with α_2-macroglobulin and albumin whereas the copper is associated with caeruloplasmin. Separation of protein-bound chromium by Sephadex G-25, followed by electrothermal atomization allows differentiation between protein-bound tervalent and sexivalent chromium, and chromium(III) formed by reduction of chromium(VI) by thiol-containing proteins [696]. Protein fractions separated by gel permeation may also be analysed by the graphite-furnace technique [697, 698]. Metallothionins, relatively small molecular-weight proteins bound to metals, have also been determined by HPLC with a gel permeation column. The outlet of the column is directly connected to the nebulizer unit of the AAS instrument [699].

12.2.2 Electrophoresis AAS Methods

Delves fractionated protein-bound copper in plasma or serum by electrophoresis on cellulose acetate at 8 mA and 30 V/cm. The protein bands were fixed by heating the electrophoretogram at 100°C and introduced directly into a graphite furnace for AAS measurement of copper at 324.7 nm [700]. About 95% of the serum copper was found to be associated with the α_2 fraction and a relation between the copper content and some disease states (Menkes' syndrome) was outlined. A similar procedure was described by Peape et al. [701] for measuring the copper content of the different fractions of serum protein. Other workers [702] fractionated the protein on cellulose acetate, digested the separated bands with 10% nitric acid in a PTFE bomb, and measured copper, lead and iron by electrothermal atomization AAS.

Combination of gel filtration and electrophoresis allows the separation and identification of individual protein fractions which can subsequently be analysed for gold [703]. By use of cellulose acetate paper or cellogel, the serum protein can be fractionated into γ, β, α_1 and α_2 globulins and albumin, the gold content of which can easily be measured by electrothermal atomization AAS. Foote and Delves [704] demonstrated that separation of protein fractions on a cellulose acetate membrane is almost complete. Albumin and non-albumin fractions give reasonably good agreement with the total zinc concentration measured by flame AAS. Ultracentrifugation of serum gives a protein-containing fraction (A) with a molecular weight > 10000 and a protein-free fraction with molecular weight < 10,000. Fraction A is further fractionated by isotachophoresis, in which the sample is treated in polyacrylamide gel at physiological pH, and beryllium is measured in each fraction by electrothermal atomization AAS [705].

12.3 PROTEIN-THIOL

It appears that one of the most specific and sensitive methods for the determination of protein-thiol is the reaction with organic mercurials, followed by AAS. p-Hydroxymercuribenzoate in 100-fold molar excess is allowed to react at $4°C$ with the protein samples in a phosphate or 'tris'-hydrochloric acid buffer at pH 7.5 and the progress of the reaction is followed by measuring the absorbance at 255 nm. After equilibrium is reached, usually within 3-24 hr, depending on the nature of the protein, the protein–p-hydroxymercuribenzoate mixtures are exhaustively dialysed against a buffer to remove the excess of unreacted reagent, and the mercury content of the dialysed fraction is measured by AAS [706]. Mercury(II) chloride or p-chloromercuribenzoate can equally well be used under the same conditions instead of p-hydroxymercuribenzoate, but they give significantly lower thiol values, probably because their binding constants are small.

Gel filtration may be used to remove the excess of reagent and has proved to be a more efficient and rapid technique. Under suitable conditions, Sephadex G-50 retards the mercury reagent, which emerges in later fractions of the eluate. The mercury–protein fractions are diluted with ethanol or acetone and directly aspirated into a hydrogen–air flame. The use of a long absorption path gives sensitive and precise results with a variety of protein samples (Table 12.5). The sensitivity of this method was later improved by digesting the protein–mercury complex before measurement by electrothermal atomization AAS [707]. The digestion was performed with $8.5M$ nitric acid at $55°C$ for 1 min, followed by heating with $18N$ sulphuric acid for 15 min and then with 5% potassium permanganate solution. The results obtained with native and denatured proteins are, in general, slightly lower than those obtained by using flame AAS with undigested protein–mercury complex (Table 12.5).

12.4 GLYCOSAMINOGLYCANS

Glycosaminoglycans (GAG) are found in most non-connective tissues at extremely concentrations. Binding of the Alcian Blue moeity of the GAG–Alcian Blue

Table 12.5. AAS determination of protein-thiol by reaction with p-chloromercuribenzoate at 4°C (reprinted by permission of the copyright holders, Academic Press.)

Protein	Buffer	Dialysis [706]		Gel filtration [707]	
		Concentration (a) (μmole/ml)	Mole–SH/mole protein	Concentration (a) (μmole/ml)	Mole–SH/mole protein
Bovine serum albumin	0.1M 'Tris'-HCl, pH 7.5	0.020 0.005	0.67 0.67	0.010 0.001	0.61 0.63
β-Lactoglobulin	0.1M 'Tris'-HCl, pH 7.5	0.030 0.0025	1.83 1.96	0.010 0.001	1.72 1.67
Ovalbumin	0.1M 'Tris'-HCl, pH 7.5	0.0500 0.0017	2.86 2.94	0.010 0.001	2.50 2.66
	0.1M Sodium phosphate, pH 7.5	–	–	0.010	3.73
Alcohol dehydrogenase	0.1M 'Tris'-HCl, pH 7.5	0.0050 0.00016	26.80 26.80	0.001 0.001	22.30 28.50
	0.1M Sodium phosphate, pH 7.5				
Haemoglobin	0.1M Sodium phosphate, pH 7.5	0.05000 0.00300	1.60 1.72	– –	– –
Papain	0.33M Acetate, pH 4.6	0.03000 0.01000	0.50 0.47	– –	– –
Lysozyme	0.1M 'Tris'-HCl, pH 7.5	–	–	0.010	0.04

(a) Final protein concentration in the incubation mixture.

complex to copper allows determination of GAG by AAS [708]. It has been demonstrated that the amount of Alcian Blue bound to purified chondroitin sulphate (CS), hyaluronic acid (HA), and heparin sulphate (HS), is linearly related to the GAG concentrations. The lower limit of detection ranges between 15 and 190 ng. Mixtures containing CS, HA and HS are measured similarly after separation by electrophoresis on cellulose acetate. Keratan sulphate and dermatan sulphate are not sufficiently resolved from each other or from hyaluronic acid. Two-dimensional electrophoresis may be used for their separation.

12.5 ENZYME ASSAY

Microsomal epoxide hydrolase enzyme (E.C.3.3.23) is known to catalyse the hydrolysis of epoxides into 1,2-diols. When *trans*-styrene oxide is used as a substrate at pH 9, styrene glycol is formed, which can be oxidized with potassium periodate; addition of lead nitrate then permits preferential precipitation of the excess of periodate and separation of the precipitate and AAS measurement of its lead content can be used for determination of the enzyme activity [709]. The hepatic microsomal activities of rabbit and fish are determined by use of 2-4 mg of microsome protein, dilution to 3.0 ml with 'tris' buffer of pH 9 containing 0.2% Tween-80, and initiation of the reaction by addition of 8 μmole of *trans*-styrene oxide in methanol. After an incubation period of 60 min, 1.0 ml of 20mM potassium periodate is added followed by 1.0 ml of 40mM lead nitrate. The precipitate is isolated and dissolved in 1M nitric acid, then the lead is measured by AAS [710]. The method is less sensitive than the radiometric of HPLC method, but it is sufficient for measuring enzyme activities in a range equivalent to 0.1-7.5 nmole of styrene glycol/min/mg of protein [709,710].

12.6 IMMUNOASSAY

The era of modern immunochemistry began in 1929 with the introduction of precise methods for quantitative measurement of antibodies and antigens. These methods are based on precipitation or complexation of the antigen with the antibody, followed by fluorimetric and enzymatic measurements. An interesting system for immunochemical studies by use of AAS has been suggested [711]. It is based on the use of organometallic complexes as labelling agents for some antigens such as oestrogens, barbiturates and cannabinoids (see Fig. 12.1). For measuring the antigen level (Ag) the following steps are used: (a) immobilization of the antibody (Ab) of the antigen on a solid support; (b) preparation of a metal-labelled antigen (Ag-M); (c) mixing of the antigen in the unknown test solution with the standard metal-labelled antigen, followed by incubation with antibody.

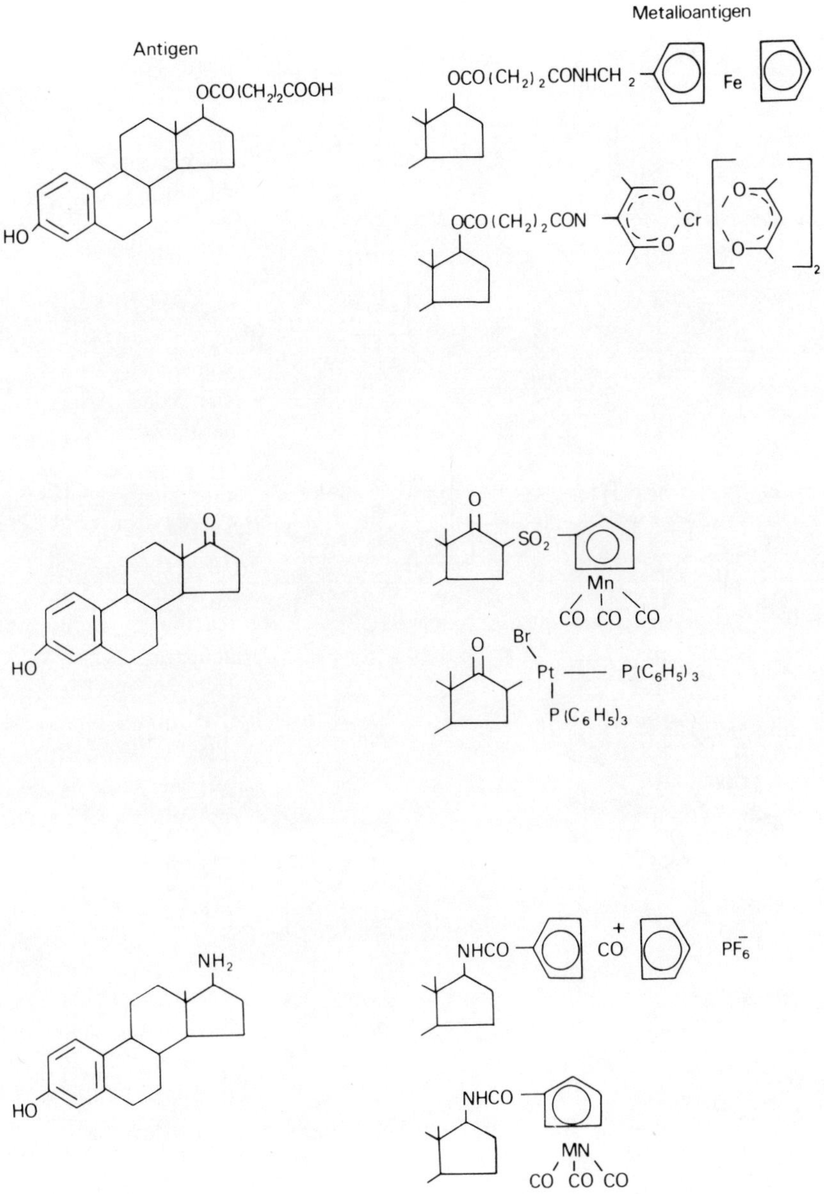

Metalioantigen

Antigen

(continued overleaf)

Fig. 12.1 Examples of some metallo-antigens [reprinted by permission from M. Cais *et al.* *Nature,* Vol. 270, pp. 534-5. Copyright © 1977 Macmillan Journals Ltd.]

Fig. 12.1–*continued*

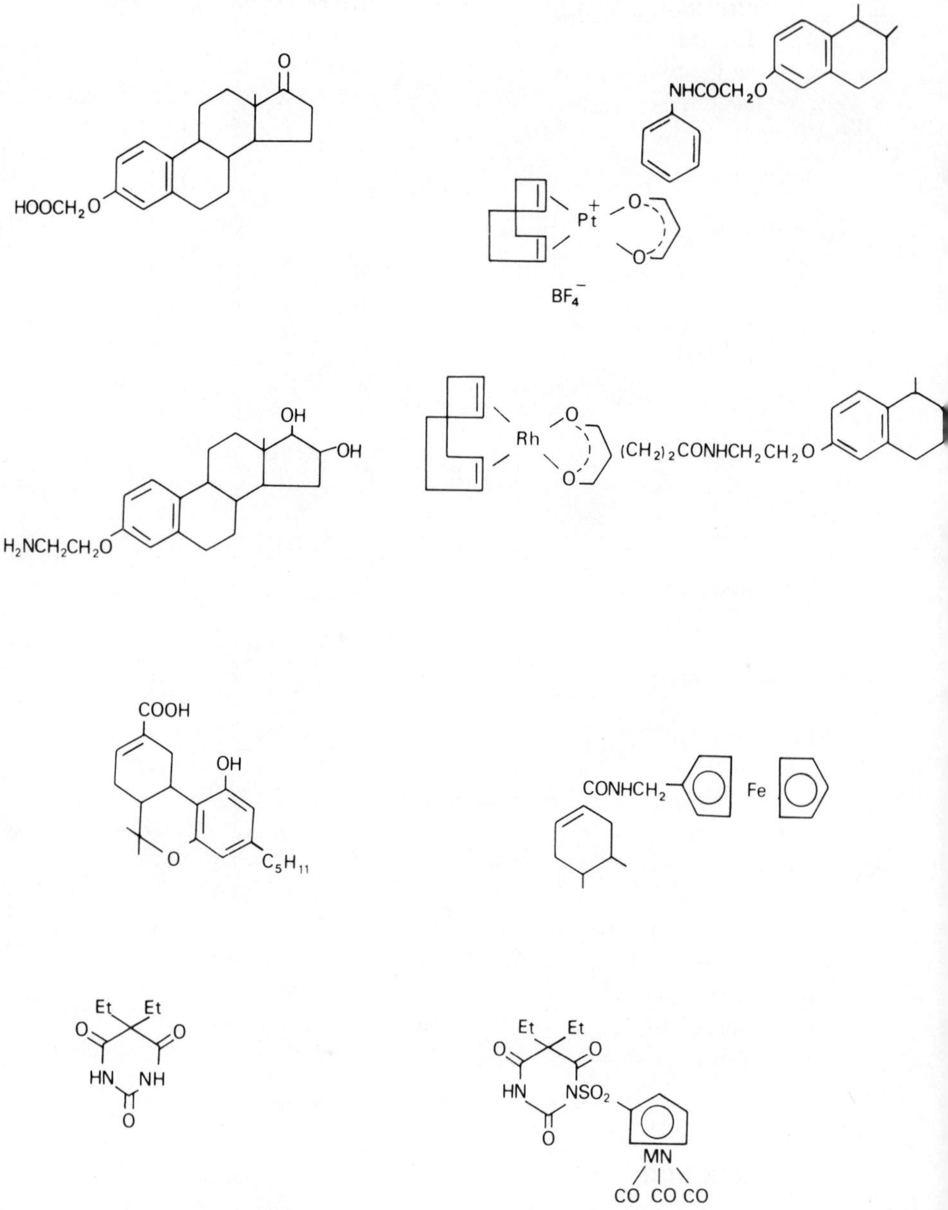

The standard metal-labelled antigen thus competes with the unlabelled form in the test solution for binding of the limited amount of antibody on the solid phase. Separation of the solid and soluble phases permits AAS measurement of the metal content in the soluble phase, as a function of the amount of antigen (Fig. 12.2). The feasibility of this approach has been demonstrated by preparation of the antibody of some steroid antigens by injecting rabbits with a bovine serum albumin (BSA) conjugated with the steroids. The antibody is isolated and immobilized on activated Sepharose 4B by reaction with cyanogen bromide.

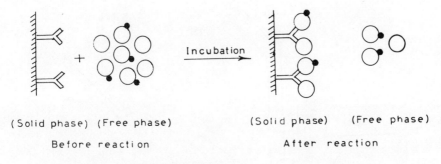

(Solid phase)   (Free phase)   (Solid phase)   (Free phase)

Before reaction   After reaction

Fig. 12.2. Competitive reaction of the antibody (⊂⟨) with antigen (◯) and metal labelled antigen (◯⬤).

Then the unknown steroid solution is mixed with the iron-antigen derivative and incubated for 30 min at room temperature with the Sepharose-bound antibody. The mixture is centrifuged for 1 hr at 3000 rpm and a 20–30 μl portion of the supernatant liquid is injected into the graphite furnace of the atomic-absorption spectrometer. The amount of free metallo-antigen is measured and compared with a calibration graph prepared from standard metallo-antigen solution in $0.1M$ phosphate buffer of pH 7.3 and containing 15% of dimethylformamide. With electrothermal atomization AAS, a calibration graph of the metal in the range 2–20 ng/ml corresponds to a 20–500 ng/ml concentration range for the unlabelled antibody. This concentration range is suitable for the determination of a relatively large number of urinary metabolites such as oestriol (in pregnancy urine), morphine, barbiturates, amphetamine and cocaine, without prior extraction or concentration steps. This concept may also be applied to other antibody-antigen systems.

12.7 HISTOCHEMISTRY

The use of AAS in histochemistry involves microdetermination of metals located in different parts of the tissue by direct survey under the microscope [712]. A spectrophotometer with an optical microscope and laser oscillator has been devised for this purpose (Fig. 12.3). In principle, the metal in a very small target area of the tissue is vaporized and atomized by irradiation with a narrow laser beam (under microscopic survey) and measured by electrothermal atomization AAS. The method has been utilized for determining cadmium in human kidney cortex. A piece of kidney cortex tissue is frozen in a slush of solid carbon

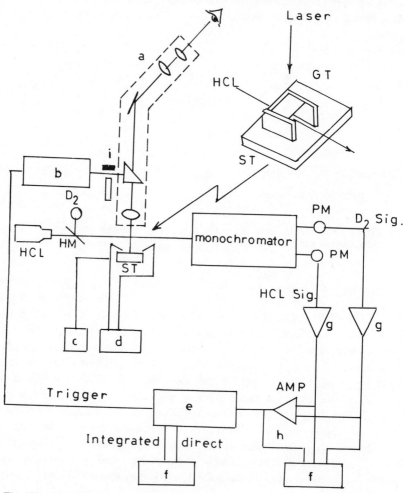

Fig. 12.3. Block diagram of laser atomic-absorption spectrometer: a, microscope; b, neodymium laser; c, metal-heater power supply; d, spark power supply; e, data memory; f, recorder; g, log amplifier; h, HCL–D_2 signal; i, diaphragm iris; ST, stage for sample or metal heater; GT, glass tunnel; HM, half mirror; HCL, hollow-cathode lamp; D_2, deuterium lamp. [Reprinted by permission from K. Sumino, R. Yamamoto, F. Hatayama, S. Kitamura and H. Itoh, *Anal. Chem.*, **52**, 1064 (1980). Copyright 1980 American Chemical Society.]

dioxide and n-pentane, and a slice 6 μm thick is sectioned in a cryostat. This section is fixed on a glass slide by dropping 0.1% phenol solution in alcohol on it and drying at room temperature, followed by staining with 0.2% Methylene Blue-0.5% gelatine solution. After drying, the thickness of the section becomes approximately 4 μm.

The tissue section at this stage is examined under the optical microscope and the crater sites for the particular material and target element are located. Then a laser beam (from a neodymium glass laser) is focused onto the selected areas in the sample. The metals in the tissue are vaporized almost completely by the laser shot. The radiation from a hollow-cathode lamp is passed through the atom cloud created by the laser shot, and then projected through the entrance slit of the spectrometer. Three signals, one from the hollow-cathode lamp, one from a deuterium lamp used for background correction, and the third the difference between them, are received by the photomultiplier, amplified, and recorded by a three-pen high-speed recorder. The intensity of the net absorption signal is directly proportional to the concentration of the target element.

This method can be used satisfactorily for measuring cadmium, mercury and zinc with high sensitivity [712]. Samples containing trace levels of these metals are covered by a small glass tunnel (1 × 2 cm, 5 mm in height) and placed on the graphite tube to maintain the atomic atmosphere as long as possible, and thus increase the sensitivity. An auxiliary attachment consisting of a pair of spark-discharge electrodes can also be used to facilitate vaporization of elements which require high temperature for atomization, such as copper, iron and chromium.

12.8 CELL PHYSIOLOGY

Sodium, potassium, magnesium, calcium and hydrogen ions are known to compete with each other for membrane binding sites. The important role of each of these ions in living systems and their binding to cellular membranes have been investigated by use of AAS. Rat-liver cell microsomes are isolated by differential centrifugation in 0.25M sucrose solution and dialysed against a citric acid medium to remove sodium, potassium, magnesium and calcium originally present. The microsomes are then suspended and equilibrated in a medium having a chosen composition of these cations, followed by washing at high speed, centrifugation and ashing. The ash is dissolved in hydrochloric acid and aliquots of the solution are diluted, treated with 15mM lanthanum to eliminate the effect of phosphate, or 4mM caesium to suppress the ionization of sodium and potassium, and analysed by AAS [713]. The effect of adenosine triphosphate (ATP) on the binding of sodium has been investigated similarly and the data demonstrate that ATP significantly decreases the binding of bivalent cations by complexing them, and that there is a remarkable increase in the binding of univalent cations.

302 **Biological Materials** [Ch.

REFERENCES

[1] B. R. Little, S. R. Platman and R. R. Fieve, *Clin. Chem.*, **14**, 1211 (1968).
[2] W. Reichel and B. G. Bleakley, *Anal. Chem.*, **46**, 59 (1974).
[3] R. Herrmann and W. Lang, *Z. Ges. Exptl. Med.*, **134**, 268 (1961).
[4] J. V. Sullivan and A. Walsh, *Spectrochim. Acta*, **22**, 1843 (1966).
[5] R. M. Forbes, *J. Nutr.*, **88**, 403 (1966).
[6] F. Bek and V. Sychra, *Chem. Listy*, **66**, 78 (1972).
[7] M. W. Gaumer, S. Sprague and W. Slavin, *Atom. Absorp. Newslett.*, **5**, 58 (1966).
[8] K. Szivós and E. Pungor, *Acta Pharm. Hung.*, **44**, 253 (1974).
[9] F. Burdevald, H. G. McCann and P. Graen, *Arch. Oral Biol.*, **13**, 877 (1968).
[10] T. L. Chang, T. A. Gover and W. W. Harrison, *Anal. Chim. Acta*, **34**, 17 (1966).
[11] R. J. Hurst and R. R. McSwiney, *Hilger J.*, **8**, 25 (1963).
[12] M. A. MacDonald and L. Watson, *Clin. Chim. Acta*, **14**, 233 (1966).
[13] C. E. Payne and H. F. Combs, *Appl. Spectrosc.*, **22**, 786 (1968).
[14] W. K. Stewart, F. Hutchinson and L. W. Fleming, *J. Lab. Clin. Med.*, **61**, 858 (1963).
[15] J. B. Willis, *Nature*, **184**, 186 (1959).
[16] J. B. Willis, *Spectrochim. Acta*, **16**, 273 (1960).
[17] D. B. Cheek, J. E. Graystone, J. B. Willis and A. B. Holt, *Clin. Sci.*, **23**, 169 (1962).
[18] J. B. Dawson and F. W. Heaton, *Biochem. J.*, **80**, 99 (1961).
[19] R. Gray and E. L. Prudeu, *Am. J. Med. Technol.*, **33**, 349 (1967).
[20] F. D. Griffith, H. E. Parker and J. C. Rogler, *J. Nutr.*, **83**, 15 (1964).
[21] J. L. Hansen and E. F. Freier, *Am. J. Med. Technol.*, **33**, 217 (1967).
[22] W. W. Harrison, J. P. Yurachek and C. A. Benson, *Clin. Chim. Acta*, **23**, 83 (1969).
[23] B. J. Hunt, *Clin. Chem.*, **15**, 979 (1969).
[24] R. Herrmann and W. Z. Lang, *Z. Ges. Exptl. Med.*, **135**, 569 (1962).
[25] D. B. Horn and A. L. Latner, *Clin. Chim. Acta*, **8**, 974 (1963).
[26] H. Husdan and A. Rapoport, *Clin. Chem.*, **15**, 669 (1969).
[27] K. H. Hyatt, L. Levy, N. Nichaman and M. Oscherwitz, *Appl. Spectrosc.*, **20**, 142 (1966).
[28] C. Iida, K. Fuwa and W. E. C. Wacker, *Anal. Biochem.*, **18**, 18 (1967).
[29] B. Klein, J. H. Kaufman and M. Oklander, *Clin. Chem.*, **13**, 788 (1967).
[30] G. E. Marks, C. E. Moore, E. L. Kanabrocki, Y. T. Oester and E. Kaplan, *Appl. Spectrosc.*, **26**, 523 (1971).
[31] J. Dubois, *Anal. Lett.*, **5**, 83 (1972).
[32] S. K. Bhattacharya and J. C. Williams, *Anal. Lett.*, **12**, 397 (1979).
[33] K. Paschen and G. Fritz, *Ärztl. Forsch.*, **24**, 202 (1970).
[34] C. Porter, *Atom. Absorp. Newslett.*, **8**, 112 (1969).
[35] D. O. Rodgerson and I. K. Moran, *Clin. Chem.*, **14**, 1206 (1968).
[36] J. Savory, J. W. Wiggins and M. G. Heintges, *Am. J. Clin. Pathol.*, **51**, 720 (1969).
[37] R. H. Seller, O. Ramirez, A. N. Brest and J. H. Moyer, *J. Am. Med. Assoc.*, **191**, 118 (1965).
[38] W. Slavin, *Occup. Health Rev.*, **17**, 9 (1965).
[39] D. J. Trent and W. Slavin, *Atom. Absorp. Newslett.*, **4**, 300 (1965).
[40] D. L. Trudeau and E. F. Freier, *Clin. Chem.*, **13**, 101 (1967).
[41] J. B. Willis, *Nature*, **186**, 249 (1960).
[42] J. B. Willis, *Spectrochim. Acta*, **16**, 259 (1960).
[43] J. B. Willis, *Anal. Chem.*, **33**, 556 (1961).
[44] A. Zettner and D. Selligson, *Clin. Chem.*, **10**, 869 (1964).
[45] J. King, Jr. and R. Buchanan, *Clin. Chem.*, **15**, 31 (1969).
[46] F. Lagerlof, *Clin. Chim. Acta*, **102**, 127 (1980).
[47] C. F. Decker, A. Aras and L. R. Decker, *Anal. Biochem.*, **8**, 344 (1964).

[48] E. G. Gimblet, A. F. Marney and R. W. Bonsnes, *Clin. Chem.*, **13**, 204 (1967).

[49] N. Gochman and H. Givelber, *Clin. Chem.*, **16**, 229 (1970).

[50] R. Hanig and M. H. Aprison, *Anal. Biochem.*, **21**, 169 (1967).

[51] H. Heinrichs and J. Lange, *Z. Anal. Chem.*, **265**, 256 (1973).

[52] J. R. K. Johnson and G. C. Riechmann, *Clin. Chem.*, **14**, 1218 (1968).

[53] B. Klein, J. H. Kaufman and S. Morgenstern, *Clin. Chem.*, **13**, 388 (1967).

[54] B. Klein, J. H. Kaufman and M. Oklander, *Clin. Chem.*, **13**, 79 (1967).

[55] C. Monder and N. Sells, *Anal. Biochem.*, **20**, 215 (1967).

[56] E. Newbrun, *Nature*, **192**, 1182 (1961).

[57] H. E. Parker, *Atom. Absorp. Newslett.*, **2**, 23 (1963).

[58] K. Paschen, *Deut. Med. Wochenschr.*, **95**, 2570 (1970).

[59] F. W. Sunderman, Jr. and J. E. Carroll, *Am. J. Clin. Pathol.*, **43**, 302 (1965).

[60] C. G. Thin and P. A. Thomson, *J. Clin. Pathol.*, **20**, 280 (1967).

[61] F. W. Sunderman, Jr. and J. R. Carroll, *Clin. Pathol. Serum Electrolytes*, 199 (1966).

[62] F. Bronner, J. P. Aubert, L. Boram, C. L. Moody, M. J. Delaney, E. S. Lang and P. J. Sammon, *Am. J. Physiol.*, **209**, 887 (1965).

[63] J. Pybus, F. J. Feldman and G. N. Bowers, Jr., *Clin. Chem.*, **16**, 998 (1970).

[64] A. H. Lichtman, G. B. Segel and M. A. Lichtman, *Clin. Chim. Acta*, **97**, 107 (1979).

[65] W. P. Tew, C. D. Malis and W. G. Walker, *Anal. Biochem.*, **112**, 346 (1981).

[66] A. Antonetti, C. Amiel, S. Couette, C. Coureau and H. Kuntziger, *Analusis*, **3**, 126 (1975).

[67] J. A. Lott, *C.R.C. Crit. Rev. Anal. Chem.*, **3**, 41 (1972).

[68] H. Kawamura, G. Tanaka and Y. Ohyogi, *Spectrochim. Acta*, **28B**, 309 (1973).

[69] D. H. Curnow, D. H. Gutteridge and E. D. Horgan, *Atom. Absorp. Newslett.*, **7**, 45 (1968).

[70] D. J. David, *Analyst*, **87**, 576 (1972).

[71] J. Descube, N. Roques, F. Rousselet and M. L. Girard, *Ann. Biol. Clin.*, **35**, 1011 (1967).

[72] B. Montford and S. C. Cribbs, *Atom. Absorp. Newslett.*, **8**, 77 (1969).

[73] J. M. Warren and H. Spencer, *Clin. Chim. Acta*, **38**, 435 (1972).

[74] F. Bek, J. Janouškova and B. Moldan, *Chem. Listy*, **66**, 867 (1972).

[75] S. van Puymbroeck and O. van der Borght, *Anal. Chim. Acta*, **57**, 441 (1971).

[76] E. Berman, *Atom. Absorp. Newslett.*, **4**, 296 (1965).

[77] M. Böhmer, E. Auer and H. Bartels, *Ärztl. Lab.*, **13**, 258 (1967).

[78] H. T. Delves, G. Shepherd and P. Vinter, *Analyst*, **96**, 260 (1971).

[79] D. B. Cheek, G. K. Powell, R. Reba and M. Feldman, *Bull. Johns Hopkins Hosp.*, **118**, 338 (1966).

[80] T. Makino and K. Takahara, *Rinsho Byori*, **28**, 478 (1980).

[81] G. Heinemann, *Z. Klin. Chem. Klin. Biochem.*, **10**, 467 (1972).

[82] R. Herrmann and W. Lang, *Z. Klin. Chem.*, **1**, 182 (1963).

[83] N. A. Holtzman, D. A. Elliott and R. H. Heller, *New England J. Med.*, **175**, 347 (1966).

[84] J. P. Matoušek and B. J. Stevens, *Clin. Chem.*, **17**, 363 (1971).

[85] M. W. Parker, F. L. Humoller and D. J. Mahler, *Clin. Chem.*, **13**, 40 (1967).

[86] A. D. Olson and W. B. Hamlin, *Atom. Absorp. Newslett.*, **7**, 69 (1968).

[87] I. Ringhardtz and B. Welz, *Z. Anal. Chem.*, **243**, 190 (1968).

[88] W. Slavin and S. Sprague, *Atom. Absorp. Newslett.*, **3**, 1 (1965).

[89] S. Sprague and W. Slavin, *Atom. Absorp. Newslett.*, **4**, 228 (1965).

[90] F. W. Sunderman, Jr. and N. O. Roszel, *Am. J. Clin. Pathol.*, **48**, 286 (1967).

[91] B. Welz and E. Wiedeking, *Z. Anal. Chem.*, **252**, 111 (1970).

[92] M. L. Girard and F. Rousselet, *Ann. Pharm. Franc.*, **25**, 353 (1967).

[93] E. Berman, *Clin. Chem.*, **9**, 459 (1963).

[94] W. W. Harrison, M. G. Netsky and M. D. Brown, *Clin. Chim. Acta*, **21**, 55 (1968).
[95] V. Bojović, L. Stojadinović and D. Djurić, *Med. Lav.*, **59**, 357 (1968).
[96] N. Weinstock and M. Uhlemann, *Clin. Chem.*, **27**, 1438 (1981).
[97] M. A. Evenson and B. L. Warren, *Clin. Chem.*, **21**, 619 (1975).
[98] B. Momčilović, B. Belonje and B. G. Shah, *Clin. Chem.*, **21**, 588 (1975).
[99] B. J. Stevens, *Clin. Chem.*, **18**, 1379 (1972).
[100] E. Berman, *Atom. Absorp. Newslett.*, **3**, 111 (1964).
[101] H. Berndt and E. Jackwerth, *Clin. Chem. Clin. Biochem.*, **17**, 489 (1979).
[102] P. D. Castilho and R. F. M. Herber, *Anal. Chim. Acta*, **94**, 269 (1977).
[103] A. D. Olson and W. B. Hamlin, *Atom. Absorp. Newslett.*, **7**, 4 (1968).
[104] N. Weinstock, *J. Clin. Chem.*, **18**, 712 (1980).
[105] S. Meret and R. I. Henkin, *Clin. Chem.*, **17**, 369 (1971).
[106] D. Kurz, J. Roach and E. J. Eyring, *Biochem. Med.*, **6**, 274 (1972).
[107] S. Levi and W. C. Purdy, *Clin. Biochem.*, **13**, 253 (1980).
[108] J. C. Meranger, B. R. Hollebone and G. A. Blanchette, *J. Anal. Toxicol.*, **5**, 33 (1981).
[109] S. B. Gross and E. S. Parkinson, *Interface*, **3**, 10 (1974).
[110] R. T. Lofberg and E. A. Levir, *Anal. Lett.*, **7**, 775 (1975).
[111] H. Kamel, J. Teape, D. H. Brown, J. M. Ottaway and W. E. Smith, *Analyst*, **103**, 921 (1978).
[112] O. Wawschinek and H. Hofler, *Atom. Absorp. Newslett.*, **18**, 97 (1979).
[113] M. Glenn, J. Savory, L. Hart, T. Glenn and J. D. Winefordner, *Anal. Chim. Acta*, **57**, 263 (1971).
[114] C. W. Fuller, *Anal. Chim. Acta*, **62**, 442 (1972).
[115] C. A. Helsby, *Talanta*, **20**, 779 (1973).
[116] F. J. Langmyhr, B. Eyde and J. Jarsen, *Anal. Chim. Acta*, **104**, 225 (1979).
[117] V. Sychra, F. Bek and J. Janoušková, *Chem. Listy*, **66**, 83 (1972).
[118] T. Maruta and T. Takeuchi, *Anal. Chim. Acta*, **62**, 253 (1972).
[119] L. Murthy, E. E. Menden, P. M. Eller and H. G. Petering, *Anal. Chem.*, **53**, 365 (1973).
[120] H. R. Imbus, J. Cholak, L. H. Miller and T. Sterling, *Arch. Environ. Health*, **6**, 28 (1959).
[121] V. Lagesson and L. Andrasko, *Clin. Chem.*, **25**, 1948 (1979).
[122] F. D. Posma, J. Balke, R. F. M. Herber and E. Stuik, *Anal. Chem.*, **47**, 834 (1975).
[123] K. G. Brodie and B. J. Stevens, *J. Anal. Toxicol.*, **1**, 282 (1977).
[124] R. T. Ross and J. G. Gonzalez, *Anal. Chim. Acta*, **70**, 443 (1974).
[125] H. T. Delves, *Analyst*, **102**, 403 (1977).
[126] G. Lundgren, *Talanta*, **23**, 309 (1976).
[127] P. Pulido, K. Fuwa and B. L. Vallee, *Anal. Biochem.*, **14**, 393 (1966).
[128] F. Lehnert, K. H. Schaller and T. Haas, *Z. Klin. Chem. Klin. Biochem.*, **6**, 174 (1968).
[129] M. Stoeppler and K. Brandt, *Z. Anal. Chem.*, **300**, 372 (1980).
[130] P. A. Ullucci and J. Y. Hwang, *Talanta*, **21**, 745 (1974).
[131] K. W. Lieberman, *Clin. Chim. Acta*, **46**, 217 (1973).
[132] E. F. Perry, S. R. Koirtyohann and H. M. Perry, Jr., *Clin. Chem.*, **21**, 626 (1975).
[133] E. Berman, *Atom. Absorp. Newslett.*, **6**, 57 (1967).
[134] A. A. Cernik, *Atom. Absorp. Newslett.*, **12**, 163 (1973).
[135] A. S. Curry and A. R. Knott, *Clin. Chim. Acta*, **30**, 115 (1970).
[136] R. D. Ediger and R. L. Coleman, *Atom. Absorp. Newslett.*, **12**, 3 (1973).
[137] T. Hauser, T. A. Hinners and J. L. Kent, *Anal. Chem.*, **44**, 1819 (1972).
[138] G. Lehnert, G. Klavis, K. H. Schaller and T. Haas, *Brit. J. Ind. Med.*, **26**, 156 (1969).
[139] P. A. Pleban and K. H. Pearson, *Clin. Chim. Acta*, **99**, 267 (1979).
[140] G. F. Carter and W. B. Yeoman, *Analyst*, **105**, 295 (1980).
[141] C. Bruhn and G. Navarrete, *Anal. Chim. Acta*, **130**, 209 (1981).
[142] P. A. Legotte, W. C. Rosa and D. C. Sutton, *Talanta*, **27**, 39 (1980).

[143] J. W. Robinson and S. Weiss, *J. Environ. Sci. Health,* **A15**, 635 (1980).

[144] E. Schumacher and F. Umland, *Z. Anal. Chem.,* **1270**, 285 (1974).

[145] G. H. Patzlaff, *Diss. Abstr. Intern. B,* **35**, 5501 (1975).

[146] N. P. Kubasik and M. T. Volosin, *Clin. Chem.,* **19**, 954 (1973).

[147] J. V. Sullivan, M. Parker and S. B. Carson, *J. Lab. Clin. Med.,* **71**, 893 (1968).

[148] J. B. Willis, *Anal. Chem.,* **34**, 614 (1962).

[149] M. Suzuki, K. Hayashi and W. E. C. Wacker, *Anal. Chim. Acta,* **104**, 389 (1979).

[150] V. V. Lidums, *Atom. Absorp. Newslett.,* **18**, 71 (1979).

[151] R. A. A. Muzzarelli and R. Rochetti, *Talanta,* **22**, 683 (1975).

[152] F. W. Sunderman, Jr., *Am. J. Clin. Pathol.,* **44**, 182 (1965).

[153] D. O. Rodgerson and R. E. Helfer, *Am. J. Clin. Pathol.,* **46**, 63 (1966).

[154] S. Sprague and W. Slavin, *Atom. Absorp. Newslett.,* **3**, 160 (1964).

[155] S. Nomoto and F. W. Sunderman, Jr., *Clin. Chem.,* **16**, 477 (1970).

[156] D. C. Hohnadel, F. W. Sunderman, Jr., M. W. Nechay and M. D. McNeely, *Clin. Chem.,* **19**, 1288 (1973).

[157] F. W. Sunderman, Jr., *Human Pathol.,* **4**, 549 (1973).

[158] S. Nomoto, *Shinshu Igaku Zasshi,* **22**, 39 (1974).

[159] H. Zachariasen, I. Anderson, C. Kostol and R. Barton, *Clin. Chem.,* **21**, 562 (1975).

[160] A. Dornemann and H. Kleist, *Z. Anal. Chem.,* **300**, 197 (1980).

[161] M. W. Routh, *Anal. Chem.,* **52**, 182 (1980).

[162] I. W. F. Davidson and W. L. Secrest, *Anal. Chem.,* **44**, 1808 (1972).

[163] K. H. Schaller, H. G. Essing, H. Valentin and G. Schäcke, *Z. Klin. Chem. Klin. Biochem.,* **10**, 434 (1972).

[164] R. T. Ross, J. G. Gonzalez and D. A. Segar, *Anal. Chim. Acta,* **63**, 205 (1973).

[165] B. E. Guthrie, W. R. Wolf and C. Veillon, *Anal. Chem.,* **50**, 1900 (1978).

[166] F. J. Kayne, G. Komar, H. Laboda and R. E. Vanderlinde, *Clin. Chem.,* **24**, 2151 (1978).

[167] R. S. Pekarek, E. C. Hauer, R. W. Wannemacher, Jr. and W. R. Beisel, *Anal. Biochem.,* **59**, 283 (1974).

[168] F. J. Feldman, E. C. Knoblock and W. C. Purdy, *Anal. Chim. Acta,* **38**, 489 (1967).

[169] J. M. Morgan, *Metabolism,* **21**, 313 (1972).

[170] G. Buttgereit, *Arbeitsmed. Sozialmed. Arbeitshyg.,* **10**, 286 (1972).

[171] G. Buttgereit, *Z. Anal. Chem.,* **267**, 81 (1973).

[172] R. S. Pekarek and E. C. Hauer, *Fed. Proc..,* **31**, 700 (1972).

[173] J. O. Pierce and J. Cholak, *Arch. Environ. Health,* **13**, 208 (1966).

[174] A. Zettner, K. Rafferty and H. J. Jarecky, *Atom. Absorp. Newslett.,* **7**, 32 (1968).

[175] E. Graf-Harsanyi and F. J. Langmyhr, *Anal. Chim. Acta,* **116**, 106 (1980).

[176] J. Kumpulainen, *Anal. Chim. Acta,* **113**, 355 (1980).

[177] H. Nomiyama, M. Yotoriyama and K. Nomiyama, *Am. Ind. Hyg. Assoc. J.,* **41**, 98 (1980).

[178] G. Nise and O. Vesterberg, *Scand. J. Work Environ. Health,* **5**, 404 (1979).

[179] M. T. Glenn, J. Savory, S. A. Fein, R. D. Reaves, C. J. Molnar and J. D. Winefordner, *Anal. Chem.,* **45**, 203 (1973).

[180] E. D. Olsen, P. I. Jatlow, F. J. Fernandez and H. L. Kahn, *Clin. Chem.,* **14**, 326 (1973).

[181] A. D. Olson and W. B. Hamlin, *Clin. Chem.,* **15**, 438 (1969).

[182] H. Spector, S. Glusman, P. I. Jatlow and D. Seligson, *Clin. Chim. Acta,* **31**, 5 (1971).

[183] F. Aldrighetti, G. Carelli, F. Ceriati, G. Cremona and M. Pomponi, *Atom. Spectrosc.,* **2**, 71 (1981).

[184] Y.-Y. Yeh and P. Zee, *Clin. Chem.,* **20**, 360 (1974).

[185] E. C. Zaino, *Atom. Absorp. Newslett.,* **6**, 93 (1967).

[186] A. Zettner and L. Mansbach, *Am. J. Clin. Pathol.,* **44**, 517 (1965).

[187] A. Zettner and A. H. Mensch, *Am. J. Clin. Pathol.,* **48**, 225 (1967).

[188] A. Zettner and A. H. Mensch, *Am. J. Clin. Pathol.,* **49**, 196 (1968).

[189] A. Zettner, L. C. Sylvia and L. Capacho-Delgado, *Am. J. Clin. Pathol.,* **45**, 533 (1966).

[190] O. W. Van Assendelft, W. G. Zijlstra, A. Buursma, E. J. Van Kampen and W. Hoek, *Clin. Chim. Acta,* **22**, 281 (1968).

[191] W. G. Zijlstra, O. W. Van Assendelft, E. J. Van Kampen and W. Hoek, *Am. J. Clin. Pathol.,* **50**, 513 (1968).

[192] A. Zettner, *Am. J. Clin. Pathol.,* **50**, 514 (1968).

[193] D. O. Rodgerson and R. E. Helfer, *Clin. Chem.,* **12**, 338 (1966).

[194] K. Nakane, K. Iguchi, A. Kosaka and H. Saito, *Rinsho Byori,* **22**, 741 (1974).

[195] K. Nakamura, H. Watanabe and H. Orii, *Anal. Chim. Acta,* **120**, 155 (1980).

[196] J. A. Ealy, N. E. Bolton, R. J. McElheny and R. W. Morrow, *Am. Ind. Hyg. Assoc. J.,* **35**, 566 (1974).

[197] M. Stoeppler, K. Brandt and T. C. Rains, *Analyst,* **103**, 714 (1978).

[198] M. A. Evenson and D. D. Pendergast, *Clin. Chem.,* **20**, 171 (1974).

[199] L. J. M. Zinterhofer, P. I. Jatlow and A. Fappiano, *J. Lab. Clin. Med.,* **78**, 664 (1971).

[200] M. P. Bratzel and J. A. Reed, *Clin. Chem.,* **20**, 217 (1974).

[201] H. T. Delves, *Analyst,* **95**, 431 (1970).

[202] J. Y. Hwang, P. A. Ullucci and C. J. Mokeler, *Anal. Chem.,* **45**, 795 (1973).

[203] M. A. Evenson and D. D. Pendergast, *Clin. Chem.,* **20**, 163 (1974).

[204] V. A. Razumov, T. P. Utkina and T. K. Ajdarov, *Zh. Analit. Khim.,* **20**, 1371 (1965).

[205] W. F. Barthel, A. L. Smrek, G. P. Angel, J. A. Liddle, P. J. Landrigan, S. H. Gehlbach and J. J. Chisolm, *J. Assoc. Off. Anal. Chem.,* **56**, 1252 (1973).

[206] E. Berman, V. Valavanis and A. Dubin, *Clin. Chem.,* **14**, 239 (1968).

[207] M. M. Joselow and J. D. Bogden, *Atom. Absorp. Newslett.,* **11**, 99 (1972).

[208] M. M. Joselow and J. D. Bogden, *Atom. Absorp. Newslett.,* **11**, 127 (1972).

[209] M. M. Joselow and N. P. Singh, *Atom. Absorp. Newslett.,* **12**, 128 (1973).

[210] H. L. Kahn, G. E. Peterson and J. E. Schallis, *Atom. Absorp. Newslett.,* **7**, 35 (1968).

[211] H. L. Kahn and J. E. Sebestyen, *Atom. Absorp. Newslett.,* **9**, 33 (1970).

[212] L. Kopito, M. A. Davis and H. Schwachman, *Clin. Chem.,* **20**, 205 (1974).

[213] D. G. Mitchell, F. J. Ryan and K. M. Aldous, *Atom. Absorp. Newslett.,* **11**, 120 (1972).

[214] G. A. Rose and E. G. Willden, *Analyst,* **98**, 243 (1973).

[215] R. J. Segal, *Clin. Chem.,* **15**, 1124 (1969).

[216] S. Selander and K. Cramer, *Brit. J. Ind. Med.,* **25**, 139 (1968).

[217] J. B. Willis, *Nature,* **191**, 381 (1961).

[218] M. D. Amos, P. A. Bennett, K. G. Brodie, P. W. Y. Lung and J. P. Matoušek, *Anal. Chem.,* **43**, 211 (1971).

[219] W. N. Anderson, P. M. G. Broughton, J. B. Dawson and G. W. Fisher, *Clin. Chim. Acta,* **50**, 129 (1974).

[220] A. A. Cernik, *Atom. Absorp. Newslett.,* **12**, 42 (1973).

[221] A. A. Cernik and M. H. P. Sayers, *Brit. J. Ind. Med.,* **28**, 392 (1971).

[222] E. Döllefeld, *Ärztl. Lab.,* **17**, 369 (1971).

[223] R. D. Ediger and R. L. Coleman, *Atom. Absorp. Newslett.,* **11**, 33 (1972).

[224] F. J. Fernandez, *Atom. Absorp. Newslett.,* **12**, 70 (1973).

[225] F. J. Fernandez and H. L. Kahn, *Atom. Absorp. Newslett.,* **10**, 1 (1971).

[226] J. M. Hicks, A. N. Gutierrez and B. E. Worthy, *Clin. Chem.,* **19**, 322 (1973).

[227] J. F. Rosen, *J. Lab. Clin. Med.,* **80**, 567 (1972).

[228] N. P. Kubasik, M. T. Volosin and M. H. Murray, *Clin. Chem.,* **18**, 410 (1972).

[229] G. Nise and O. Vesterberg, *Clin. Chim. Acta,* **84**, 129 (1978).

[230] J. Y. Hwang, P. A. Ullucci, S. B. Smith, Jr. and A. L. Malenfant, *Anal. Chem.,* **43**, 1319 (1971).

[231] F. J. Fernandez, *Clin. Chem.,* **21**, 558 (1975).

[232] D. W. Hessel, *Atom. Absorp. Newslett.,* **7**, 55 (1968).

[233] O. Einarrson and G. Lindstedt, *Scand. J. Clin. Lab. Invest.*, **23**, 367 (1969).

[234] D. Roosels and J. V. Vanderkeel, *Atom. Absorp. Newslett.*, **7**, 9 (1968).

[235] C. Kopito and H. Schwachman, *J. Lab. Clin. Med.*, **70**, 326 (1967).

[236] K. Selander and K. Cramer, *Brit. J. Ind. Med.*, **25**, 209 (1968).

[237] R. O. Farrelly and J. N. Pybus, *Clin. Chem.*, **15**, 566 (1969).

[238] F. J. Fernandez and H. L. Kahn, *Atom. Absorp. Newslett.*, **11**, 33 (1972).

[239] E. D. Olsen and P. J. Jatlow, *Clin. Chem.*, **18**, 1312 (1972).

[240] M. Plechaty, B. Noll and F. W. Sunderman, *Ann. Clin. Lab. Sci.*, **7**, 515 (1977).

[241] J. Ebert and H. Jungmann, *Z. Anal. Chem.*, **272**, 287 (1974).

[242] G. R. Sirota and J. F. Uthe, *Anal. Chem.*, **49**, 823 (1977).

[243] S. G. Eltoaum, R. Juliano, R. E. MacFarland and D. L. Pfeil, *Clin. Chem.*, **18**, 316 (1974).

[244] L. Kopito, H. Schwachman and L. A. Williams, *Std. Methods Clin. Chem.*, **7**, 151 (1972).

[245] T. A. Kilroe-Smith, *Clin. Chem.*, **21**, 630 (1975).

[246] D. G. Mitchell, K. M. Aldous and F. J. Ryan, *N.Y. State J. Med.*, **74**, 1599 (1974).

[247] E. J. M. De Haas and F. A. De Wolff, *Clin. Chem.*, **27**, 205 (1981).

[248] L. E. Wittmers, A. Alich and A. C. Aufderhelde, *Am. J. Clin. Pathol.*, **75**, 80 (1981).

[249] E. Norval and L. R. P. Butler, *Anal. Chim. Acta*, **58**, 47 (1972).

[250] F. Bek, J. Janoušková and B. Moldan, *Atom. Absorp. Newslett.*, **13**, 47 (1974).

[251] V. B. Grafflage, G. Buttgereit, W. Kübler and H.-M. Mertenz, *Z. Klin. Chem. Klin. Biochem.*, **12**, 287 (1974).

[252] R. T. Ross and J. G. Gonzalez, *Bull. Environ. Contam. Toxicol.*, **12**, 470 (1974).

[253] J. P. Mahoney, K. Sargent, M. Greland and W. Small, *Clin. Chem.*, **15**, 312 (1969).

[254] D. J. D'Amico and H. L. Klawans, *Anal. Chem.*, **48**, 1469 (1976).

[255] R. S. Ajemian and N. E. Whitman, *Ann. Ind. Hyg. Assoc. J.*, **30**, 52 (1969).

[256] D. G. Van Ormer and W. C. Purdy, *Anal. Chim. Acta*, **64**, 93 (1973).

[257] H. Yasuda, T. Sasaki and K. Saito, *Bunseki Kagaku*, **30**, 538 (1981).

[258] M. Mishima, *Bunseki Kagaku*, **24**, 433 (1975).

[259] M. Suzuki and W. E. C. Wacker, *Anal. Biochem.*, **57**, 605 (1974).

[260] P. A. Pleban, J. Kerkay and K. H. Pearson, *Clin. Chem.*, **27**, 68 (1981).

[261] D. I. Paynter, *Anal. Chem.*, **51**, 2086 (1979).

[262] K. Fuwa, P. Pulido, R. McKay and B. L. Vallee, *Anal. Chem.*, **36**, 2407 (1964).

[263] B. M. Hackley, J. C. Smith, Jr. and J. A. Halsted, *Clin. Chem.*, **14**, 1 (1968).

[264] M. D. Stevens, W. F. Mackenzie and V. D. Anand, *Biochem. Med.*, **18**, 158 (1977).

[265] J. C. Smith, Jr., G. P. Butrimovitz and W. C. Purdy, *Clin. Chem.*, **25**, 1487 (1979).

[266] D. Kurz, J. Roach and E. J. Eyring, *Anal. Biochem.*, **53**, 586 (1973).

[267] R. S. Pekarek, W. R. Beisel, P. J. Bartelloni and K. A. Bostian, *Am. J. Clin. Pathol.*, **57**, 506 (1972).

[268] J. H. Freeland-Graves, P. J. Hendrickson, M. L. Ebangit and J. Y. Snowden, *Am. J. Clin. Nutr.*, **34**, 312 (1981).

[269] N. E. Vieira and J. W. Hansen, *Clin. Chem.*, **27**, 73 (1981).

[270] M. K. Chooi, J. K. Todd and N. D. Boyd, *Clin. Chem.*, **21**, 632 (1975).

[271] J. R. Kelson and R. J. Shamberger, *Clin. Chem.*, **24**, 240 (1978).

[272] H. Haas, G. Lehnert and K. M. Schaller, *Z. Klin. Chem. Klin. Biochem.*, **5**, 27 (1967).

[273] G. Hauck, *Z. Anal. Chem.*, **267**, 337 (1973).

[274] G. Machata and R. Binder, *Z. Rechtsmed.*, **73**, 29 (1973).

[275] A. S. Prasad, D. Oberleas and J. A. Halsted, *J. Lab. Clin. Med.*, **66**, 508 (1965).

[276] J. G. Reinhold, E. Pascoe and G. A. Kfoury, *Anal. Biochem.*, **25**, 557 (1968).

[277] P. Gonick, D. Oberleas, T. Knechtges and A. S. Prasad, *Invest. Urol.*, **6**, 345c (1969).

[278] K. H. Hu and R. L. Friede, *J. Neurochem.*, **15**, 677 (1968).

[279] N. Honegger, *Ärztl. Lab.*, **2**, 41 (1963).

[280] M. Arroyo and E. Palenque, *Rev. Clin. Esp.*, **134**, 227 (1974).

[281] B. Bergman, R. Sjostrom and K. R. Wing, *Acta Physiol. Scand.*, **92**, 440 (1974).

[282] R. I. Henkin, C. W. Mueller and R. O. Wolf, *J. Lab. Clin. Med.*, **86**, 175 (1975).

[283] E. Machata, Y. Kanohda and H. Naka, *Eisei Kensa*, **23**, 687 (1974).

[284] E. W. Reimold and D. J. Besch, *Clin. Chem.*, **24**, 675 (1978).

[285] A. Mazzucotelli, M. Galli, E. Benassi, C. Loeb, G. A. Ottonello and P. T. Tanganelli, *Analyst*, **103**, 863 (1978).

[286] D. Clark, R. M. Dagnall and T. S. West, *Anal. Chim. Acta*, **63**, 11 (1973).

[287] P. E. Gardiner, J. M. Ottaway, G. S. Fell and R. R. Burns, *Anal. Chim. Acta*, **124**, 281 (1981).

[288] H. T. Delves, *Prog. Anal. Atom. Spectrosc.*, **4**, 1 (1981).

[289] J. R. Kelson, *Clin. Chem.*, **26**, 349 (1980).

[290] S. Kiilerich, M. S. Christensen, J. Naestoft and C. Christiansen, *Clin. Chim. Acta*, **105**, 231 (1980).

[291] S. Kiilerich, M. S. Christensen, J. Naestoft and C. Christiansen, *Clin. Chim. Acta*, **114**, 117 (1981).

[292] K. Ladefoged, *Clin. Chim. Acta*, **100**, 149 (1980).

[293] S. Levi, R. C. Fortin and W. C. Purdy, *Anal. Chim. Acta*, **127**, 103 (1981).

[294] J. C. Smith, G. P. Butrimovitz and W. C. Purdy, *Clin. Chem.*, **26**, 350 (1980).

[295] D. M. Williams, *Clin. Chim. Acta*, **99**, 23 (1979).

[296] D. J. Johnson, T. S. West and R. M. Dagnall, *Anal. Chim. Acta*, **66**, 171 (1973).

[297] Ch. Fuchs, M. Brasche, K. Paschen, H. Nordbeck and E. Quellhorst, *Clin. Chim. Acta*, **52**, 71 (1974).

[298] S. S. Krishnan, K. A. Gillespie and D. R. Crapper, *Anal. Chem.*, **44**, 1469 (1972).

[299] Y. Pegon, *Anal. Chim. Acta*, **101**, 385 (1978).

[300] S. W. King, M. R. Wills and J. Savory, *Anal. Chim. Acta*, **128**, 221 (1981).

[301] F. Buneaux and P. Fabiani, *Ann. Biol. Clin.*, **28**, 273 (1970).

[302] R. C. Rooney, *Analyst*, **100**, 471 (1975).

[303] T. H. Rosenquist and J. W. Rosenquist, *J. Histochem. Cytochem.*, **22**, 104 (1974).

[304] M. Ishizaki and S. Ueno, *Talanta*, **26**, 523 (1979).

[305] E. Chrenekova' and N. Rusinova', *Chem. Listy*, **72**, 990 (1978).

[306] N. Thiex, *J. Assoc. Off. Anal. Chem.*, **63**, 496 (1980).

[307] S. Dupire and M. Hoenig, *Analusis*, **8**, 153 (1980).

[308] M. Ishazaki, *Bunseki Kagaku*, **26**, 667 (1977).

[309] A. Yasui, C. Tsutsumi and S. Toda, *Agric. Biol. Chem.*, **42**, 2139 (1978).

[310] W. L. Hoover, J. R. Melton, P. A. Howard and J. W. Bassett, Jr., *J. Assoc. Off. Anal. Chem.*, **57**, 18 (1974).

[311] R. R. Lauwerys, J. P. Buchet and H. Roels, *Arch. Toxicol.*, **41**, 239 (1979).

[312] I. Paralescu and A. Tasca-Danescu, *Rev. Chim. (Bucharest)*, **26**, 507 (1975).

[313] R. G. Smith, J. C. Van Loon, J. R. Knechtel, J. L. Fraser, A. E. Pitts and A. S. Hodges, *Anal. Chim. Acta*, **93**, 61 (1977).

[314] B. W. Haynes, *Atom. Absorp. Newslett.*, **17**, 49 (1978).

[315] F. LaVilla and F. Queraud, *Rev. Inst. Fr. Pet.*, **32**, 413 (1977).

[316] Y. Odanaka, O. Matano and S. Goto, *Bunseki Kagaku*, **28**, 517 (1979).

[317] H. Woidich and W. Pfannhauser, *Z. Anal. Chem.*, **276**, 61 (1975).

[318] J. Holm, *Fleischwirtschaft*, **58**, 745 (1978).

[319] S. Peats, *Atom. Absorp. Newslett.*, **18**, 118 (1979).

[320] R. M. Orheim and H. H. Bovee, *Anal. Chem.*, **46**, 921 (1974).

[321] D. B. Lo and R. L. Coleman, *Atom. Absorp. Newslett.*, **18**, 10 (1979).

[322] B. Welz and M. Melcher, *Atom. Absorp. Newslett.*, **18**, 121 (1979).

[323] L. Pozzoli and C. Minoia, *Ann. Ist. Super. Sanita*, **13**, 377 (1977).

[324] G. De Grott, A. Van. Dijk and R. A. A. Maes, *Pharm. Weekblad*, **112**, 949 (1977).

[325] T. J. Kneip, *Health Lab. Sci.*, **14**, 53 (1977).

[326] F. Peters, G. Growcock and G. Strunc, *Anal. Chim. Acta*, **104**, 177 (1979).

[327] J. Y. Hwang, P. A. Ullucci, C. J. Mokeler and S. B. Smith, *Am. Lab.*, **5**, No. 3, 43 (1973).

[328] D. G. Van Ormer, *J. Forensic Sci.*, **20**, 595 (1975).

[329] F. J. Amore, *Anal. Chem.*, **46**, 1597 (1974).

[330] J. Nève and M. Hanocq, *Anal. Chim. Acta*, **93**, 85 (1977).

[331] B. C. Severne and R. R. Brooks, *Talanta*, **19**, 1467 (1972).

[332] M. J. Siren, *Sci. Tools*, **11**, 37 (1964).

[333] M. Ihnat and R. J. Westerby, *Anal. Lett.*, **7**, 257 (1974).

[334] O. E. Clinton, *Analyst*, **102**, 187 (1977).

[335] M. Ishizaki, *Talanta*, **25**, 167 (1978).

[336] K. Saeed, Y. Thomassen and F. T. Langmyhr, *Anal. Chim. Acta*, **110**, 285 (1979).

[337] D. H. Cox and A. E. Bibb, *J. Assoc. Off. Anal. Chem.*, **64**, 265 (1981).

[338] B. C. Severne and R. R. Brooks, *Anal. Chim. Acta*, **58**, 216 (1972).

[339] A. S. Curry, J. F. Read and A. R. Knott, *Analyst*, **94**, 744 (1969).

[340] J. Savory, N. O. Rozel, P. Mushak and F. W. Sunderman, *Am. J. Clin. Pathol.*, **50**, 505 (1968).

[341] G. M. Shkolnik and R. F. Bevill, *Atom. Absorp. Newslett.*, **12**, 112 (1973).

[342] R. E. Kinser, *Am. Ind. Hyg. Assoc. J.*, **27**, 260 (1966).

[343] H. Bader and H. Brandenberger, *Atom. Absorp. Newslett.*, **7**, 1 (1968).

[344] D. L. Bokowski, *Am. Ind. Hyg. Assoc. J.*, **29**, 474 (1968).

[345] T. Stiefel, K. Schulze, G. Tölg and H. Zorn, *Anal. Chim. Acta*, **87**, 67 (1976).

[346] G. Devoto, *Boll. Soc. Ital. Biol. Sper.*, **44**, 1253 (1968).

[347] R. Boudon, M. Galliot, F. Prouillet, *Ann. Biol. Clin. (Paris)*, **32**, 413 (1974).

[348] M. L. Denniston, L. A. Sternson and A. J. Repta, *Anal. Lett.*, **14**, 451 (1981).

[349] D. A. Hull, N. Muhammad, L. G. Lanese, S. D. Reich, T. T. Finkelstein and S. Fandrich, *J. Pharm. Sci.*, **70**, 500 (1981).

[350] J. Smeyers-Verbeke, M. R. Detaevernier, L. Denis and D. L. Massart, *Clin. Chim. Acta*, **113**, 329 (1981).

[351] J. Aggett, *Anal. Chim. Acta*, **63**, 473 (1973).

[352] R. M. Turkall and J. R. Bianchine, *Analyst*, **106**, 1096 (1981).

[353] A. Bouchard, *Atom. Absorp. Newslett.*, **12**, 115 (1973).

[354] M. B. Jacobs, L. J. Goldwater and H. Gilbert, *Am. Ind. Hyg. Assoc. J.*, **22**, 276 (1961).

[355] M. B. Jacobs and A. Singerman, *J. Lab. Clin. Med.*, **59**, 871 (1962).

[356] M. B. Jacobs, S. Yamaguchi, L. J. Goldwater and H. Gilbert, *Am. Ind. Hyg. Assoc. J.*, **21**, 475 (1960).

[357] L. A. Kauser, R. Henderson, H. P. Shotwell and D. A. Culp, *Am. Ind. Hyg. Assoc. J.*, **32**, 331 (1971).

[358] N. P. Kubasik, H. E. Sine and M. T. Volosin, *Clin. Chem.*, **18**, 1326 (1972).

[359] C. J. Least, Jr., T. A. Rejent and H. Lees, *Atom. Absorp. Newslett.*, **13**, 4 (1974).

[360] G. Lindstedt, *Analyst*, **95**, 264 (1970).

[361] L. Magos and A. A. Cernik, *Brit. J. Ind. Med.*, **26**, 144 (1969).

[362] B. B. Mesman and B. S. Smith, *Atom. Absorp. Newslett.*, **9**, 81 (1970).

[363] B. B. Mesman, B. S. Smith and J. O. Pierce, *Am. Ind. Hyg. Assoc. J.*, **31**, 701 (1970).

[364] A. Moffitt, Jr. and R. E. Kupel, *Atom. Absorp. Newslett.*, **9**, 113 (1970).

[365] P. J. Nord, M. P. Kadaba and J. R. J. Sorenson, *Arch. Environ. Health*, **27**, 40 (1973).

[366] M. Olivier, *Z. Anal. Chem.*, **257**, 187 (1971).

[367] U. Ulfvarson, *Acta Chem. Scand.*, **21**, 641 (1967).

[368] T. Szprengier, *Med. Wet.*, **33**, 182 (1977).

[369] J. P. Farant, D. Brissette, L. Moncion, L. Bigras and A. Chartrand, *J. Anal. Toxicol.*, **5**, 147 (1981).

[370] Y. Sadin and P. Deldime, *Anal. Lett.*, **12**, 563 (1979).
[371] K. Matsunaga, T. Takahashi and M. Nishimura, *Mizu Shori Gijutsu*, **15**, 1053 (1974).
[372] J. Toffaletti and J. Savory, *Anal. Chem.*, **47**, 2091 (1975).
[373] J. Y. Hwang, P. A. Ullucci and A. L. Malenfant, *Can. Spectrosc.*, **16**, 2 (1971).
[374] G. Hallerer and J. Hoffmann, *Z. Lebensm.-Unters.-Forsch.*, **150**, 277 (1973).
[375] G. Cumont, *Chim. Anal. (Paris)*, **53**, 634 (1971).
[376] L. Magos, *Analyst*, **96**, 847 (1971).
[377] A. E. Moffitt, Jr. and R. E. Kupel, *Am. Ind. Hyg. Assoc. J.*, **32**, 614 (1971).
[378] S. H. Omang, *Anal. Chim. Acta*, **63**, 247 (1973).
[379] V. A. Thorpe, *J. Assoc. Off. Anal. Chem.*, **54**, 206 (1971).
[380] M. A. Evenson and T. Anderson, Jr., *Clin. Chem.*, **21**, 537 (1975).
[381] L. T. Wetzel and J. U. John, *Clin. Chem.*, **26**, 1796 (1980).
[382] F. R. Alderman and H. J. Gitelman, *Clin. Chem.*, **26**, 258 (1980).
[383] J. E. Gorsky and A. A. Dietz, *Clin. Chem.*, **24**, 1485 (1978).
[384] A. Taylor and T. N. Bryant, *Clin. Chim. Acta*, **110**, 83 (1981).
[385] S. Arpadyan and I. Kachov, *Zentralbl. Pharm. Pharmakother. Laboratoriumsdiagn.*, **117**, 237 (1978).
[386] P. E. Gardiner, J. M. Ottaway, G. S. Fell and D. J. Halls, *Anal. Chim. Acta*, **128**, 57 (1981).
[387] D. J. Halls and G. S. Fell, *Anal. Chim. Acta*, **129**, 205 (1981).
[388] T. Makino and K. Takahara, *Clin. Chem.*, **27**, 1445 (1981).
[389] F. Alt, *Z. Anal. Chem.*, **308**, 137 (1981).
[390] W. Toda, J. Lux and J. Van Loon, *Anal. Lett.*, **13**, 1105 (1980).
[391] U. Voellkopf, Z. Grobenski and B. Welz, *Atom. Spectrosc.*, **2**, 68 (1981).
[392] O. Vesterberg and T. Bergstrom, *Clin. Chem.*, **23**, 555 (1977).
[393] B. N. Noller and H. Bloom, *Aust. J. Med. Technol.*, **7**, 22 (1976).
[394] D. C. Paschal and C. J. Bell, *Atom. Spectrosc.*, **2**, 146 (1981).
[395] O. Vesterberg and K. Wrangskogh, *Clin. Chem.*, **24**, 681 (1978).
[396] J. W. Robinson and S. Weiss, *Spectrosc. Lett.*, **11**, 715 (1978).
[397] S. K. Bhattacharya, *Anal. Lett.*, **10**, 817 (1977).
[398] C. H. McMurray and W. J. Blanchflower, *Clin. Chem.*, **24**, 344 (1978).
[399] P. A. Sebesta and L. A. Danzer, *Clin. Chim. Acta*, **68**, 309 (1976).
[400] H. T. Delves and J. Woodward, *Atom. Spectrosc.*, **2**, 65 (1981).
[401] M. Speich, B. Bousquet and G. Nicolas, *Clin. Chem.*, **27**, 246 (1981).
[402] M. Stiefel, K. Schulze, G. Tölg and H. Zorn, *Z. Anal. Chem.*, **300**, 189 (1980).
[403] M. C. Lois Gonzalez, A. Gonzalez Portal and C. Baluja Santos, *Quim. Anal.*, **30**, 307 (1976).
[404] M. Prosbová and G. Kovac, *Chem. Listy*, **71**, 978 (1977).
[405] V. Krcma and J. Komárek, *Chem. Listy*, **74**, 770 (1980).
[406] K.-R. Sperling and B. Bahr, *Z. Anal. Chem.*, **301**, 29 (1980).
[407] P. Allain and Y. Mauras, *Clin. Chim. Acta*, **91**, 41 (1979).
[408] S. Levi and W. C. Purdy, *Anal. Chim. Acta*, **116**, 375 (1980).
[409] C. Minoia, M. Colli and L. Pozzoli, *Atom. Spectrosc.*, **2**, 163 (1981).
[410] D. J. Halls, G. S. Fell and P. M. Dunbar, *Clin. Chim. Acta*, **114**, 21 (1981).
[411] M. Ishizaki, N. Oyamada, M. Fujiki and S. Yamaguchi, *Sangyo Igaku*, **20**, 174 (1978).
[412] N. Oyamada and M. Ishizaki, *Bunseki Kagaku*, **28**, 289 (1979).
[413] International Union of Pure and Applied Chemistry, Clinical Chemistry Division, Commission on Toxicology, Sub-Committee on Environmental and Occupational Toxicology of Nickel, *Pure Appl. Chem.*, **53**, 773 (1981).
[414] D. Ader and M. Stoeppler, *J. Anal. Toxicol.*, **1**, 252 (1977).
[415] A. Dornemann and H. Kleist, *Zentralbl. Arbeitsmed., Arbeitsschutz Prophyl.*, **28**, 165 (1978).

[416] K.-R. Sperling and B. Bahr, *Z. Anal. Chem.*, **301**, 31 (1980).

[417] H. L. Boiteau and C. Meteyer, *Analusis*, **6**, 350 (1978).

[418] S. S. Chao and E. E. Pickett, *Anal. Chem.*, **52**, 335 (1980).

[419] A. L. Gelman, *J. Sci. Fd. Agric.*, **27**, 520 (1976).

[420] U. Harms and J. Kunze, *Z. Lebensm.-Unters. Forsch.*, **164**, 204 (1977).

[421] C. Schmidt, *Am. Ind. Hyg. Assoc. J.*, **40**, 1085 (1979).

[422] F. F. Farris, A. Poklis and G. E. Griesmann, *J. Assoc. Off. Anal. Chem.*, **61**, 660 (1978).

[423] J. P. Buchet, R. Lauwerys, H. Roels and C. De Vos, *Clin. Chim. Acta*, **73**, 481 (1976).

[424] S. U. Khan, R. O. Cloutier and M. Hidiroglou, *J. Assoc. Off. Anal. Chem.*, **62**, 1062 (1979).

[425] W. Torjussen, I. Anderson and H. Zachariasen, *Clin. Chem.*, **23**, 1018 (1977).

[426] D. Mikac-Devic, F. W. Sunderman, Jr. and S. Nomoto, *Clin. Chem.*, **23**, 948 (1977).

[427] M. Aihara and M. Kiboku, *Bunseki Kagaku*, **26**, 559 (1977).

[428] J. Nève, M. Hanocq and L. Molle, *Anal. Chim. Acta*, **115**, 133 (1980).

[429] K. Yasuda, M. Taguchi, S. Tamura and S. Toda, *Bunseki Kagaku*, **26**, 442 (1977).

[430] M. Ishizaki, *Bunseki Kagaku*, **26**, 206 (1977).

[431] S. Ueno and M. Ishizaki, *Nippon Kagaku Kaishi*, 217 (1979).

[432] M. Ishizaki, S. Ueno. M. Fujiki and S. Yamaguchi, *Sangyo Iagku*, **20**, 30 (1978).

[433] E. Bonilla, *Clin. Chem.*, **24**, 471 (1978).

[434] D. A. Shearer, R. O. Cloutier and M. Hidiroglou, *J. Assoc. Off. Anal. Chem.*, **60**, 155 (1977).

[435] J. Smeyers-Verbeke, G. Segebarth and D. L. Massart, *Atom. Absorp. Newslett.*, **14**, 153 (1975).

[436] W. B. Robbins and J. A. Caruso, *Spectrosc. Lett.*, **11**, 333 (1978).

[437] C. Iida, T. Uchida and I. Kojima, *Anal. Chim. Acta*, **113**, 365 (1980).

[438] K. Julshamn, K. J. Anderson, Y. Willassen and O. R. Braekkan, *Anal. Biochem.*, **88**, 552 (1978).

[439] D. Seifert, *Z. Anal. Chem.*, **287**, 317 (1977).

[440] S. L. Gaffin, *Clin. Toxicol.*, **15**, 293 (1979).

[441] K. Yanagi and M. Ambe, *Bunseki Kagaku*, **30**, 209 (1981).

[442] T. Shimizu, T. Hiyama, Y. Shijo and K. Sakai, *Bunseki Kagaku*, **29**, 680 (1980).

[443] T. Nishijima, Y. Dokiya, H. Iizuka and S. Toda, *Bunseki Kagaku*, **26**, 349 (1977).

[444] D. Gardner and G. Dal Pont, *Anal. Chim. Acta*, **108**, 13 (1979).

[445] S. M. Pederson and P. M. Graabaek, *Scand. J. Clin. Lab. Invest.*, **37**, 91 (1977).

[446] D. W. Bolin and O. E. Stamberg, *Ind. Eng. Chem.*, *Anal. Ed.*, **16**, 345 (1944).

[447] V. Korunová and J. Dědina, *Analyst*, **105**, 48 (1980).

[448] J. Zmudzki, *Bromatol. Chem. Toksykol.*, **13**, 77 (1980).

[449] E. E. Menden, D. Brockman and H. G. Petering, *Anal. Chem.*, **49**, 1644 (1977).

[450] T. Kawaraya, M. Kawasaki, K. Haruki, K. Tomita, M. Oka and S. Horiguchi, *Bunseki Kagaku*, **25**, 464 (1976).

[451] J. Locke, *Anal. Chim. Acta*, **104**, 225 (1979).

[452] R. Djudzman, E. Van den Eeckhout and P. De Moerloose, *Analyst*, **102**, 688 (1977).

[453] C. E. Gleit and W. D. Holland, *Anal. Chem.*, **34**, 1454 (1962).

[454] C. E. Gleit and W. D. Holland, *Nature*, **200**, 69 (1963).

[455] C. E. Gleit, *Anal. Chem.*, **37**, 314 (1965).

[456] B. J. Aungst, J. Dolce and H.-L. Fung, *Anal. Lett.*, **13**, 347 (1980).

[457] H. L. Trachman and A. Tyberg, *Anal. Chem.*, **49**, 1090 (1977).

[458] C. Pechery, R. Depraitere, D. De Lauture, N. Roullane and C. Durlach, *Feuill. Biol.*, **21**, 75 (1980).

[459] W. J. Goldberg and N. Allen, *Clin. Chem.*, **27**, 562 (1981).

[460] S. J. Horsky, *Atom. Spectrosc.*, **1**, 129 (1980).

[461] D. I. Paynter, *Anal. Chem.*, **51**, 2086 (1979).

[462] R. C. Carpenter, *Anal. Chim. Acta*, **125**, 209 (1981).
[463] S. R. Koirtyohann and C. A. Hopkins, *Analyst*, **101**, 870 (1976).
[464] M. Feinberg and C. Ducauze, *Bull. Soc. Chim. France*, 419 (1978).
[465] H. I. Issaq, *Anal. Chem.*, **51**, 657 (1979).
[466] N. Mohamed and R. C. Fry, *Anal. Chem.*, **53**, 450 (1981).
[467] J. W. Robinson, D. K. Wolcott and L. Rhodes, *Anal. Chim. Acta*, **78**, 285 (1975).
[468] C. Howlett and A. Taylor, *Analyst*, **103**, 916 (1978).
[469] B. L. Therrell, Jr., J. M. Drosche and T. W. Dziuk, *Clin. Chem.*, **24**, 1182 (1978).
[470] J. Smeyers-Verbeke, D. Verbeelen and D. L. Massart, *Clin. Chim. Acta*, **108**, (1980).
[471] J. V. Dunckley and F. A. Staynes, *Ann. Clin. Biochem.*, **14**, 53 (1977).
[472] E. J. Hinderberger, M. L. Kaiser and S. R. Koirtyohann, *Atom. Spectrosc.*, **2**, 1 (1981).
[473] P. A. Pleban and K. H. Pearson, *Anal. Lett.*, **12**, 935 (1979).
[474] P. Frigieri and R. Trucco, *Spectrochim. Acta*, **35B**, 113 (1980).
[475] K. Dittrich, *Zentralbl. Pharm. Pharmakother. Laboratoriumsdiagn.*, **117**, 761 (1978).
[476] G. D. Christian and J. J. Feldman, *Atomic Absorption Spectroscopy: Applications in Agriculture, Biology and Medicine*, Wiley, New York, 1970.
[477] R. Berman, *Appl. Spectrosc.*, **29**, 1 (1975).
[478] J. B. Dawson and F. W. Heaton, *Spectrochemical Analysis of Clinical Materials*, Thomes, Springfield, USA (1967).
[479] W. J. Price, *Med. Lab. World*, **2**, 135 (1978).
[480] G. J. Waughman and T. Brett, *Environ. Res.*, **21**, 385 (1980).
[480a] *Talanta*, 1982, **29**, No. 11B.
[480b] L. J. Hinks, M. Colmsee and H. T. Delves, *Analyst*, **107**, 815 (1982).
[481] F. J. Langmyhr and D. Tsalev, *Anal. Chim. Acta*, **92**, 79 (1977).
[482] S. W. King, M. R. Wills and J. Savory, *Anal. Chim. Acta*, **128**, 221 (1981).
[483] J. E. Gorsky and A. A. Dietz, *Clin. Chem.*, **24**, 169 (1978).
[484] O. Oster, *Clin. Chim. Acta*, **114**, 53 (1981).
[485] D. L. Collett, D. E. Fleming and G. A. Taylor, *Analyst*, **103**, 1074 (1978).
[486] K. Morita and M. Mishima, *Bunseki Kagaku*, **30**, 170 (1981).
[487] T. J. Kneip, *Health Lab. Sci.*, **14**, 53 (1977).
[488] E. Chreneková and N. Rusinová, *Chem. Listy*, **72**, 990 (1978).
[489] D. H. Cox, *J. Anal. Toxicol.*, **4**, 207 (1980).
[490] G. Devoto, *Boll. Soc. Ital. Biol. Sper.*, **44**, 425 (1968).
[491] G. Drasch, L. von Meyer and G. Kauert, *Z. Anal. Chem.*, **304**, 141 (1980).
[492] D. S. Grewal and F. X. Kearns, *Atom. Absorp. Newslett.*, **16**, 131 (1977).
[493] N. Rombach and K. Kock, *Z. Anal. Chem.*, **292**, 365 (1978).
[494] R. C. Rooney, *Analyst*, **101**, 749 (1976).
[495] M. Palliere and G. Gernez, *Ann. Pharm. Franc.*, **34**, 183 (1976).
[496] P. Allain, *Clin. Chim. Acta*, **64**, 281 (1975).
[497] P. Del Castilho and R. E. M. Herber, *Anal. Chim. Acta*, **94**, 269 (1977).
[498] V. Lagesson and L. Andrasko, *Clin. Chem.*, **25**, 1948 (1979).
[499] F. J. Langmyhr, B. Eyde and J. Jonsen, *Anal. Chim. Acta*, **107**, 211 (1979).
[500] P. E. Gardiner, J. M. Ottaway and G. S. Fell, *Talanta*, **26**, 841 (1979).
[501] N. Lekehal, M. Hanocq and M. Helson-Cambier, *J. Pharm. Belg.*, **32**, 76 (1977).
[502] D. J. Bell, J. L. Braidwood, J. G. Campbell, J. Culbert, J. Filshie and P. E. Lake, *Comp. Biochem. Physiol.*, **19**, 133 (1966).
[503] H. J. Hendriks and W. Klazinga, *Tijdschr. Diergeneesk.*, **90**, 155 (1965).
[504] E. Newbrun, *Nature*, **192**, 1182 (1961).
[505] F. J. Fernandez and H. L. Kahn, *Atom. Absorp. Newslett.*, **3**, 24 (1971).
[506] B. E. Capeland, D. W. Grisley, J. Casella and H. Bailey, *Am. J. Clin. Pathol.*, **66**, 619 (1976).
[507] J. Kocian and I. Rubeška, *Cesk. Gastroenterol. Vyziva*, **22**, 188 (1968).

[508] M. L. Girard and F. Rousselet, *Ann. Pharm. Franc.*, **25**, 271 (1967).

[509] A. M. Briscoe and C. Rogan, *J. Appl. Physiol.*, **20**, 453 (1965).

[510] G. Devoto, *Boll. Soc. Ital. Biol. Sper.*, **44**, 1251 (1968).

[511] W. Seeling, A. Grünert, K. H. Kienle, R. Opferkuch and M. Swobodnik, *Z. Anal. Chem.*, **299**, 368 (1979).

[512] Y. Hayashi, K. Hunakawa, N. Yoshida, M. Ishizawa and R. Tsujino, *Bunseki Kagaku*, **25**, 409 (1976).

[513] M. F. Moster, O. Riebel and J. Komárek, *Cesk. Oftalmol.*, **36**, 190 (1980).

[514] K. Murakami, Y. Ito, K. Taguchi, K. Ogata and T. Imanari, *Bunseki Kagaku*, **30**, 200 (1981).

[515] W. T. Binnerts and T. Achterop, *Tijdschr. Diergeneesk.*, **92**, 639 (1967).

[516] M. L. Girard, *Clin. Chim. Acta*, **20**, 243 (1968).

[517] J. B. Dawson, D. J. Ellis and H. Newton-John, *Clin. Chim. Acta*, **21**, 33 (1968).

[518] C. B. Lawrence and M. Phillippo, *Anal. Chim. Acta*, **118**, 153 (1980).

[519] W. Schmidt, *Z. Anal. Chem.*, **243**, 198 (1968).

[520] O. Wawschinek, *Mikrochim. Acta*, 111 (1979 **II**).

[521] G. Devoto, *Boll. Soc. Ital. Biol. Sper.*, **44**, 1249 (1968).

[522] K. Furuno, *Okayama Daigaku Onsen Kenkyusho Hokoku*, 13 (1979).

[523] D. G. Berge and R. T. Pflaum, *Am. J. Med. Technol.*, **34**, 725 (1968).

[524] S. Melethil, A. Poklis and V. A. Sagar, *J. Pharm. Sci.*, **69**, 585 (1980).

[525] A. Lorber, R. L. Cohen, C. C. Chang and H. E. Anderson, *Arthritis Rheumat.*, **11**, 170 (1968).

[526] S. M. Pederson and P. M. Graabaek, *Scand. J. Clin. Lab. Invest.*, **37**, 91 (1977).

[527] P. Tavenier and H. B. A. Hellendoorn, *Clin. Chim. Acta*, **23**, 47 (1969).

[528] P. Baily, H. B. Rollin and T. A. Kilroe-Smith, *Microchem. J.*, **26**, 250 (1981).

[529] G. Devoto, *Rass. Med. Sarda*, **71**, 357 (1968).

[530] M. E. Tatro, W. L. Raynolds and F. M. Costa, *Atom. Absorp. Newslett.*, **16**, 143 (1977).

[530a] S. K. Giri, C. K. Shields, D. Littlejohn and J. M. Ottaway, *Analyst*, **108**, 244 (1983).

[531] P. A. Pleban and K. H. Pearson, *Anal. Lett.*, **12**, 935 (1979).

[532] E. Auermann, G. Heidel, J. Cumbrowski, J. Jacobi and U. Meckel, *Dtsch. Gesundheitswes.*, **33**, 1769 (1978).

[533] P. N. Vijan and G. R. Wood, *Analyst*, **101**, 966 (1976).

[534] F. Alt and H. Massmann, *Spectrochim. Acta*, **33B**, 337 (1978).

[535] G. Garelli, V. Rimatori and B. Sperduto, *Med. Lav.*, **70**, 313 (1979).

[536] C. Minoia, G. Catenacci, A. Baruffini and A. Prestinoni, *Ann. Ist. Super. Sanita*, **14**, 753 (1978).

[537] P. Baily, E. Norval, T. A. Kilroe-Smith, M. I. Shikne and H. B. Roellin, *Microchem. J.*, **24**, 107 (1979).

[538] R. Knutti, C. Balsiger and C. Schlatter, *Mitt. Geb. Lebensmittelunters. Hyg.*, **68**, 78 (1977).

[539] D. J. Hodges, *Analyst*, **102**, 66 (1977).

[540] G. H. Hisayasu, J. L. Cohen and R. W. Nelson, *Clin. Chem.*, **23**, 41 (1977).

[541] V. Lehmann, *Clin. Chim. Acta*, **20**, 523 (1968).

[542] A. E. Woods, R. D. Crowder, J. T. Coates and J. J. Wittrig, *Atom. Absorp. Newslett.*, **7**, 85 (1968).

[543] R. Giese, *Ärtzl. Lab.*, **12**, 285 (1966).

[544] W. K. Stewart, F. Hutchinson and L. W. Fleming, *J. Lab. Clin. Med.*, **61**, 858 (1963).

[545] L. W. Fleming and W. K. Stewart, *Clin. Chim. Acta*, **14**, 131 (1966).

[546] A. D. Care, *Nature*, **199**, 818 (1963).

[547] W. E. C. Wacker, C. Iida and K. Fuwa, *Nature*, **202**, 659 (1964).

[548] F. Rousselet, *Afinidad*, **22**, 185 (1965).

[549] W. Scholl, *Landwirt. Forsch.*, **19**, 131 (1965).

[550] J. L. Hansen and E. F. Freier, *Am. J. Med. Technol.*, **33**, 158 (1967).

[551] D. L. Tsalev, F. J. Langmyhr and B. Gunderson, *Bull. Environ. Contam. Toxicol.*, **17**, 660 (1977).

[552] T. Watanabe, R. Tokunaga, T. Iwahana, M. Tati and M. Ikeda, *Br. J. Ind. Med.*, **35**, 73 (1978).

[553] D. C. Sharma and P. S. Davis, *Clin. Chem.*, **25**, 769 (1979).

[554] R. C. Rooney, *Analyst*, **101**, 678 (1976).

[555] T. R. Collier, *U.K. Atom. Energy Auth., Rept.*, M 2930 (1978).

[556] R. A. Richardson, *Clin. Chem.*, **22**, 1604 (1976).

[557] M. R. Greenwood, P. Dhahir, T. W. Clarkson, J. P. Farant, A. Chartrand and A. Khayat, *J. Anal. Toxicol.*, **1**, 265 (1977).

[558] S. Dogan and W. Haerdi, *Anal. Chim. Acta*, **84**, 89 (1976).

[559] T. Ohkawa, H. Uenoyama, K. Tanida and T. Ohmae, *Eisei Kagaku*, **23**, 13 (1977).

[560] S. Dogan and W. Haerdi, *Intern. J. Environ. Anal. Chem.*, **6**, 327 (1979).

[561] P. Coyle and T. Hartley, *Anal. Chem.*, **53**, 354 (1981).

[562] R. Knutti and C. Balsiger, *GIT Labor. Med.*, 201 (1978).

[563] K. H. Schaller, A. Kuehner and G. Lehnert, *Blut*, **17**, 155 (1968).

[564] D. B. Adams, S. S. Brown, F. W. Sunderman, Jr. and H. Zachariasen, *Clin. Chem.*, **24**, 862 (1978).

[565] A. H. Jones, *Anal. Chem.*, **48**, 1472 (1976).

[566] L. J. Dillon, D. C. Hilderbrand and K. S. Groon, *Atom. Spectrosc.*, **3**, 5 (1982).

[567] Y. Tada, T. Yonemoto, A. Iwasa and K. Nakagawa, *Bunseki Kagaku*, **29**, 248 (1980).

[568] D. B. Lo and G. D. Christian, *Microchem. J.*, **23**, 481 (1978).

[569] J. B. Willis, *Spectrochim. Acta*, **16**, 551 (1960).

[570] D. C. Curnow, D. H. Gutteridge and E. D. Horgan, *Atom. Absorp. Newslett.*, **7**, 45 (1968).

[571] C. D. Wall, *Clin. Chim. Acta*, **76**, 259 (1977).

[572] J. P. Buchet, E. Knepper and R. Lauwerys, *Anal. Chim. Acta*, **136**, 243 (1982).

[573] M. L. Girard and F. Rousselet, *Ann. Pharm. Franc.*, **25**, 353 (1967).

[574] A. Saleh, J. N. Udall and N. W. Solomons, *Clin. Chem.*, **27**, 338 (1981).

[575] G. P. Butrimovitz and W. C. Purdy, *Anal. Chim. Acta*, **94**, 63 (1977).

[576] J. R. Kelson, *Clin. Chem.*, **26**, 349 (1980).

[577] A. S. Prasad, D. Oberleas and J. A. Halsted, *J. Lab. Clin. Med.*, **66**, 508 (1965).

[578] T. Haas, G. Lehnert and K. H. Schaller, *Beckman Rept.*, 3 (1967).

[579] T. Haas, G. Lehnert and K. H. Schaller, *Z. Klin. Chem. Klin. Biochem.*, **5**, 218 (1967).

[580] H. Matsumoto, K. Tsummatsu and T. Shiraishi, *Bunseki Kagaku*, **17**, 703 (1968).

[581] K. Oiwa, T. Kimuro, M. Mikino and M. Okuda, *Bunseki Kagaku*, **17**, 810 (1968).

[582] F. L. Humoller and D. J. Mahler, *Clin. Chem.*, **13**, 40 (1967).

[583] N. Honegger, *Ärtzl. Lab.*, **9**, 41 (1963).

[584] K. Matsumiya, T. Yoshinaga and K. Omori, *Rinsho Byori*, **16**, 103 (1968).

[585] K. Fuwa, P. Pulido, R. McKay and B. L. Vallee, *Anal. Chem.*, **36**, 2407 (1964).

[586] J. R. McDermott and I. Whitehill, *Anal. Chim. Acta*, **85**, 195 (1976).

[587] S. H. Weissman and D. B. DeNicola, *Bull. Environ. Contam. Toxicol.*, **27**, 139 (1981).

[588] H. Woidich and W. Pfannhauser, *Nährung*, **24**, 367 (1980).

[589] A. Dornemann and H. Kleist, *Z. Anal. Chem.*, **294**, 402 (1979).

[590] S. E. Raptis, W. Wegscheider and G. Knapp, *Mikrochim. Acta*, 93 (1981 I).

[591] P.-O. Berggren, *Anal. Chim. Acta*, **119**, 161 (1980).

[592] H. Koizumi and K. Yasuda, *Anal. Chem.*, **48**, 1178 (1976).

[593] M. Blanusa and D. Breski, *Talanta*, **28**, 681 (1981).

[594] G. Ramelow, S. Tugrul, M. A. Ozkan, G. Tuncel, C. Saydam and T. I. Balkas, *Intern. J. Environ. Anal. Chem.*, **5**, 125 (1978).

[595] K.-R. Sperling, *Z. Anal. Chem.*, **299**, 103 (1979).

[596] S. K. Syamal, J. C. Williams and G. M. A. Palmieri, *Anal. Lett.*, **12**, 1451 (1979).

[597] P.-O. Berggren, O. Berglund and B. Hellman, *Anal. Biochem.*, **84**, 393 (1978).

[598] P. Schramel, *Anal. Chim. Acta*, **67**, 69 (1973).

[599] J. J. Christensen, P. A. Hearty and R. M. Izatt, *J. Agr. Food Chem.*, **24**, 811 (1976).

[600] U. Pfueller, V. Fuchs, S. Golbs, E. Ebert and D. Pfeifer, *Arch. Exp. Veterinaermed.*, **34**, 367 (1980).

[601] U. Harms and J. Kunze, *Z. Lebensm.-Unters. Forsch.*, **164**, 204 (1977).

[602] L. Wuyts, J. Smeyers-Verbeke and D. L. Massart, *Clin. Chim. Acta*, **72**, 405 (1976).

[603] F. H. Nielson, M. L. Sunde and W. G. Hoekstro, *J. Nutr.*, **89**, 24 (1966).

[604] P. Baily, T. A. Kilroe-Smith and H. B. Roellin, *Lab. Pract.*, **29**, 141 (1980).

[605] S. R. Koirtyohann, G. Wallace and E. Hinderberger, *Can. J. Spectrosc.*, **21**, 61 (1976).

[606] K. Nakamura, M. Fujimori, H. Tsuchiya and H. Orii, *Anal. Chim. Acta*, **138**, 129 (1982).

[607] H. Kamel, D. H. Brown, J. M. Ottaway and W. E. Smith, *Talanta*, **24**, 309 (1977).

[608] R. L. Deter, H. Martinez and D. Cantu, *Tech. Rept. Biol. Med.*, **33**, 407 (1975).

[609] R. J. Bittel and P. P. Graham, *J. Assoc. Off. Anal. Chem.*, **60**, 63 (1977).

[610] K. W. Jackson, E. Marczak and D. G. Mitchell, *Anal. Chim. Acta*, **97**, 37 (1978).

[611] K. Stegavik, G. Mikalsen, E. M. Ophus and E. A. Mylius, *Bull. Environ. Contam. Toxicol.*, **15**, 734 (1976).

[612] P. J. Barlow and A. K. Khera, *Atom. Absorp. Newslett.*, **14**, 149 (1975).

[613] W. Matthes, R. Flucht and M. Stoeppler, *Z. Anal. Chem.*, **291**, 20 (1978).

[614] D. R. Boureier and R. P. Sharma, *J. Anal. Toxicol.*, **5**, 65 (1981).

[615] Y. Horimoto and S. Nishi, *Bunseki Kagaku*, **27**, 69 (1978).

[616] N. P. Elakhovskaya, E. K. P. Ershova and A. I. Itskova, *Gig. Sanit.*, **64** (1978).

[617] M. F. Pera, Jr. and H. C. Harder, *Clin. Chem.*, **23**, 1245 (1977).

[618] G. Tanaka, A. Tomikawa and H. Kawamura, *Bull. Chem. Soc. Japan*, **50**, 2310 (1977).

[619] D. C. Reamer and C. Veillon, *Anal. Chem.*, **53**, 1192 (1981).

[620] S. Dogan and W. Haerdi, *Intern. J. Environ. Anal. Chem.*, **8**, 249 (1980).

[621] A. M. Ure and M. C. Mitchell, *Anal. Chim. Acta*, **87**, 283 (1976).

[622] J. Zvara, *Bodenkultur*, **27**, 361 (1976).

[623] Y. Yano, N. Odaka, S. Takei and K. Nagashima, *Bunseki Kagaku*, **27**, T 25 (1978).

[624] R. C. Daniel, *Mikrochim. Acta*, 289 (1977 II).

[625] M. Shamsipoor and F. Wahdat, *Z. Anal. Chem.*, **288**, 191 (1977).

[626] H.-U. Meisch and W. Reinle, *Mikrochim. Acta*, 505 (1977 I).

[627] R. Albert, E. Beigl, H. Kinzel and G. M. Steiner, *Z. Pflanzenphysiol.*, **80**, 43 (1976).

[628] T. P. Gaines and G. A. Mitchell, *J. Assoc. Off. Anal. Chem.*, **61**, 1179 (1978).

[629] M. Wettern, *Z. Anal. Chem.*, **292**, 279 (1978).

[630] S. Nakamura, N. Fudagawa and A. Kawase, *Bunseki Kagaku*, **29**, 477 (1980).

[631] R. R. Eltou-Bott, *Anal. Chim. Acta*, **86**, 281 (1976).

[632] T. C. Woodis, Jr., G. B. Hunter and F. J. Johnson, *Anal. Chim. Acta*, **90**, 127 (1977).

[633] D. R. Boline and W. G. Schrenk, *Appl. Spectrosc.*, **30**, 607 (1976).

[634] A. M. Ure, M. P. Hernandez-Artiga and M. C. Mitchell, *Anal. Chim. Acta*, **96**, 37 (1978).

[635] D. J. David, *Commun. Soil. Sci. Plant Anal.*, **11**, 189 (1980).

[636] O. C. Bataglia, *Cinc. Cult.*, **29**, 71 (1977).

[637] R. J. Zasoski and R. G. Burau, *Commun. Soil. Sci. Plant Anal.*, **8**, 425 (1977).

[638] W. J. Arian and M. L. Stevens, *Analyst*, **102**, 446 (1977).

[639] W. J. Simmons, *Anal. Chem.*, **50**, 870 (1978).

[640] W. Griebenow, B. Werthmann and H.-J. Sembritzki, *Material Prüfung*, **18**, 51 (1976).

[641] A. Clement, *Bull. Ec. Natl. Super. Agron. Ind. Aliment.*, **22**, 75 (1980).

[642] V. D. Mattera, Jr., V. A. Arbige, Jr., S. A. Tomellini, D. A. Erbe, M. M. Doxtader and R. K. Forcé, *Anal. Chim. Acta*, **124**, 409 (1981).

[643] S. Salmela, E. Vuori and J. O. Kilpio, *Anal. Chim. Acta*, **125**, 131 (1981).

[644] W. W. Harrison, J. P. Yurachek and C. A. Benson, *Clin. Chim. Acta*, **23**, 83 (1969).

[645] J. F. Chapman and B. E. Leadbeatter, *Anal. Lett.*, **13**, 439 (1980).

[646] T. Tanaka, Y. Hayashi, K. Funakawa and M. Ishizawa, *Nippon Kagaku Kaishi*, 169 (1981).

[647] D. Gardner and G. Dal Pont, *Anal. Chim. Acta*, **108**, 13 (1979).

[648] L. Fiková, *Chem. Listy*, **74**, 533 (1980).

[649] G. Bagliano, F. Benischek and I. Huber, *Anal. Chim. Acta*, **123**, 45 (1981).

[650] J. A. Hurlbut, *Atom. Absorp. Newslett.*, **17**, 121 (1978).

[651] R. Nakashima, *Bunseki Kagaku*, **27**, 185 (1978).

[652] J. F. Alder, A. J. Samuel and T. S. West, *Anal. Chim. Acta*, **92**, 217 (1977).

[653] G. D. Renshaw, C. A. Pounds and E. F. Pearson, *Nature*, **238**, 162 (1972).

[654] G. D. Renshaw, C. A. Pounds and E. F. Pearson, *J. Forensic Sci.*, **18**, 143 (1973).

[655] J. F. Alder, A. J. Samuel and T. S. West, *Anal. Chim. Acta*, **87**, 313 (1976).

[656] J. F. Alder, A. J. Samuel and T. S. West, *Anal. Chim. Acta*, **94**, 187 (1977).

[657] J. W. Robinson and S. Weiss, *J. Environ. Sci. Health*, **15A**, 663 (1980).

[658] K. Katsuzawa, *Nagano-Ken Eisei Kogai Kenkyusho Kenkyu Hokoku*, 15 (1979).

[659] L. M. Klevay, *Am. J. Clin. Nutr.*, **23**, 284 (1970).

[660] H. G. Petering and D. W. Yeager, *Proc. West. Hemisphere Nutr. Congr.*, **II**, 38 (1969).

[661] H. A. Schroeder and A. P. Nason, *J. Invest. Dermatol.*, **53**, 71 (1969).

[662] J. G. Reinhold, G. A. Kfoury and M. Arslanian, *J. Nutr.*, **96**, 519 (1968).

[663] U. G. Oleru, *Am. Ind. Hyg. Assoc. J.*, 229 (1975).

[664] L. Kopito, R. K. Byers and H. Schwachman, *N. Engl. J. Med.*, **276**, 949 (1967).

[665] J. P. Creason, T. A. Hinners, J. E. Bumgarner and C. Pinkerton, *Clin. Chem.*, **21**, 603 (1975).

[666] H. K. Y. Lau and H. Ashmead, *Anal. Lett.*, **8**, 815 (1975).

[667] J. R. J. Sorenson, E. G. Melby, P. J. Nord and H. G. Petering, *Arch. Environ. Health*, **27**, 36 (1973).

[668] H. Hagedorn-Götz, G. Kuppers and M. Stoeppler, *Arch. Toxicol.*, **38**, 275 (1977).

[669] J. F. Alder, D. Alger, A. J. Samuel and T. S. West, *Anal. Chim. Acta*, **87**, 301 (1976).

[670] Ch. A. Helsby, *Anal. Chim. Acta*, **82**, 427 (1976).

[671] J. A. Hurlbut, *U.S. Energy Res. Dev. Adm. Rept.*, PEP-2443 (1976).

[672] D. C. Wigfield, S. M. Croteau and S. L. Perkins, *J. Anal. Toxicol.*, **5**, 52 (1981).

[673] J. F. Alder, C. A. Pankhurst, A. J. Samuel and T. S. West, *Anal. Chim. Acta*, **91**, 407 (1977).

[674] S. M. De Antonio, S. A. Katz, D. M. Scheiner and J. D. Wood, *Anal. Proc.*, **18**, 162 (1981).

[675] R. S. Farag, S. T. El-Aassar and M. M. Soliman, *J. Drug Res.*, **12**, 227 (1980).

[676] R. S. Farag, S. T. El-Aassar, M. A. Mostafa and E. A. Abdel-Rahim, *J. Drug Res.*, **12**, 217 (1980).

[677] J. F. McMullin, J. G. Pritchard and A. H. Sikondari, *Analyst*, **107**, 803 (1982).

[678] A. Sohler, P. Wolcott and C. C. Pfeiffer, *Clin. Chim. Acta*, **70**, 391 (1976).

[679] C. A. Helsby, *Anal. Chim. Acta*, **69**, 259 (1974).

[680] C. A. Helsby, *Talanta*, **24**, 46 (1977).

[681] P. Quint and H. J. Höhling, *Z. Anal. Chem.*, **296**, 411 (1979).

[682] H. Malissa, K. Maly and T. Till, *Z. Anal. Chem.*, **293**, 141 (1979).

[683] A. Reda, B. N. Srinvasan and F. Brudevold, *J. Dent. Res.*, **52** (1974), Supplement IADR Abstract 126.

[684] F. J. Langmyhr and A. Sundli, *Anal. Chim. Acta*, **73**, 81 (1974).

[685] F. J. Langmyhr, T. Lin and J. Jonsen, *Anal. Chim. Acta*, **80**, 297 (1975).

[686] F. Dolinšek, J. Štupar and M. Špenko, *Analyst*, **100**, 884 (1975).

[687] T. Okano, K. Ikebe, T. Ichikawa and M. Kondo, *Eisei Kagaku*, **24**, 159 (1978).

[688] T. Okano, T. Ichikawa, M. Kondo and K. Ikebe, *Eisei Kagaku*, **24**, 231 (1978).
[689] R. Nakashima, *Bunseki Kagaku*, **27**, 185 (1978).
[690] F. Ishino, H. Matsumae, K. Shibata, N. Ariga and F. Goshima, *Bunseki Kagaku*, **27**, 232 (1978).
[691] E. Norval, *Anal. Chim. Acta*, **97**, 399 (1978).
[692] F. J. Langmyhr and I. Kjuus, *Anal. Chim. Acta*, **100**, 139 (1978).
[693] C. Selden and T. J. Peters, *Clin. Chim. Acta*, **98**, 47 (1979).
[694] S. R. Himmelhoch, H. A. Sober, B. L. Vallee, E. A. Peterson and K. Fuwa, *Biochem.*, **5**, 2523 (1966).
[695] J. B. Dawson, M. H. Bahreyni-Toosi, D. J. Ellis and A. Hodgkinson, *Analyst*, **106**, 153 (1981).
[696] H. Sakurai, T. Tachikawa and S. Shimomura, *Anal. Lett.*, **11**, 879 (1978).
[697] K. H. Falchuk and N. Engle, *J. Med.*, **296**, 1129 (1977).
[698] E. Graf and F. J. Langmyhr, *Magy. Kem. Foly.*, **86**, 412 (1980).
[699] K. T. Suzuki, *Anal. Biochem.*, **102**, 31 (1980).
[700] H. T. Delves, *Clin. Chim. Acta*, **71**, 495 (1976).
[701] J. Teape, H. Kamel, D. H. Brown, J. M. Ottaway and W. Smith, *Clin. Chim. Acta*, **94**, 1 (1979).
[702] M. Pinta, D. Baron, C. Riandey and W. Ghidalia, *Spectrochim. Acta*, **33B**, 489 (1978).
[703] H. Kamel, D. H. Brown, J. M. Ottaway and W. E. Smith, *Analyst*, **102**, 645 (1977).
[704] J. W. Foote and H. T. Delves, *Analyst*, **107**, 121 (1982).
[705] T. Stiefel, K. Schulze, G. Tölg and H. Zorn, *Z. Anal. Chem.*, **300**, 189 (1980).
[706] M. Suzuki, T. L. Coombs and B. L. Vallee, *Anal. Biochem.*, **32**, 106 (1969).
[707] J. B. Carlsen, *Anal. Biochem.*, **64**, 53 (1975).
[708] J. F. O'Brien and M. E. Emmerling, *Anal. Biochem.*, **85**, 377 (1978).
[709] B. Tan and P. Melius, *Anal. Proc.*, **18**, 384 (1981).
[710] B. Tan and P. Melius, *Anal. Lett.*, **14**, 311 (1981).
[711] M. Cais, S. Dani, Y. Eden, O. Gandolfi, M. Horn, E. E. Isaacs, Y. Josephy, Y. Saar, E. Slovin and L. Snarsky, *Nature*, **270**, 534 (1977).
[712] K. Sumino, R. Yamamoto, F. Hatayama, S. Kitamura and H. Itoh, *Anal. Chem.*, **52**, 1064 (1980).
[713] H. Sanui and N. Pace, *Appl. Spectrosc.*, **20**, 135 (1966).

13

Miscellaneous applications

13.1 FOOD PRODUCTS

Analysis of foodstuffs is of great importance, because of possible contamination during processing and packaging. Atomic-absorption spectrometry is widely used for metal analysis by food manufacturers and government inspectors [1-4]. Most foodstuffs, however, require a preparation step involving wet digestion with strong acids, dry ashing or dilution, depending on the nature of the sample. In analysis for heavy toxic metals, the mineralization step is the main source of difficulty. Dry ashing seems best suited for this type of problem as it is safer, can be run automatically and needs smaller quantities of reagents [5,6]. Although dry ashing at temperatures below 500°C is commonly used when volatile toxic metals are present [7,8], addition of sulphuric acid as an ashing aid for some metals such as lead and cadmium permits increase in the mineralization temperature up to 980°C [9]. This technique has proved to be successful with vegetables, dairy products, meat and fish.

A comparison of some methods used for wet digestion of foodstuffs [10-13] and the accuracy of results at levels in the region of the detection limit [14] have been discussed. It should be noted that most of the inconsistencies arise from differences in the nature of the matrices. Thus, it is essential to treat samples individually rather than to accept a standard procedure for all matrices. Contamination from reagents, losses due to volatilization at the ashing temperature, and losses by retention when the element of interest reacts with the ashing vessel or solids within the ash, should all be avoided.

13.1.1 Grains and Cereals

In determination of metals in grains, the sample is first dried at 40-105°C, then ground to a fine powder and subjected to wet or dry decomposition. Lead and cadmium in wheat grain are determined by digestion with 5:1 v/v nitric-perchloric acid mixture in a Kjeldahl flask and injection of a portion of the diluted digest into a graphite furnace [15]. Potassium, magnesium, calcium, sodium, manganese and iron are similarly determined in wheat grain and flours by prior ashing of either the ground grain with HNO_3-$HClO_4$-H_2SO_4 mixture or of the unground samples with HNO_3-$HClO_4$ in a pressure vessel for 3 hr at 70°C [16]. Determination of arsenic in rice by wet digestion with H_2SO_4-HNO_3-$HClO_4$ at 380°C and arsine generation has been adapted by Yasui and Tsutsumi [17]. The precision is 5-8% for 0.1 μg of As_2O_3 and 1-4% for 0.4 μg.

Talc on rice is determined by removing the coating by washing with hot 5% H_2O_2 solution and concentrated ammonia solution, followed by acidification of the suspension with hydrochloric acid, heating to boiling, filtration, ashing at 550°C and measurement of magnesium in the ash [18]. The talc content is obtained by multiplying the magnesium content by the factor 5.5. The error of the method is ±5%. The ash content of flour-mill streams of hard red spring wheat has been estimated by determination of the manganese content of the water extract [19]. Mercury in some cereals and cereal products is determined by grinding the samples, mixing with sodium carbonate and heating at 550°C in a stream of oxygen in a silica tube [20]. The combustion products are passed through silver gauze to trap the mercury liberated, and this mercury is then released by heating at 500°C, absorbed in potassium permanganate solution and determined by cold vapour AAS. The limit of detection is 3 ng/g but the method gives values lower than those reported by other workers. Soya beans, wheat and lucerne containing down to 2 ppm of copper and down to 16 ppm of iron are decomposed by wet or dry digestion, the residue is dissolved in 2:3 v/v hydrochloric acid-methanol mixture and aspirated into an air-acetylene flame [21].

Low levels of metals are determined by using electrothermal atomization and/or prior extraction of the metal. Tin in potatoes has been determined with a precision better than ±5% by digestion with nitric acid under reflux, followed by evaporation in the presence of sulphuric acid and atomization in a graphite tube furnace [22]. When the tin content is low (< 1 ppm) the tin neocupferron complex is extracted into chloroform and the organic extract is atomized. The chromium content of the whole grain or flour of wheat and rye is determined by ashing at 480°C, oxidation to Cr(VI) and extraction into methyl isobutyl ketone (MIBK). However, a comparative study of the results obtained by 6 laboratories for three standard flours showed wide and unexplained differences [23]. Cadmium in polished rice is determined by low-temperature ashing, extraction with diethyldithiocarbamate into **MIBK**, and aspiration into an air-acetylene flame [24].

13.1.2 Beverages

13.1.2.1 Fruit and vegetable juices. Fruit juices stored in glass or tin-plated containers are commonly analysed for tin. The samples are digested with hydrochloric acid to eliminate interferences from organic materials, diluted and atomized. The tin content in the juice ranges from 0 to 497 ppm, depending on the condition of the containers [25]. A long-path absorption cell (60 cm × 12 mm) fitted with a hydride inlet and a nitrogen-hydrogen flame system has been used to enhance the sensitivity and reliability of measurement of tin in canned fruit juices [26]. Low levels of tin can be determined by co-precipitation with zirconium hydroxide and electrothermal AAS measurement [27]. Although Al, Fe, Zn, Cu, Ph, Ni, Cd and As are also co-precipitated, they do not interfere in up to 100-fold amounts relative to tin.

The methods used for AAS determination of Al, Sb, As, B, Cd, Ca, Cr, Co, Cu, Fe, Pb, Mg, Mn, Ni, K, Se, Na, Sn and Zn in beverages, including sample preparation and applications of flame and furnace atomization techniques, have been reviewed [28]. The general procedure is centrifugation or filtration to remove suspended matter and pulp, followed by dilution or acidification before atomization. The effect of several analytical factors on the results of simultaneous determination of Pb, Zn, Cu and Cd can be minimized by using the standard addition technique [29]. Florida orange juice has been analysed for Ca, Cu, Fe, K, Mg, Mn and Na after hydrolytic preparation [30]. Lemon, orange or mandarin juice is mixed with equal volumes of 50% ammonium nitrate solution and injected into the graphite furnace for the determination of Pb, Cu, Fe and Zn [31]. The results obtained agree satisfactorily with those obtained by wet digestion with nitric acid or dry ashing with magnesium nitrate at $\sim 450°C$. Lead in grape juice is determined after concentration with immobilized ED3A on a glass support [32]. Canned vegetable juices are analysed for lead by dry ashing at $\sim 450°C$ and extraction [33]. Many other cations have been determined in pineapple and orange juice [34,35]. Chloride in Italian citrus fruit juice has been determined by precipitation with silver nitrate and measurement of the unconsumed silver by flame AAS [36]. Arsenic is determined by arsine generation and atomization in a hydrogen–oxygen flame [37].

13.1.2.2 Beers and wines. The concentration of some metals in beers affects the stability and quantity of foam and is the subject of legislation by some countries. The results of a collaborative study of the AAS procedures of the American Society of Brewery Chemists for the determination of Cu, Fe and Ca (adopted by the European Brewery convention) have been summarized and statistically evaluated [38]. A collaborative study of analysis for Mg and Zn in wort and beer has also been evaluated [39]. Helin and Slaughter [40] determined Mn, Mg, Ca, Fe, Cu, Al, Ni, Co, and Pb in malt, hops, wort and beer, and traced the fate of these metals in the brewing process. Several trace elements in beer have been determined by AAS after either prior dilution or extraction [41–43].

Metals in wines are commonly determined by direct atomization of a diluted solution. For samples containing high alcohol or sugar content the standards should be carefully matrix-matched with respect to ethanol or glucose. The results obtained for the determination of sodium and potassium in 65 wine samples from the Mancha region (Spain) showed that pretreatment with hydrochloric acid to remove ethanol and other organic matter gave better precision than direct injection did [44]. The effect of viticultural conditions on the manganese content of Galician wine (Spain) [45] and the aluminium content and precipitate caused by aluminium in Tokay [46] have been evaluated on dry-ashed samples. Wines may also be analysed by evaporation to dryness with sulphuric acid and then ashing at 500°C [47]. Many of these methods have been used for the determination of a variety of trace metals [48–51].

13.1.3 Edible Fats and Oils

Atomic-absorption spectrometry has been extensively used for the determination of metals in fats and oils [52,53] since the presence of small quantities of metals is known to have serious deleterious effects on the quality and stability of these products. A quite general procedure for the analysis of these products is up to tenfold dilution with an appropriate organic solvent followed by direct atomization into the flame or graphite furnace. Phosphorus in edible oil is determined by 1:1 dilution with either a 0.5% chloroform solution of lanthanum acetylacetonate [54] or a 1% MIBK solution of lanthanum 4-cyclohexylbutyrate [55] and injection into the graphite furnace. The sensitivity is 8 ng of phosphorus for 1% absorption and the detection limit is 0.5 μg/g. The reproducibility is 2% at phosphorus concentration > 20 μg/g and 5% at 3 μg/g. Vegetable oils are dissolved in either MIBK–ethanol mixture (8:3) or propionic acid and analysed directly for iron at levels < 5 ppm, with a coefficient of variation of < 5% [56]. Many other metals, such as Na, Ca, Mg, Zn, Fe, Pb, Cu, Al, K, Ni, Mn and Sn, can also be determined by direct atomization [57,58].

Trace levels of Cu, Fe, Ni and Pb in finished margarines are determined by dry ashing at 450°C and electrothermal atomization [59]. Iron in fats and oils is similarly determined by use of an air–acetylene flame [60]. Chiricosta and Bruno [61] showed that dry ashing of essential oils at 550°C leads to losses of some metals by volatilization, whereas digestion with H_2SO_4–HNO_3 mixture leads to difficulties in the destruction of hydrocarbons. They recommended extraction with hydrochloric acid, acetic acid or citric acid to ensure complete recoveries of such metals. When the concentration of the metals is less than 1 μg/g, an enrichment procedure is required. This is usually dry ashing [62–64] or wet digestion [65–68], followed by extraction of the metals and atomization.

13.1.4 Fish and Meat

13.1.4.1 Fish. Mercury is one of the most important and widely determined metals in sea-food. The analysis is commonly done by homogenization of the tissues by digestion, and atomization of the digest electrothermally or by the cold-vapour technique. Wet digestion of fish products with a mixture of concentrated nitric and sulphuric acids for 30 min at 140°C in an autoclave [69] has been combined with measurement of mercury by electrothermal atomization in a closed recirculating system [70] or by a stationary cold-vapour technique [71]. Digestion with HNO_3–H_2SO_4 mixture in the presence of permanganate

[72] or vanadium pentoxide [73] gives results similar to those obtained by the more hazardous and less convenient A.O.A.C. (25.103) method. Shrimp, crab and cod-liver can be similarly digested and their mercury and selenium contents measured by the cold-vapour and hydride generation techniques respectively [74]. Hydrogen peroxide alone or in admixutre with acids may also be used for digestion of fish tissues. The homogenized, freeze-dried and ground tissues are allowed to react with H_2O_2 [75] at 80°C or with HNO_3-H_2O_2 [76,77], followed by cold-vapour measurement of mercury. Digestion with H_2O_2-H_2SO_4-$KMnO_4$-$(NH_4)_2S_2O_8$ mixture [78] followed by exchanging the mercury in the tissue for copper [79] will permit measurement of mercury at levels as low as 0.14 ppm in fish tissues. Organomercury compounds in canned tuna have been measured by extraction into benzene–cysteine mixture, separation by gas-liquid chromatography and analysis of each fraction by electrothermal AAS [80]. Methylmercury in homogenized fresh or freeze-dried fish can be extracted with concentrated hydrobromic acid and toluene, the organic phase treated with cysteine, and the elemental mercury generated measured by the cold-vapour technique [81].

Nanogram levels (as low as 0.25 ng) of mercury in fish tissues can be determined by using a dual-channel instrument and the cold-vapour technique, at the rate of 60 samples per hour [82]. Fish gonads and liver have been digested with nitric acid under pressure and the mercury released enriched by collection on gold, and measured [83]. The limit of detection is 0.4 ng. A comparison has been made of the combustion–gold-trap technique and the wet-digestion-reduction method [84]; higher values were obtained with the first method. In another comparative study, however, no significant difference was found between the results obtained by using (a) dry ashing followed by amalgamation on gold, (b) digestion with H_2SO_4-HNO_3 mixture at 97°C for 3 hr, and (c) digestion with H_2SO_4-HNO_3-HF in a PTFE bomb at 150°C [85]. Fish tissue has also been dry-ashed by heating with MgO, SiO_2 and V_2O_5 at 800°C for 5 min in a stream of oxygen, this being followed by absorption of mercury in iodine, reduction, and measurement by the cold-vapour technique [86].

Other elements in sea-food have also been determined by wet or dry decomposition and flame or electrothermal AAS measurements. Arsenic and selenium in fish tissues have been determined by digestion with HNO_3-$HClO_4$-H_2SO_4 (9:2:1 v/v) for 3 hr and hydride generation [87]. The limit of detection was ~ 5 ng for both elements. Total inorganic arsenic has been determined either by distillation as $AsCl_3$ (b.p. 130°C) from $6M$ hydrochloric acid in the presence of iron(III) sulphate, or by chelation and extraction into an MIBK solution of ammonium pyrrolidinedithiocarbamate, followed by AAS [88]. Total arsenic in tuna and shrimp has been determined by wet digestion with HNO_3-$HClO_4$-H_2SO_4 (in the presence of nickel nitrate) at 190°C for 90 min, followed by electrothermal atomization [89]. The wet digestion procedure has been used for the determination of Cd, Pb, Cu, Cr, Ni and Zn. The digest is electrothermally atomized directly or after extraction of these metals into organic solvents

[90-94]. Dry ashing by heating at ∼ 500°C [95,96], combustion in an oxygen-filled flask [97] and by using a low-temperature asher [98] has also been used for the determination of various heavy metals in fish and tuna tissues. A glass nebulizer for the injection of a discrete volume (up to 500 μl) of the digest into an air–acetylene flame has been described [99].

13.1.4.2 Oysters, lobsters and clams. Oysters, lobsters and clams are analysed by similar methods. Wet digestion with HNO_3-$HClO_4$ [100] and direct injection of the digest into a graphite furnace allows measurement of lead and cadmium with recoveries of 90-110% and detection limits of 4 pg for lead and 0.2 pg for cadmium. Cadmium may also be extracted from the digest before atomization [101]. Clams and oysters can be analysed for Pb, Cd, Zn and Cu by mineralization by heating with sulphuric acid and then dry ashing at 475°C, followed by extraction of the metals and aspiration into a lean air–acetylene flame [102]. Recoveries are 94-100%, except for Pb and Cd in the presence of high levels of Cu and Zn (the extraction is then incomplete). A simpler method for the determination of cadmium in oysters and lobsters has been described, involving homogenization with $1M$ ammonium sulphate at pH 4, extraction with chloroform–hexane mixture and electrothermal AAS [103]. The detection limit is ∼ 50 ng per g of raw tissue.

13.1.4.3 Meat. A comparison has been made of atomic-absorption spectrometry, neutron-activation analysis (NAA) and X-ray fluorescence (XRF) methods for the determination of Zn, Fe, Rb and Cu in dried turkey breast [104]. For Zn, AAS and NAA gave similar results but the XRF values were significantly higher than the AAS values. For Rb, NAA and XRF gave similar results. Copper and iron were not determined by NAA, and copper was not determined by XRF. In general, the AAS methods were the most precise. Sodium and potassium in broiler-chicken breast muscles have been determined by digestion with nitric acid and with 70% w/v perchloric acid at 170°C [105]. The K:Na ratio is ∼ 6.5 in untreated chicken, but approaches 1.2 for chicken injected with aqueous sodium salt solutions. The iron content of pork can be determined by dry ashing and AAS measurement [106]. Slavin and Peterson [107] determined heavy metals in beef liver, beef muscles and turkey muscles by dry ashing and flame AAS for Cr, Fe and Zn, and electrothermal AAS for Cd, Co, Pb and Ni.

13.1.5 Fruits and Vegetables

Wehrer *et al.* [108] studied the methods used for the determination of tin in canned fruits and vegetables. This involved investigation of the sample preparation procedures (acid digestion and dry ashing), type of flame, and sources of interference. The authors recommended decomposition of the sample with hydrochloric acid and, in case of doubt, use of dry ashing for comparison.

Digestion with 3:1 v/v HNO_3-H_2SO_4 mixture may also be used for green beans, apple and fruits [109]. The digest is mixed with saturated ammonium chloride solution and methanol, and aspirated into an acetylene–nitrous oxide flame. Insecticides containing tin (e.g. cyhexatin, tricyclohexyltin and fenbutatin oxide) and present in apples, oranges and other crops can be determined by extraction with a mixture of diethyl ether and acetic acid or of hexane and acetic acid and clean-up on a basic alumina or silica gel column, followed by measurement of tin by electrothermal atomization [110–112].

Arsenic in fruit and vegetables has been determined by wet digestion with HNO_3-H_2SO_4 mixture or dry ashing at 550°C in the presence of magnesium nitrate, followed by arsine generation and atomization in the flame [113]. Total Cr, Co and Ag in spinach [114], Pb, Cd, Zn and Cu in fresh potatoes, carrots, cabbage, canned peas, french beans, cucumbers and tomato [115], Cu and Fe in peas, french beans and asparagus [116] and Na and K in cabbage, squash and potato [117] have been measured by dry ashing at $\sim$ 500°C and atomization in an air–acetylene flame.

13.1.6 Milk and Dairy Products

Milk can be introduced directly, after simple dilution or after protein precipitation, into the flame or graphite furnace. Cadmium and lead have been determined by diluting milk with Soluene-350 and toluene as solubilizing agent [118]. The sodium content of cheese has been determined by solubilizing the samples with a mixture of 8% aqueous ammonia solution and MIBK [119] and aspiration into an air–acetylene flame. The results compared favourably with those obtained by wet ashing with perchloric acid. A modified nebulizer has been devised to provide stable aerosols to feed the aspirate to a high-solids single-slot burner and allow analysis of powdered milk with solid content > 50% [120]. Cadmium and copper in powdered milk can be determined by treating the sample with 10% ammonium citrate solution, adjusting the pH to 8 and extracting with sodium diethyldithiocarbamate into MIBK [121].

Casein-free milk can be used for the determination of the major cations in milk, such as Ca, Mg, Na and K, at concentrations of 1.2, 0.1, 0.5 and 1.5 g/l., respectively [122]. Trichloroacetic acid is used for protein precipitation [123]. Similar results are obtained with ashed milk samples except for potassium, for which the ashing procedure gives much lower values. It has been shown that many of the metals in milk can be determined by AAS, with ashed samples. Bruhn and Franke [124] determined lead and cadmium in California raw milk by dry ashing of a sample of the whole milk at a temperature < 325°C and extraction of the metals from the ash with ammonium pyrrolidine dithiocarbamate into isoamyl acetate. A comparative study on the determination of lead in evaporated milk by extraction–AAS, anodic stripping voltammetry and direct plating, gave similar results for two samples containing 0.15 and 0.4 ppm Pb [125]. Oxygen-flask combustion, bomb [126] and oxygen plasma ashing techniques have been utilized for the determination of selenium [126] and chromium [127]. The effect of several analytical factors on the results of simultaneous determination of Pb, Zn, Cu and Cd, has been reported [128]. The presence of 10% sugar in evaporated milk interferes in the determination of Zn and Mg [129].

Wet digestion of milk samples has also been used in determination of some metals. Cadmium and lead in pasteurized liquid milk have been determined by electrothermal AAS [130]. The freeze-dried sample is digested with 70% nitric acid for 3 hr at 150°C in a closed stainless-steel bomb. Manning [131] has described a method for the determination of lead in evaporated milk by digestion with nitric acid, chelation, extraction into butyl acetate, and either flame (Delves cup) or electrothermal atomization. A similar method has been used for the determination of nickel in condensed milk [132]. The content of Cu, Fe and Mn in Swedish market milk has been surveyed by wet digestion with either 1N sulphuric acid (for Cu and Fe) or 1M hydrochloric acid (for Mn) [133]. Copper and iron in milk and cream [134] have been determined by treatment of the samples with a mixture of water, concentrated aqueous ammonia and 1,4-dioxan (1:2:1 v/v). Butter and cheese samples are digested with concentrated nitric acid at 100°C. The separated fat is extracted with light petroleum and the aqueous phase is diluted and injected into the graphite furnace [134].

13.1.7 Tea and Coffee

Tea leaves and coffee are dried, crushed to pass through a 0.5–0.7 mm sieve and then oven-dried at 105°C before dry ashing or wet digestion. Tea leaves containing naturally occurring tin and tricyclohexyltin hydroxide and its degradation products are mixed with Na_4EDTA solution, dried, and ashed at 550°C for 6 hr, then the tin is extracted into an MIBK solution of ammonium pyrrolidine dithiocarbamate. The detection limit is 0.01 ppm Sn [135]. If a silica or aluminium vessel (open or closed) is used for the ashing, the temperature should not exceed 550°C, since significant amounts of some metals (e.g. Cu, Zn) are retained[136] and other metals (e.g. Ni) are lost at higher temperatures [137]. Extraction of Pb, Cd and Cu after ashing increases the reproducibility of the results. Average values for Cu, Pb and Cd in 54 tea samples were 11.9, 0.31 and 0.023 $\mu g/g$, respectively [138].

Wet digestion with 1:10 v/v H_2SO_4-HNO_3 or HNO_3-$HClO_4$ mixture, with or without subsequent extraction, has been used in conjunction with flame or electrothermal AAS measurements for determining Mn and Co in tea [139,140]. Tsushida and Takeo [141] determined the concentration of some trace metals and the infusion rates of a range of teas. Tea infusions were prepared by steeping green or black tea in boiling water for 5 min, then filtered and injected into a graphite furnace. The copper level of an infusion is found to be strongly influenced by the copper content of the tea itself. Lead is not detected in a second

Table 13.1. Bibliographical survey of some AAS methods used for the determination of trace elements in foodstuffs

Element	Grains and cereals	Beverages	Fats and oils	Fish and meat	Fruits and vegetables	Milk and dairy products	Tea and coffee	Miscellaneous foodstuffs
Aluminium		28,40,46	57,61					
Antimony		28						222
Arsenic	17	28,37,144		78,87,89	113		142	211–228,230, 231,239
Boron		28		185				
Cadmium	15,24	28,29	63,146	90–92,94,95, 100–103,175	115	118,121,124,128, 130	138,141, 142	218,227,228, 238–242,245
Calcium	16	28,30,34,38,40, 42,159,164	57,170,171	182,197		156,157,204		248,250
Chloride		36,161						
Chromium	23	28,160,164		94,95,175	114	127		
Cobalt		28,40,160		191	114		139	
Copper	21	28–31,38,40,42, 48,51,158–160, 164,166,167	57–59,61, 174	94,99,102,104, 150,175,179, 182–184,192,193	115,116	121,122,128,133, 134,155–157, 208,210	136,138, 141	218,228,232, 233,246–250
Iron	16,21	28,30,31,34,38, 40,42,49,51,143, 159,160,164,166	56–61	99,104,106,148, 182,192	116	133,134		243,244, 248–250,253
Lead	15	28,29,31–33,40, 41,47,48,143, 165,168	57,59,61, 146	91–95,100,102, 147,150,178, 186,190,198	115	118,124,125,128, 130,205,206	138,141, 142	218,227–229, 234–244
Lithium		164		152				

Magnesium	16,18	28,30,39,40,51, 144,160,163	57,171	182,192		129		248,250
Manganese	16,19	28,30,40,45, 160,164	57,58,61, 63,174	147,182,192		133	140	246,248–250
Mercury	20	64,68,146		69–86,92,149, 151,176,177, 187–189,194– 196,199,200		153	142	227,229,239, 254–256
Molybdenum						209		
Nickel		28,40,158,160	58,59,65, 173,174	94,98		132	137	240
Phosphorus	16		54,55,174					
Potassium		28,30,34,44, 159,162,164	57,172	183,201		204		248,250–252
Rhodium			66					
Rubidium		164		104				
Selenium			67	74,87		126,154		218,224,229, 257–259
Silver	16				114			
Sodium	16	28,30,44,159, 162,164,169	57,172	105,183		119,204		248,250–252
Strontium				180,202		207,208		
Tin	22	25–28,34,145	57	181	108–112		135	221,243,244, 260–262
Vanadium				98				
Zinc		28–31,47,51, 160,164	57,63	94,99,102,104, 175,182,192, 197,203	115	122,128,129	136	218,228,246, 249,250

infusion of either green or black tea, owing to its low content. Mercury, arsenic, cadmium and lead at the nanogram level in raw coffee can be determined by heating the samples at 190°C with nitric acid in a pressurized vessel for 3-4 hr. For the determination of lead and cadmium it is recommended to extract both metals into xylene with diethylammonium diethyldithiocarbamate [142].

A bibliographical survey of some methods used for the determination of trace elements in various food products is given in Table 13.1.

13.2 INDUSTRIAL PRODUCTS

13.2.1 Petroleum Products
Trace metals in crude oil and its products are commonly determined by AAS. The presence of some metals (e.g. Ni, Cu, Fe) in the crude oil causes poisoning of the catalyst used in the cracking process, and other metals (e.g. V) cause corrosion problems, thus reducing the efficiency of the refining plant [263-265]. On the other hand, the nature and level of metals in the oil can be used for identification of the source of the oil. The presence of copper in the refined products promotes formation of gums and lacquers during heating or storage and thus accelerates the deterioration of these products and reduces the storage stability. Wear metals in lubricating oil are also monitored to provide early detection and correction of major trouble in the various machine parts. For example, the presence of Pb, Sn or Sb can arise from wear in a bearing, whereas high levels of Fe, Cr and Ni indicate piston wear. A leak in the cooling system is associated with the presence of Na and B.

Most of the metallic species in petroleum products are present as volatile organometallic compounds, salts, fine metallic particles and colloidal suspensions. The general pretreatment procedures used before flame or electrothermal atomization are: (a) dilution with a suitable organic solvent or mixture of solvents (e.g. xylene, MIBK, dioxan, iso-octane, heptane); (b) dilution with organic solvents in the presence of an emulsifier, and atomization of the emulsion; (c) treatment with mixtures of acids followed by dilution with an organic solvent; (d) extraction of the metallic species; (e) dry decomposition at 500-600°C in the presence of an ashing aid.

13.2.1.1 Crude oils. Before analysis, crude oils are warmed to ~ 60°C in a water-bath. This causes mixing before direct introduction of a portion of the homogeneous solution into the flame or graphite furnace directly or after up to tenfold dilution with dioxan [266], *p*-xylene [263,267], white spirit (b.p. 140-197°C) [268], or n-heptane [269]. Grizzle *et al.* [270] evaluated four techniques for the determination of V and Ni in crude oil. These authors found that dilution with organic solvents followed by atomization in the flame gives erratic and inaccurate results, whereas reliable and reproducible results are obtainable by wet digestion.

Dry and wet ashing of crude oils before atomization have both been investigated. Lead in petroleum is determined by heating with sulphuric acid, ashing in a muffle furnace at 550°C, co-precipitation with thorium hydroxide to minimize the interference from V, Ni, Na and Fe, and finally electrothermal atomization [271]. Lead levels down to 10 ng/g can be satisfactorily measured. Vanadium is determined by a similar ashing procedure [272]. The ash is dissolved in nitric acid, followed by dilution, treatment with permanganate to oxidize all forms of V to V(V), and extraction of this as its complex with 4-(2-pyridylazo)resorcinol, into chloroform. *Trans*-1,2-diaminocyclohexane-*N,N,N',N'*-tetra-acetic acid is added to mask the interference of other metals, although Na, Mg, Al and Fe still interfere when present in 5000-fold amounts relative to vanadium. A technique for continuous monitoring of mercury in oil-shale gases has been described, involving prior heating to 900°C in a tube furnace and use of Zeeman-AAS [273]. Sychra *et al.* [274] studied the effect of interferences in the AAS determination of V, Ni and Fe in petroleum by mineralization, addition of K^+ and Al^{3+} as ionization suppressors, and use of an oxidizing flame. They compared their results with those obtained by spectrophotometry and square-wave polarography.

13.2.1.2 Fuels. Copper and nickel [266] and zinc [275] in petrol have been determined by direct aspiration of the undiluted samples. Nickel in petroleum samples may be determined by dilution in xylene and atomization in an air-acetylene or nitrous oxide–acetylene flame [276]. Eight different organonickel compounds of the type encountered in petroleum have been tested for calibration. However, the results do not agree with those obtained after prior mineralization.

A significant improvement can be ensured by using the standard-addition technique. Low levels of copper in gasoline can be determined by 1:2 dilution with iso-octane before injection into the graphite furnace [277]. Manganese, added as methylcyclopentadienylmanganesetricarbonyl to jet fuel, is determined by dilution of the fuel with organic solvent and aspiration into the flame [278]. Beryllium in petroleum products has been similarly determined [279,280].

Vanadium at the ppm level in fuel gas is determined by 1:9 dilution of the sample with MIBK and direct atomization [281]. Lower concentrations require either prior separation and concentration or direct electrothermal atomization [282]. Copper, nickel and lead in petroleum products are determined by extraction and concentration, followed by electrothermal atomization of the organic extract [283]. Vigler and Gaylor [284] described an analytical procedure for the determination of 23 elements in petrol products. These (i.e. Al, Sb, Ba, Be, Bi, Ca, Cd, Co, Cr, Cu, Fe, K, Pb, Mg, Mn, Mo, Ni, Na, Si, Sn, Ti, V, Zn) are determined by ashing the samples at 650°C with "magnesium sulphonate" as an ashing aid, followed by flame AAS.

Determination of the total lead content [275,285–295] and speciation of various alkyl-leads [296-302] in fuel have been thoroughly investigated. These methods have been discussed in detail in Chapter 10. The official (and most commonly used) method for the determination of total lead in gasoline (ASTM) is based on dilution of the sample with MIBK, addition of iodine and a quaternary ammonium salt, and flame AAS [303]. Solvents other than MIBK may also be used [304]. The difference between results obtained with toluene and MIBK as diluent is < 4%. Petrol samples can also be analysed by reaction with iodine in benzene solution, followed by addition of an emulsifier (Emulsogen) and atomization of the emulsion [305]. For lead levels in the range 0.2-0.6 g/l., the results do not differ by more than 2% from those obtained by the ASTM method. Extraction of lead into diethylammonium diethyldithiocarbamate in xylene and electrothermal atomization of the organic extract has also been reported [306]. Determination and speciation of various lead alkyls in gasoline are commonly achieved by prior chromatographic separation followed by AAS measurement [296-302]; see Chapter 10.

13.2.1.3 Lubricating oils. Analysis of lubricating oils for metals is of great importance since many of the commonly used colour improvers, antifoam agents, oxidation inhibitors, corrosion inhibitors, detergents, pour-point depressents and viscosity index improvers contain metallic components. On the other hand, the degree of component wear is ascertained by determination of the concentration of metallic species present in the used lubricating oil. Several techniques [307-311] for AAS determination of wear metals in used lubricating oils have been described. Automated procedures for the determination of Fe, Ni, Cr, Pb, Cu, Ag, Sn, Mg and Al in aircraft lubricating oil have been reported [312]. The application of electrothermal atomization [313,314], the effect of ionization suppressors [315], and the interference of common elements in the additives used in lubricating oil (e.g. Ba, P, Zn), in the AAS determination of the common wear metals (e.g. Fe, Cr, Ag, Mg, Cu, Ni, Pb) [316] have all been studied.

Additives in fresh lubricating oil and wear metals in used oil can be determined by dilution with organic solvents and atomization. The most commonly used solvents are aromatic hydrocarbons, ketones, alcohols and acids. n-Heptane [317] and p-xylene [318-321] are used for the determination of Ca, Ba, Zn and Ni. p-Xylene has also been recommended for dilution of jet engine lubricants for the flame AAS determination of Fe, Ag, Cu, Mg, Cr, Sn, Pb and Ni [321]. A mixture of 2-methylpropan-2-ol and toluene (3:2 v/v) has been used as diluent for lubricants analysed for Ba either alone or in the presence of Ca [322]. A mixed solvent of toluene and acetic acid (1:5 v/v) may be utilized for the determination of Ca, Mg and Zn [323]. Lubricating oils may also be aspirated into the flame as an emulsion. Zinc [324] and lead [325] are determined by heating the lubricant to 50-60°C, filtering, treating with 10% solution of Hoechst MS12 lipophilic surfactant in benzene, aqueous 4% polysorbate-20 and water. The mixture is ultrasonically agitated and the resulting emulsion is diluted with water and aspirated into an air-acetylene flame.

MIBK is widely used as diluent in the AAS determination of many metals. Antimony dialkyldithiocarbamate and dialkylphosphorodithionate, which are added as anti-oxidant and anti-wear reagents can be determined in an MIBK solution of the oil [326]. Lead diamyldithiocarbamate is similarly determined [327]. MIBK has been used for dilution of both railway engine oils [328] and lubricants based on lithium sebacate [329]. Isobutyric acid is also shown to provide a suitable medium for the determination of Ca, Zn, Cu, Fe and Ba in used and fresh lubricating oil. The results are in good agreement with those obtained by using xylene or MIBK as solvents [330].

Although wear metals are present in the form of fine metallic particles or colloidal suspensions and can be atomized completely with hot flames, organic diluents containing acid have been recommended to effect complete solubilization of the metallic species. A mixture consisting of cyclohexanone-butan-1-ol-ethanol-hydrochloric acid-water (10:6:4:1:1 v/v) has been used for the determination of Ca and Zn [331]. Methyl hexyl ketone-ethanol-water-hydrochloric acid mixture (85:13.5:0.5:1 v/v) may also be used [332]. Dilution with either MIBK and a 16:1:3 methanol-water-hydrochloric acid mixture [333,334] or MIBK containing small amounts of HCl-HF mixture [335] has also been utilized.

Several procedures for the determination of Al, Cu, Fe, Mg, Mo, Ni, Sn and Ti in lubricating oil have been published which incorporate prior acid dissolution of the metal particles followed by dilution with organic solvents [333,336,337]. The oil is shaken with a 1:7 v/v mixture of concentrated hydrofluoric acid and nitric acid, then the mixture is diluted with MIBK and atomized in a nitrous oxide-acetylene flame. Application of this method to samples containing particles of molybdenum powder as large as 200 mesh gives results with an accuracy of 95.5 ± 2.6% [338]. A substantial improvement of this method has recently been described by Kauffman et al. [339] for the determination of various metallic wear species in lubricating oils and hydraulic fluids. The oil sample is treated with HF-HCl-HNO$_3$ mixture (1:8:1 v/v), ultrasonically agitated at 65°C, and diluted with a mixture of MIBK and Neodol 91-6. The latter solvent is an ethoxylated primary alcohol (Shell Chemical Co.), having the molecular formula RO(CH$_2$CH$_2$O)$_6$H where R is a C$_9$H$_{19}$, C$_{10}$H$_{21}$ or C$_{11}$H$_{23}$ linear chain. The recoveries of various metals range from 89% to 102%, with relative standard deviation of 2-12%.

Lubricating oils may also be decomposed before atomization. An improved Wickbold apparatus has been suggested for combustion of MIBK solutions of used lubricants, followed by electrothermal atomization [340]. A comparison has been made of decomposition methods used in the determination of traces of Ca, V, and Zn in spent lubricating oil, including dry ashing, wet digestion, oxygen-flask combustion, cold ashing in an oxygen-plasma asher, and heating under pressure with concentrated nitric acid at 160°C [341]. A comparison of the results obtained by these methods with those obtained by dilution of the oil with toluene–acetic acid mixture (7:3 v/v) and direct atomization shows the superiority of high-pressure acid treatment.

13.2.1.4 Greases and waxes. Metals in greases can be determined after treatment by one of two methods: dissolution in a suitable organic solvent [342] or dry ashing [317] followed by dissolution of the residue in hydrochloric acid. The first approach has been used for the determination of Li, Na and Ca in grease by homogenizing 10–200-mg samples with butanol and $1M$ hydrochloric acid, followed by atomization of the acid extract [342]. The second method has been used for the determination of molybdenum [343].

Lead in wax crayons is determined by inserting the solid samples (~ 0.15 mg) into the graphite furnace, drying at 100°C for 60 sec, charring at 500° for 120 sec and atomizing at 2300°C for 10 sec. Alternatively, the sample is dry ashed at 500°C, then the residue is dissolved in nitric acid and atomized. The two procedures give closely similar results which compare favourably with those obtained by differential pulse polarography [344].

Table 13.2 summarizes some selected methods for the AAS determination of various elements in petroleum products.

13.2.2 Surfactants

Organic compounds functioning as wetting, foaming, dispersing, emulsifying and/or penetrating agents are known as detergents or surfactants, the latter name being a convenient contraction of 'surface active agents'. These compounds are commonly classified according to whether or not they are ionized in solution, and the nature of the ionic or electrical charge. The use of these compounds is steadily increasing, as indicated by the world production. However, these compounds cause pollution of surface water. Monitoring of low levels of surfactants in surface water and effluents is based chiefly on their reaction with metal-containing compounds to form adducts that can be extracted and measured by AAS.

13.2.2.1 Non-ionic surfactants. Sheridan *et al.* [345] determined some non-ionic surfactants (e.g. Tween-80, carbowax and emulphor) by precipitation with excess of phosphomolybdic acid in the presence of barium chloride, and digestion at 85°C for 1 hr, followed by AAS measurement of the molybdenum content of the solution. The reaction was done in acid medium (1.25% v/v hydrochloric acid) and the precipitate separated by either centrifugation or filtration. This method (without the hydrochloric acid) has been used by Chlebicki and Garncarz [346,347] for the determination of 0.4–1.2 mg/l. levels of non-ionic surfactants in surface water, effluents and sewage treatment plants, by prior extraction into ethyl acetate. All types of non-surfactants obtained by

Table 13.2. AAS determination of some elements in petroleum products

Element	Petroleum product	Pretreatment step	Atomization system	Reference
Aluminium	Lubricating oils, hydraulic fluids,	Mixing with HF–HNO₃–HCl, MIBK–Neodol 96-6	F	339
	Kerosine, naphtha, furnace oil, heating gas oil	Dry ashing at 650°C	F	284
Antimony	Kerosine, naphtha, furnace oil, heating gas oil	Dry ashing at 650°C	F	284
Barium	Lubricating oils	Dilution with isobutyric acid	F	330
		Dilution with MIBK	F	310
		Dilution with xylene	F	318
Beryllium	Kerosine, naphtha, furnace oil, heating gas oil	Dry ashing at 650°C	F	284
Bismuth	Kerosine, naphtha, furnace oil, heating gas oil	Dry ashing at 650°C	F	284
Cadmium and cobalt	Kerosine, naphtha, furnace oil, heating gas oil	Dry ashing at 650°C	F	284
Calcium	Lubricating oils	Dilution with toluene–acetic anhydride	F	323
		Dilution with isobutyric acid	F	330
		Dilution with xylene	F	318
		Dilution with white spirit	F	315
		Dilution with MIBK	F	310
		Combustion in oxygen-filled flask	F	341
Chromium	Kerosine, naphtha, furnace oil, heating gas oil	Dry ashing at 650°C	F	284
	Lubricating oils	Dilution with organic solvents	F	316
		Mixing with HF–HNO₃–HCl, MIBK–Neodol 96-6	F	339
		Dilution with xylene	F	308
Copper	Kerosine, naphtha, furnace oil, heating gas oil	Dry ashing at 650°C	F	284
	Gasoline	Dilution with 2,2,4-trimethylpentane	GF	283
		Dilution with iso-octane	GF	277
	Kerosine, naphtha, furnace oil, heating gas oil	Dry ashing at 650°C	F	284
	Lubricating oils	Dilution with isobutyric acid	F	330
		Dilution with organic solvents	F	316
		Dilution with xylene	F	308
		Mixing with HF–HNO₃–HCl, MIBK–Neodol 96-6	F	339
Iron	Kerosine, naphtha, furnace oil, heating gas oil	Dry ashing at 650°C	F	284
	Lubricating oils	Dilution with organic solvents	F	316
		Dilution with isobutyric acid	F	330
		Dilution with xylene	F	308
		Mixing with HF–HNO₃–HCl, MIBK–Neodol 96-6	F	339
		Mineralization with H₂SO₄	F	274

Table 13.2. *continued*

Element	Petroleum product	Pretreatment step	Atomization system	Reference
Lead	Gasoline	Dilution with methyl ethyl ketone	F	286
		Dilution with MIBK–I$_2$	CR	292
		Dilution with toluene–I$_2$	F	304
	Gasoline, bunker heating oil, jet engine lubricating oils, crude oil	Extraction	CR	293
	Kerosine, naphtha, furnace oil, heating gas oil	Dry ashing at 650°C	F	284
	Crude oils	Mineralization with H$_2$SO$_4$ at 500°C	GF	271
	Lubricating oils	Dilution with organic solvents	F	316
		Dilution with emulsifier	F	325
		Mixing with HF–HNO$_3$–HCl, MIBK–Neodol 96-6	F	339
		Dilution with xylene	F	308
	Petrol	Dilution with benzene–I$_2$	F	305
	Naphtha	Extraction	GF	306
	Wax	Dry ashing at 500°C	GF	344
Magnesium	Kerosine, naphtha, furnace oil, heating gas oil	Dry ashing at 650°C	F	284
	Lubricating oils	Dilution with organic solvents	F	316
		Dilution with toluene–acetic anhydride	F	323
		Dilution with xylene	F	308
Manganese	Kerosine, naphtha, furnace oil, heating gas oil	Mixing with HF–HNO$_3$–HCl, MIBK–Neodol 96-6	F	339
Mercury	Kerosine, naphtha, furnace oil, heating gas oil	Dry ashing at 650°C	F	284
Molybdenum	Oil shale	Heating at 900°C	ZGF	273
	Kerosine, naphtha, furnace oil, heating gas oil	Dry ashing at 650°C	F	284
	Lubricating oils	Mixing with HF–HNO$_3$–HCl, MIBK–Neodol 96-6	F	339
		Mixing with HF–HNO$_3$–MIBK	F	338
Nickel	Crude oil	Dilution with organic solvents or wet ashing	F,GF	270
	Gasoline	Dilution with xylene	F	276
	Kerosine, naphtha, furnace oil, heating gas oil	Dry ashing at 650°C	F	284
	Lubrication oils	Dilution with organic solvents	F	316
		Mixing with HF–HNO$_3$–HCl, MIBK–Neodol 96-6	F	339
		Dilution with xylene	F	308

Element	Material	Method	Technique	Page
Potassium	Petroleum products	Mineralization with H_2SO_4	F	274
	Kerosine, naphtha, furnace oil, heating gas oil	Dry ashing at 650°C	F	284
Silicon	Lubricating oils	Mixing with HF–HNO_3–HCl, MIBK–Neodol 96-6	F	339
	Kerosine, naphtha, furnace oil, heating gas oil	Dry ashing at 650°C	F	284
Silver	Lubricating oils	Dilution with organic solvents	F	316
	Lubricating oils	Dilution with xylene	F	308
Sodium	Kerosine, naphtha, furnace oil, heating gas oil	Dry ashing at 650°C	F	284
Tin	Kerosine, naphtha, furnace oil, heating gas oil	Dry ashing at 650°C	F	284
	Lubricating oils	Dilution with xylene	F	308
Titanium	Kerosine, naphtha, furnace oil, heating gas oil	Dry ashing at 650°C	F	284
	Lubricating oils	Dilution with MIBK, HCl–HF	F	336
		Mixing with HF–HNO_3–HCl, MIBK–Neodol 96-6	F	339
Vanadium	Crude oil	Dilution with organic solvents or wet ashing	F,GF	270
	Kerosine, naphtha, furnace oil, heating gas oil	Dry ashing at 650°C	F	284
	Fuel oil	Dilution with MIBK	F	281
	Petroleum products	Dry ashing at 600°C, extraction	GF	272
		Dilution with xylene	F	267
		Mineralization with H_2SO_4	F	274
	Lubricating oils	Combustion	GF	340
		Combustion or dilution with toluene–acetic anhydride	F	341
Zinc	Kerosine, naphtha, furnace oil, heating gas oil	Dry ashing at 650°C	F	284
	Lubricating oils	Dilution with xylene	F	318
		Dilution with isobutyric acid	F	330
		Dilution with n-heptane	F	343
		Dilution with toluene–acetic anhydride	F	323
		Mixing with HF–HNO_3–HCl, MIBK–Neodol 96-6	F	339
		Combustion or dilution with toluene–acetic anhydride	F	341
		Dilution with emulsifier	F	324
Mineral oils		Dilution with white spirit	F	268

F = flame,   GF = graphite tube furnace,   CR = carbon rod,   ZGF = Zeeman graphite tube furnace oxyethylation of alkylphenols, alcohols and fatty amines are extracted practically quantitatively (apparent recovery 97.3-102.1%) into the organic phase. However, block co-polymers of the pluronic and tetronic types cannot be extracted under these conditions. No interferences are caused from all metal salts and protein substances which are known to affect the AAS measurement of molybdenum. Cationic surfactants cannot be extracted under these conditions and anionic surfactants (e.g. Manoxol OT) are only 10% extracted.

Non-ionic surfactants have been determined by extraction into benzene at pH 7.5 as their cobaltothiocyanate adducts $[Co_n.RO(CH_2OCH_2)_{6n}H]^{2n+}$ $[Co(SCN)_4^{2-}]_n$. The benzene extract is evaporated to dryness, the residue is dissolved in MIBK, and cobalt is measured by flame [348] or electrothermal AAS [349]. Triton X-100, Mergital ST 30, Mergital LM11 and Sinnopal PN 17 at the concentration level of 0.02-0.5 mg/l. have been satisfactorily determined with minimal interference from anionic surfactants. Polyoxyethylene oxide may also be precipitated with a modified Dragendorff's reagent. The precipitate is washed, dried, and dissolved in ammonium tartrate solution, and its bismuth content is measured and compared with a calibration graph prepared from ethoxylated nonylphenol (10 ethylene oxide units) [350]. It is possible to determine polyoxyethylene surfactants down to the 20-µg/l. level by prior concentration and purification steps [351]. This is done by extracting the surfactants with ethyl acetate, evaporating the extract to dryness, dissolving the precipitate in 75% ethanol, passing the solution through a column of Amberlite IR-120 B plus IRA410, and measuring of the bismuth content of the complex. The sensitivity of this method is 10 times that of the tetraisothiocyanatocobaltate method and is applicable to the determination of polyethylene nonionic surfactants in river water and sea-water [352].

Triton X-100 at concentrations of 0.05-2 mg/l. has been determined by extraction as a neutral adduct with potassium tetrathiocyanatozincate at pH 6-8 into 1,2-dichlorobenzene, followed by AAS measurement of the zinc content [353]. This method requires a single phase-separation step, is most suitable for surfactants containing between 8 and 40 ethoxy units, and is relatively free from interference by soaps (sodium stearate) and anionic surfactants. Cationic surfactants, however, form extractable ion-association complexes and thus interfere.

13.2.2.2 Anionic surfactants. Linear alkylbenzene sulphonates (LAS) are the most widely used anionic surfactants. Le Bihan and Courtot-Coupez [354] described procedures for the determination of this class of surfactants in seawater [354] and fresh water [355]. The anionic surfactants Manoxol OT and dodecylbenzene sulphate are extracted into MIBK as ion-association compounds with tris(1,10-phenanthroline)copper(II) and the copper content of the organic phase is measured by AAS [354-356]. The limit of detection is 1-3 µg/l. To overcome interference by other anions in water, a multiple extraction procedure is used.

Extraction of anionic surfactants with bis(ethylenediamine)copper(II) into chloroform has also been suggested [357]. The chloroform extract containing $Cu(en)_2(LAS)_2$ is shaken with dilute nitric acid and the ethylenediaminecopper(II) stripped into the aqueous phase is directly measured in an air–acetylene flame. The method can be used for fresh, estuarine and sea waters. The limit of detection is ~ 60 $\mu g/l$. (LAS). Lower concentrations are determined by a similar procedure entailing electrothermal atomization AAS [358]. With a 750-ml water sample, the limit of detection for LAS is 2 $\mu g/l$. The only ions that interfere at the concentrations likely to be found in polluted water are sulphide and iron(III). These interferences, however, can be eliminated by treating the water sample with 30% H_2O_2 and 2% EDTA before addition of the bis(ethylenediamine)copper(II) reagent. Non-ionic surfactants (e.g. Triton X-100) do not interfere.

An alternative counter-ion is the thiourea complex of copper(I). Matsueda and Morimoto [359] showed that addition of thiourea to an aqueous medium containing copper(II) results in the formation of various thiourea–copper(I) complexes which are extractable as ion-pairs with LAS into MIBK. This permits measurement of the LAS (0.01–20 $\mu g/l$.) by means of the copper concentration. Thiocyanate, nitrite and iodide interfere, but their effect can be removed by washing the organic layer with 0.1M sodium borate or phosphate–hydrochloric acid buffer of pH 3. Cationic surfactants seriously interfere. Matsueda [360] determined LAS and di(2-ethylhexyl)sulphosuccinate by extraction of their sodium salts at pH > 5 into MIBK. The organic layer was analysed for sodium by flame photometry. The method was applied to tap, river and sewage waters. Cationic (e.g. benzethonium chloride) and amphoteric surfactants (e.g. lauryl betain), if present, interfere seriously.

The total phosphate content in liquid and powder detergents (2–17% P_2O_5) has been determined by measuring the emission of the HPO species at 528 nm in a cool hydrogen–nitrogen diffusion flame [361]. Alternatively the samples, e.g. Dash (13.8% P) or Axion (1.87% P), are digested with an HNO_3–$HClO_4$ mixture, the residue is treated with hydrofluoric acid and the phosphorus is determined by AAS and the method of standard-addition [362].

13.2.3 Polymers

Atomic-absorption spectrometry is one of the most satisfactory techniques for trace-metal analysis of natural and synthetic fibres, textiles, plastics, rubber and paper. The application of AAS extends from raw materials to finished products. Tonini [363] has recently reviewed some of the applications in the textile industry. In general, determination of various elements in textiles, fibres, plastics, rubber and paper is based on one of the following pretreatment techniques: (a) grinding and direct electrothermal atomization of the solid samples or slurries; (b) extraction of the metals into dilute acids; (c) dissolution of the polymeric substance in organic solvents; (d) dry ashing at 550°C and digestion of the ash with suitable acids; (e) wet ashing with concentrated HNO_3, $HClO_4$ or H_2O_2, alone or in admixture.

The use of solid sampling provides distinct advantages in speed and convenience and eliminates the necessity of preparing a concentrated solution of the polymeric materials. Henn [364] determined Cu, Cr and Fe in polymers by direct atomization in the graphite furnace, from a titanium spoon. The largest practical sample size used was 5 mg. Similarly Cd, Cu, Pb and Mn have been determined in paper by introduction of solid or slurried samples [365]. A simple vacuum-actuated device for complete transfer of a weighed polymer sample into the graphite atomizer has been described [366]. This device permits transfer of samples weighing 0.02-2 mg in ~ 3 sec and removal of the decomposition products from the atomizer at the same time. In general, the results obtained by the solid sampling technique are somewhat lower than those obtained by the solution method.

Plastic utensils made from polystyrene, polyethylene, polypropene and polyamide and coloured with cadmium red pigment are analysed for cadmium after leaching with water or 3-4% acetic acid [367]. The extract is evaporated and a solution of the residue in $1M$ nitric acid is atomized in the flame. Tributyltin compounds in textiles are extracted into methanol containing hydrochloric acid, the extract is filtered, and the tin is measured by use of the standard-addition technique [368]. Copper in textiles is determined after acid extraction [369].

Dissolution of the polymeric substance in a suitable solvent and direct atomization of the solution into the flame or graphite furnace may also be used. This technique allows preparation of a concentrated solution of the polymer. However, there is a limit to how concentrated a polymer solution may be before the viscosity of the solution prevents efficient aspiration. Polystyrene and cellulose acetate are easily dissolved in MIBK, whereas polyacrylonitrile and polyimide are appreciably dissolved in dimethylformamide [370-373]. Dimethylacetamide, cyclohexanone, formic acid and ethanol are used to dissolve polycarbonate, poly(vinyl chloride), polyamide and polyether polymers, respectively [370-372,374]. Wool is solubilized by heating with 5% sodium hydroxide solution or hydrochloric acid [370,375], and cellulose is dissolved in 72% sulphuric acid [370]. Mixtures of metal-containing macromolecules (e.g. tripropyltin methacrylate and methyl methacrylate) are dissolved in a suitable solvent and separated by size-exclusion chromatography, the eluate being injected into a graphite furnace [376]. Polymers which do not dissolve in organic solvents must be wet or dry ashed.

Slavin [377] and Druckman [378] have reported a method in which the polymer is ashed and the residue is taken up in acid solution for AAS measurement. Dry ashing is done by heating 0.5-1 g of sample in a muffle furnace at temperatures ranging from 260°C for paper to 800°C for rubber. The ash is digested in dilute acid and the solution is atomized. Oxygen-flask decomposition [379] may also be used. Antimony [380] and manganese [374] in polyester fibres are determined by dry ashing at 525°C for ~ 3 hr, and flame atomization of the solution of the ash in 6M hydrochloric acid. Iron in nylon 66 has been determined, with use of a similar procedure [381]. Dry ashing of polypropylene and rubber requires high temperatures (650-800°C) and the residue may be further subjected to fusion with 2:1 v/v sodium carbonate-borax mixture before dissolution in hydrochloric acid. This procedure has been used for the determination of Al, Ti, Fe, Cu, Ni, Co, Zn, Mn, Pb, Mg, K, Na and Cd in rubber [378,382,383]. Cellulose materials and paper are also analysed for trace metals by dry ashing under mild conditions [384]. It has been reported that different types of paper can be categorized by the level of trace metals present, by use of a pattern recognition technique [385]. Specimens of paper are dried *in vacuo* at 70°C, then ignited at 260°C, and the residue is dissolved in nitric acid and atomized in a graphite cup. This method was applied to 19 paper samples representing 11 types from 7 manufacturers. The elements most successfully used in paper classification were Cu, Mn, Sb, Cr and Co [385].

Various polymeric substances are decomposed by wet digestion with strong acids. Mixtures of HNO_3, $HClO_4$ and H_2SO_4 are used for decomposing poly(vinyl chloride) [386-388] and acrylic fibres [389] for the determination of calcium and barium [386], lead [387], cadmium and zinc [388], and arsenic [389]. Traces of arsenic, barium and mercury in polycarbonate, polyethylene, polypropene and poly(vinyl chloride) are determined by wet digestion with 65% nitric acid at 240°C for 8 hr [390], followed by determination by the hydride-generation and cold-vapour techniques for arsenic and mercury, respectively. Nylon and polyester, however, are not easily solubilized with pure acids. Price [391] described a rapid decomposition method for dissolution and digestion of rayon, nylon and polyester fibres with a mixture of H_2SO_4 and H_2O_2. Elements encountered as normal contaminants, light stabilizers, catalysts or tracers for identification, have been determined by use of air–acetylene (with a 3-slot burner) or nitrous oxide-acetylene flames. This digestion procedure has been used successfully for the determination of Ca and Ba [386], Pb [387] and Sn [392] in poly(vinyl chloride) composites. References to the determination of a number of elements in various polymeric substances are shown in Table 13.3.

13.2.4 Paints

A paint is a suspension of a pigment in an oil vehicle. Most white pigments are oxides, sulphates, carbonates and sulphides, of Ti, Ca, Zn, Pb, Ba, Sb, Si, Al and Mg (as appropriate). Inorganic coloured pigments are mainly compounds of Cd, Fe, Cr, Ni and Mo. Characterization and identification of the paint is an important forensic and industrial task. Problems of lead poisoning have led to the establishment of stringent regulations in some countries to define the level of toxic metal content permitted in toys and appliances. In the U.S.A., the Federal

Table 13.3. AAS determination of trace metals in some polymers

Polymer	Elements measured	Pretreatment step	Atomization technique	Reference
Acrylic fibres	As	Wet digestion HNO_3–$HClO_4$–H_2SO_4	F	389
Cellulose materials	Ca,Cu,Fe,Mn,Mg,Na	Dry ashing	F	384
	Cu,Fe,In,Pb,Mn,Sb,Sn,Zn,Au,Cr	Wet digestion H_2SO_4–H_2O_2	F	391
	Ca,Cu,Fe,Na,K,Mg,Zn		F	395
Fluorinated polyhydrocarbons	Cu,Fe	None	GF	394
Nylon	Au,Cr,Cu,Fe,In,Pb,Mn,Sb,Sn,Zn	Wet digestion H_2SO_4–H_2O_2	F	391
	Fe			381
Paper	Cd,Cu,Pb,Mn	Slurry	GF	365
	Ag,Cd,Cr,Cu,Co,Fe,Mn,Sb,Pb	Dry ashing at 260°C	GC	385
	Cu,Fe,Mn,Si		GF,F	394
	Si	Extraction	F	396
Polyacrylamide	Cr,Cu,Fe	None	GF	364
Polyamide	Cd	Extraction	F	367
	Mn	Solubilization with formic acid	GF	374
Polycarbonate	As,Ba,Hg	Wet digestion HNO_3 at 240°C	GF	390
Polyester	Sb	Dry ashing at 525°C	F	380
	Au,Cr,Cu,Fe,In,Pb,Mn,Sb,Sn,Zn	Wet digestion H_2SO_4–H_2O_2	F	391
	Au	None	GF	393
Polyethylene	Cd	Extraction	F	367
	As,Ba,Hg	Wet digestion HNO_3 at 240°C	GF	390
Polyethylene oxide	Cr,Cu,Fe	None	GF	364
Polyimide	Cu,Fe,Mn	Solubilization with DMF	GF	373
Polypropene	Cd	Extraction	F	367
	As,Ba,Hg	Wet digestion HNO_3 at 240°C	GF	390
	Al,Fe,Ti			378
Polystyrene	Al,Fe	None	GF	394
	Cd	Extraction	F	367
Poly(vinyl chloride)	Cd,Zn	Wet digestion HNO_3, $HClO_4$–H_2SO_4	F	388
	Pb	Wet digestion H_2SO_4–H_2O_2	F	387
	Ca,Ba	Wet digestion H_2SO_4–H_2O_2	F	386
	As,Ba,Hg	Wet digestion HNO_3 at 240°C	GF	390
	Sn	Wet digestion H_2SO_4–H_2O_2	F	392
Rayon	Au,Cr,Cu,Fe,In,Pb,Mn,Sb,Sn,Zn	Wet digestion H_2SO_4–H_2O_2	F	391
Rubber	Cu,Co,Fe,Ni,Mn,Zn	Dry ashing at 800°C	F	382
	Cd,Cu,Fe,Mg,Pb,Na,K,Zn			383
Wool	Al	Wet digestion HCl	GF	375

F = flame,   GF = graphite tube furnace,   GC = graphite cup

Hazardous Substances Act requires that no household paint has a lead content in excess of 0.06% by weight of the dried film. Similar restrictions have been set up in U.K. Compliance with such regulations requires the availability of an accurate method for determination of metals in paints.

Determination of metals in paints by AAS involves prior dry ashing, fusion, wet digestion or extraction procedures. In the dry ashing methods, the paint samples are ignited, and the ash is dissolved in dilute acids and atomized [397-399]. Scott [400] reported the results obtained by four laboratories for the determination of heavy metals in 98 different paints and related products. The general procedure is dry ashing at $\sim 500°C$, or digestion with HNO_3, $HClO_4$ and H_2SO_4 in open vessels. The Delves sampling cup technique has also been used. The samples are dissolved in water (for water-base paints) and MIBK [401,402] or toluene–dimethylformamide mixture [403] (for oil-base paints) followed by preliminary ashing at $\sim 500°C$ and direct atomization of the ash. Analysis for lead in different types of paints by the Delves cup technique and Scott's method [402] shows very good agreement between both sets of results (Table 13.4). Lead scrapes of pencil paint may be similarly determined by direct transfer of a solid portion to the Delves cup [404]. Fusion with either carbonate–borax mixture [405] or sodium peroxide–sodium carbonate–potassium carbonate mixture [406] may also be used, followed by acid extraction of the metals and atomization.

Table 13.4. Comparison of ashing method with Delves cup method for the determination of lead in paints [401]. (By permission of the copyright holders, Perkin-Elmer Ltd.)

Paint type	Base	Lead (ppm)	
		Ashing	Delves cup
Acrylic	water	10	<10
Alkyd	oil	<5	<10
Alkyd enamel	oil	1750	1680
Alkyd modified acrylic	water	12	12
Glass enamel	oil	1420	1540
Latex enamel	water	11	12
Latex enamel	water	9	9
Latex enamel	water	<5	<5
Poly(vinyl acetate)	water	<5	<5
Quick-drying enamel	oil	29	32
Vinyl acrylic	water	5	10
Waterspar enamel	oil	3220	3400

Antifouling paints and powder paints containing organometallics are best treated by wet digestion. Digestion with nitric acid at $\sim 150°C$ under pressure has been used for alkyd and latex paints [407]. A mixture of H_2SO_4 and $KMnO_4$

under pressure may also be used for digestion [408]. Holak [409] compared the results obtained for the determination of lead in 14 paint samples by means of wet digestion in closed and open vessels. Both methods gave almost identical results. Acid extraction of toxic metals (e.g. Pb, Cr, Ba) from magazine colours with $0.1M$ hydrochloric acid at $\sim 37°C$ for 4 hr, followed by atomization, has been described [410]. Metals such as Al, Fe, Cu and Mn in TiO_2 pigment are determined after acid extraction [411]. In general, some polymers and additives [412] and the main pigment elements in paints, including Cr [406,408,410, 413–415], Co [413,416], Pb [399,401–404,409,410,416,417], Cu [411,414], Mn [411,414], Fe [411,414], Al [411], Mg [416], Ba [410] and Cd [399, 403] have been determined by flame or electrothermal atomization procedures after treatment by one of the above-mentioned techniques.

13.3 GUNSHOT RESIDUES

Forensic and criminal research laboratories are often required to analyse and compare the constituents of bullets or bullet fragments. This usually involves analysis for Sb, Ba, Pb, Bi, Cu and Ag. Copper is the major element used in cartridge cases and is also used as a coating for projectiles. Most primers, except those in rimfire and some shot shells, and centre fire ammunition contain antimony. Barium is used as an oxidizing agent in the primer composition and lead is used as fuel, igniter and detonator, and as a primary element in the projectile. Thus AAS analysis of gunshot residues for these elements is quite feasible. The analysis is done by dissolving the bullet residue in nitric acid or swabbing the inside of the spent cartridge with a cloth moistened in dilute acid, then making the AAS measurement by flame or electrothermal atomization [418–420]. The variations observed from lot to lot of bullets indicate the possibility of distinguishing between different types and batches [420]. Analysis for trace elements in gunshot residues deposited around the bullet hole can also be used for determination of the firing distance. The amount and distribution of Pb and Sb around bullet holes is a function of the firing distance [421].

    AAS can also be used to establish whether or not a suspect used a gun. This is based on the fact that on detonation of the cartridge, gas leakage around loose-fitting gun parts and the blow-back from the muzzle carry Ba, Sb, Pb and Cu from the primer to the shooter's hand. These elements can be collected from the skin of the back of the shooter's hand with cotton-tipped swabs moistened with 5% nitric acid [422]. The palm of the hand may also be swabbed, but these swabs should be collected and analysed separately because some metals may be detected if the person in question has held a gun but not fired it [420]. The test swabs are treated with a small volume of 5% nitric acid and agitated in an ultrasonic vibrator for 15-20 min. Then three aliquots of the test solution are used for the measurement of Sb, Cu and Pb in an air–acetylene or nitrous oxide–acetylene flame. Barium is measured in a fourth aliquot after the addition of sodium chloride to enhance the absorption [420,423].

Gunshot residues may also be removed from the hand by swabbing with a small cloth moistened with dilute hydrochloric acid. The residue is dissolved in 3*M* hydrochloric acid, and the metals are subsequently extracted and injected into a graphite tube furnace [424]. Lead is extracted into MIBK at pH 2-3 as the pyrrolidine dithiocarbamate complex, barium is complexed at pH 8-9 with thenoyltrifluoroacetate and extracted into MIBK, and antimony is extracted as $SbCl_5$ into MIBK.

The applicability of these methods has been demonstrated in several reports. Absorption signals for Sb in gunshot residues removed from the back of the firing hand are shown in Fig. 13.1 [424]. The signals represent the Sb level on the hand before firing the weapon (blank) and after 1-4 shots. Significant levels of Sb are detected after firing, even on hands washed with soap and water before sample collection. It should be noted that the amount of gunshot residues deposited on the hand is controlled by several factors such as the condition, age, type and calibre of the weapon. It is believed by some workers that detection of these metals at low levels, though not strong evidence of the suspect having a weapon, certainly indicates the presence of a firearm.

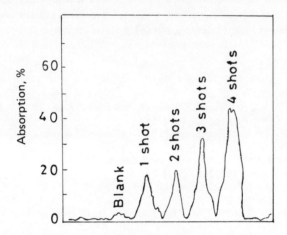

Fig. 13.1. Absorption signals for antimony deposited on the back of the hand during firing of a Walther PP semi-automatic pistol; the hand was washed between each firing. [According to G. D. Renshaw, C. A. Pounds and E. F. Pearson, *Atom. Absorp. Newslett.*, **12**, 55 (1973). By permission of the copyright holders, Perkin-Elmer Corporation.]

13.4 CARBON, HYDROGEN AND HYDROCARBONS

An AAS method for the determination of carbon in organic compounds has been described [425]. The sample (1-5 mg) is weighed in a tin-foil boat, wrapped in aluminium foil, and inserted into the platinum coil of a Schöniger flask.

Decomposition is initiated electrically and the carbon dioxide produced is absorbed in calcium hydroxide solution. The excess of calcium in the solution (or alternatively in a solution of the precipitate in $0.1M$ hydrochloric acid) is determined by flame AAS. The results obtained by the AAS method have been compared with those obtained by EDTA titration of the calcium.

Lede and Villermaux [426] described a method for AAS determination of atomic hydrogen in concentrations as low as $5 \times 10^{-10}M$. The hydrogen is allowed to react with a mercury(II) oxide target and the free mercury released is measured by the cold-vapour technique.

Ethylene is determined by AAS by prior trapping as an organomercury compound, separation, conversion into $HgCl_2$ and measurement of the mercury. The method has been used for determination of ethylene and its derivatives in air [427]. Air samples are passed through 0.1% methanolic mercury(II) acetate, the solution is evaporated at $35°C$ to 1 ml, and an aliquot is applied to a thin layer of silica gel. The chromatogram is developed with methanol–aqueous ammonia (1:1) and is placed in an atmosphere of hydrogen chloride before being sprayed with 0.1% dithizone solution in carbon tetrachloride. The band at R_f 0.20–0.24 is removed, treated with 1.5% sulphuric acid, and reduced with $NaBH_4$, and the mercury is measured at 243.7 nm. The average recovery is 90–95% and the lower limit of detection is 0.1 ppm of ethylene.

REFERENCES

[1] S. Lamm, B. Cole, K. Glynn and W. Ullmann, *New Engl. J. Med.*, **289**, 574 (1973).

[2] P. A. Martin, *J. Inst. Brewing*, 79, 289 (1973).

[3] L. Favretto, G. Pertoldi Marletta and L. Favretto Gabrielli, *Atom. Absorp. Newslett.*, **12**, 101 (1973).

[4] B. Krinitz and W. Holak, *J. Assoc. Off. Anal. Chem.*, 57, 568 (1974).

[5] N. T. Crosby, *Analyst*, **102**, 225 (1977).

[6] W. Holak, *J. Assoc. Off. Anal. Chem.*, **60**, 239 (1977).

[7] T. T. Gorsuch, *Analyst*, 84, 135 (1959).

[8] E. E. Menden, D. Brockman, H. Choudhury and H. G. Petering, *Anal. Chem.*, **49**, 1644 (1977).

[9] M. Feinberg and C. Ducauze, *Anal. Chem.*, **52**, 207 (1980).

[10] J. Maurer, *Z. Lebensm.-Unters. Forsch.*, **165**, 1 (1977).

[11] U. Rutkowska and I. Trzebska-Jeske, *Rocz. Panstw. Zakl. Hig.*, **29**, 33 (1978).

[12] A. Kuentz, R. E. Simard, J. A. Zee and M. Desmarais, *Can. Inst. Food Sci. Technol. J.*, **9**, 147 (1976).

[13] M. Heckman, *J. Assoc. Off. Anal. Chem.*, **50**, 45 (1967).

[14] W. H. Evans, *Analyst*, **103**, 452 (1978).

[15] O. Horak, *Landwirtsch. Forsch.*, **29**, 289 (1976).

[16] K. Lorenz, G. MacFarland and J. Maga, *Cereal Chem.*, **54**, 281 (1977).

[17] A. Yasui and C. Tsutsumi, *Bunseki Kagaku*, **26**, 809 (1977).

[18] H. S. Grunwaldt and H. Heigener, *Landwirtsch. Forsch.*, **28**, 298 (1975).

[19] C. A. Watson, W. C. Shuey and K. J. Sprick, *Cereal Chem.*, **54**, 189 (1977).

[20] S. A. Rakuns and L. E. Smythe, *Food Technol. Aust.*, **30**, 271 (1978).

[21] D. R. Boline and W. G. Schrenk, *J. Assoc. Off. Anal. Chem.*, **60**, 1170 (1977).

[22] P. Hocquellet and N. Labeyrie, *Atom. Absorp. Newslett.*, **16**, 124 (1977).

[23] J. Kumpulainen and P. Koivistoinen, *Acta Agr. Scand.*, **27**, 35 (1977).

[24] H. Narasaki, *Anal. Chim. Acta*, **104**, 393 (1979).

[25] R. Garcia Olmendo, P. Garcia Puertas, M. Salesa Perez, T. A. Masoud and F. J. Liso Rubio, *Ann. Bromat.*, **28**, 1 (1976).

[26] S. Nakashima, *Ber. Ohara Inst. Landwirtsch. Biol., Okayama Univ.*, **17**, 187 (1979).

[27] E. Amakawa, K. Ohnishi, N. Taguchi and H. Seki, *Bunseki Kagaku*, **27**, 81 (1978).

[28] F. L. Fricke, W. B. Robbins and J. Caruso, *Prog. Anal. At. Spectrosc.*, **2**, 185 (1979).

[29] B. Brzozowska, *Rocz. Panstw. Zakl. Hig.*, **28**, 355 (1977).

[30] J. A. McHard, J. D. Winefordner and S.-V. Ting, *J. Agr. Food Chem.*, **24**, 950 (1976).

[31] C. Caristi, G. Cimino and M. Ziino, *Essenze Deriv. Agrum.*, **50**, 165 (1980).

[32] M. M. Guedes da Mota and B. Griepink, *Z. Anal. Chem.*, **291**, 128 (1978).

[33] K. Yamaoto, Y. Mogi, H. Watanabe, E. Ninomiya, T. Horio, H. Toyoda, S. Yoshimoto, T. Murata and S. Masuda, *Shokuhin Eiseigaku Zasshi*, **20**, 317 (1979).

[34] W. J. Price and J. T. H. Roos, *J. Sci. Food Agr.*, **20**, 437 (1969).

[35] J. T. H. Roos and W. J. Price, *J. Sci. Food Agr.*, **21**, 51 (1970).

[36] E. Postorino, R. Franzi, S. Bonanno and A. Di Giacomo, *Essenze Deriv. Agrum.*, **48**, 316 (1978).

[37] D. D. Siemer, R. K. Vitek, P. Koteel and W. C. Houser, *Anal. Lett.*, **10**, 357 (1977).

[38] M. Moll, *Brauwissenschaft*, **30**, 347 (1977).

[39] M. Lepage, *J. Am. Soc. Brew. Chem.*, **38**, 116 (1980).

[40] T. R. M. Helin and J. C. Slaughter, *J. Inst. Brewing*, **83**, 15 (1977).

[41] Analysis Committee of the Institute of Brewing, *J. Inst. Brewing*, **83**, 82 (1977).

[42] Analysis Committee of the European Brewing Convention, *J. Inst. Brewing*, **84**, 156 (1978).

[43] S. W. Frey, W. G. DeWitt and B. R. Bellomy, *Am. Soc. Brewing Chemists Proc.*, 199 (1967).

[44] R. Garcia Olmendo, P. Garcia Puertas, T. A. Masoud and C. Diez Marques, *Ann. Bromatol.*, **29**, 281 (1977).

[45] C. Baluja Santos, M. L. Gil de la Pena and A. Gonzalez Portal, *Quim. Anal.*, **30**, 295 (1976).

[46] A. Laszlo, M. Varju and S. Ferenczi, *Z. Lebensm.-Unters. Forsch.*, **167**, 333 (1978).

[47] P. B. Zeeman and L. R. P. Butler, *Appl. Spectrosc.*, **16**, 120 (1962).

[48] D. Mack, *Dtsch. Lebensm.-Rundsch.*, **71**, 71 (1975).

[49] M. K. Meredith, S. Baldwin and A. A. Andreasen, *J. Assoc. Off. Anal. Chem.*, **53**, 12 (1970).

[50] D. H. Strunk and A. A. Andreasen, *J. Assoc. Off. Anal. Chem.*, **50**, 339 (1967).

[51] J. P. Weiner and L. Taylor, *J. Inst. Brewing*, **75**, 195 (1969).

[52] G. R. List, C. D. Evans and W. F. Kwolek, *J. Am. Oil Chem. Soc.*, **48**, 438 (1971).

[53] C. D. Evans, G. R. List and L. T. Black, *J. Am. Oil Chem. Soc.*, **48**, 480 (1971).

[54] F. J. Slikkerveer, A. A. Braad and P. W. Hendrikse, *Atom. Spectrosc.*, **1**, 30 (1980).

[55] M. Gente-Jauniaux and A. Prevot, *Rev. Fr. Corps Gras*, **26**, 325 (1979).

[56] P. K. Hon, O. W. Lau, S. F. Luk and C. S. Mok, *Anal. Chim. Acta*, **113**, 175 (1980).

[57] S. Chiricosta, G. Calabro and E. Bruno, *Essenze Deriv. Agrum.*, **48**, 107 (1979).

[58] J. T. Olejko, *J. Am. Oil Chem. Soc.*, **53**, 480 (1976).

[59] H. Pihlaja, *Fette, Seifen, Anstrichm.*, **83**, 294 (1981).

[60] H. Hirayama, Z. Shimoda, A. Kobayashi, T. Shige, T. Takahashi, K. Tsuji, S. Nakasato, M. Higuchi and Y. Machida, *Yukagaku*, **26**, 216 (1977).

[61] S. Chiricosta and E. Bruno, *Essenze Deriv. Agrum.*, **46**, 310 (1976).

[62] U. Persmark and B. Toregard, *J. Am. Oil Chem. Soc.*, **48**, 650 (1971).

[63] W. C. Tsai, C.-P. Lin, L.-J. Shiau and S.-D. Pan, *J. Am. Oil Chem. Soc.*, **55**, 695 (1978).

[64] W.-C. Tsai and L.-J. Shiau, *Anal. Chem.*, **49**, 1641 (1977).
[65] A. Nakamura and T. Kashimoto, *Bull. Environ. Contam. Toxicol.*, **22**, 345 (1979).
[66] E. J. Dufek and G. R. List, *J. Am. Oil Chem. Soc.*, **54**, 271 (1977).
[67] F. J. Szydlowski, *Atom. Absorp. Newslett.*, **16**, 60 (1977).
[68] I. Schuetze and W. Mueller, *Nahrung*, **23**, 867 (1969).
[69] H. J. Auslitz, *Arch. Lebensmittelhyg.*, **27**, 68 (1976).
[70] M. D. K. Abo-Rady, *Z. Anal. Chem.*, **299**, 187 (1979).
[71] S.-L. Tong, *Anal. Chem.*, **50**, 412 (1978).
[72] L. J. Dusci and L. P. Hackett, *J. Assoc. Off. Anal. Chem.*, **59**, 1183 (1976).
[73] R. K. Munns and D. C. Holland, *J. Assoc. Off. Anal. Chem.*, **60**, 833 (1977).
[74] E. Egaas and K. Julshamn, *Atom. Absorp. Newslett.*, **17**, 135 (1978).
[75] J. W. Davidson, *Analyst*, **104**, 683 (1979).
[76] M. Taguchi, K. Yasuda, M. Hashimoto and S. Toda, *Bunseki Kagaku*, **28**, T 33 (1979).
[77] M. Taguchi, Y. Kenji, Y. Dokiya, M. Shimizu and S. Toda, *Bunseki Kagaku*, **26**, 438 (1977).
[78] H. Agemian and V. Cheam, *Anal. Chim. Acta*, **101**, 193 (1978).
[79] K. Matsunaga, T. Ishida and T. Oda, *Anal. Chem.*, **48**, 1421 (1976).
[80] G. Castello and S. Kanitz, *Boll. Chim. Lab. Prov.*, **4**, 57 (1978).
[81] R. Capelli, C. Fezia, A. Franchi and G. Zanicchi, *Analyst*, **104**, 1197 (1979).
[82] W. R. Simpson and G. Nickless, *Analyst*, **102**, 86 (1977).
[83] U. Harms, *Z. Lebensm.-Unters. Forsch.*, **172**, 118 (1981).
[84] M. Taguchi, K. Yasuda, Y. Dokiya, M. Shimizu and S. Toda, *Bunseki Kagaku*, **26**, 497 (1977).
[85] M. Ambe and N. Niikura, *Bunseki Kagaku*, **29**, T 5 (1980).
[86] Yu. V. Zelyukova, R. A. Vitkun, T. O. Didorenko and N. S. Poluektov, *Zh. Analit. Khim.*, **36**, 454 (1981).
[87] J. Flanjak, *J. Assoc. Off. Anal. Chem.*, **61**, 1299 (1978).
[88] P. J. Brooke and W. H. Evans, *Analyst*, **106**, 514 (1981).
[89] T. Hirayama, Y. Sakagami, N. Motoshi and S. Fukui, *Bunseki Kagaku*, **30**, 278 (1981).
[90] G. Parolari and G. Pezzani, *Ind. Conserve*, **52**, 130 (1977).
[91] R. Kruse, *Z. Lebensm.-Unters. Forsch.*, **171**, 261 (1980).
[92] M. T. Friend, C. A. Smith and D. Wishart, *Atom. Absorp. Newslett.*, **16**, 46 (1977).
[93] L. R. Hageman, J. A. Nichols, P. Viswanadham and R. Woodriff, *Anal. Chem.*, **51**, 1406 (1979).
[94] H. Agemian, D. P. Sturtevant and K. D. Austen, *Analyst*, **105**, 125 (1980).
[95] I. Okuno, J. A. Whitehead and R. E. White, *J. Assoc. Off. Anal. Chem.*, **61**, 664 (1978).
[96] D. T. Dennis, *J. Assoc. Off. Anal. Chem.*, **61**, 715 (1978).
[97] H. Mohadjerani and E. Hauser, *Fleischwirtschaft*, **56**, 258 (1976).
[98] K. Ikebe and R. Tanaka, *Bull. Environ. Contam. Toxicol.*, **21**, 526 (1979).
[99] G. J. Ramelow and T. I. Balkas, *Anal. Lett.*, **10**, 733 (1977).
[100] J. E. Poldoski, *Anal. Chem.*, **52**, 1147 (1980).
[101] T. Nagahiro, O. Fujino, M. Matsui and T. Shigematsu, *Bull. Inst. Chem. Res., Kyoto Univ.*, **56**, 274 (1978).
[102] S. G. Capar, *J. Assoc. Off. Anal. Chem.*, **60**, 1400 (1977).
[103] C. L. Chou, J. F. Uthe and E. G. Zook, *J. Fish. Res. Board Can.*, **35**, 409 (1978).
[104] O. C. Zenoble, J. A. Bowers, G. Seaman and S. D. Howe, *J. Food Sci.*, **42**, 847 (1977).
[105] T. C. Grey, D. Robinson and J. M. Jones, *J. Sci. Food Agric.*, **28**, 822 (1977).
[106] R. Tamato and F. Ohtaka, *Rakuno Kagaku Shokuhin No Kenkyu*, **27**, A 67 (1978).
[107] S. Slavin, G. E. Peterson and P. C. Lindahl, *Atom. Absorp. Newslett.*, **14**, 57 (1975).
[108] C. Wehrer, J. Thiersault and P. Laugel, *Ind. Aliment. Agric.*, **93**, 1439 (1976).
[109] E. R. Elkins and A. Sulek, *J. Assoc. Off. Anal. Chem.*, **62**, 1050 (1979).
[110] J. L. Love and J. E. Patterson, *J. Assoc. Off. Anal. Chem.*, **61**, 627 (1978).

[111] N. Shiga, O. Matano and S. Goto, *Nippon Nayaku Gakkaishi*, **5**, 255 (1980).
[112] M. Sano, M. Furukawa, M. Kourai and I. Tomita, *J. Assoc. Off. Anal. Chem.*, **62**, 764 (1979).
[113] S. Gherardi, G. Dall'Aglio and A. Versitano, *Ind. Conserve*, **50**, 284 (1975).
[114] F. J. Jackson, J. I. Read and B. Lucas, *Analyst*, **105**, 359 (1980).
[115] B. Brozozowska and T. Zawadzka, *Rocz. Panstw. Zakl. Hig.*, **32**, 9 (1981).
[116] M. Del Carmen Martinez Para, T. A. Masoud and M. Esperanza Torija Isasa, *Ann. Bromatol.*, **31**, 189 (1979).
[117] C. A. Rowan, O. T. Zajicek and E. J. Calabrese, *Anal. Chem.*, **54**, 149 (1982).
[118] P. J. Barlow, *J. Dairy Res.*, **44**, 377 (1977).
[119] L. Maurer, *J. Milk Food Technol.*, **38**, 693 (1975).
[120] G. K. Murthy and M. Rhea, *J. Dairy Sci.*, **50**, 313 (1967).
[121] T. Suzuki, M. Takedea and M. Uchiyama, *Eisei Shikensho Hokoku*, 102 (1980).
[122] R. C. Fry and M. B. Denton, *Anal. Chem.*, **49**, 1413 (1977).
[123] I. B. Brooks, G. A. Luster and D. G. Easterly, *Atom. Absorp. Newslett.*, **9**, 93 (1970).
[124] J. C. Bruhn and A. A. Franke, *J. Dairy Sci.*, **59**, 1711 (1976).
[125] A. M. Sulek, E. R. Elkins and E. W. Zink, *J. Assoc. Off. Anal. Chem.*, **61**, 931 (1978).
[126] The Chemical Society, Analytical Methods Committee, Metallic Impurities in Organic Matter Sub-Committee, *Analyst*, **104**, 778 (1979).
[127] J. Kumpulainen, *Anal. Chim. Acta*, **113**, 355 (1980).
[128] B. Brzozowska, *Rocz. Panstw. Zakl. Hig.*, **28**, 441 (1977).
[129] A. O. Lustre and C. H. Lacebal, *Philipp. J. Nutr.*, **29**, 25 (1976).
[130] J. Koops and D. Westerbeek, *Neth. Milk Dairy J.*, **32**, 149 (1978).
[131] D. C. Manning, *Am. Lab.*, **5**, No. 8, 37 (1973).
[132] Report prepared by the Metallic Impurities in Organic Matter Sub-Committee of the Analytical Methods Committee of the Analytical Division of the Chemical Society, *Analyst*, **104**, 1070 (1979).
[133] H. Jonsson, *Milchwissenschaft*, **31**, 210 (1976).
[134] H. Haenni, J. Hulstkamp and A. Rothenbuehler, *Mitt. Geb. Lebensmittelunters. u. Hyg.*, **67**, 448 (1976).
[135] O. Nishijima and H. Nakamura, *Noyaku Kensasho Hokoku*, 54 (1980).
[136] N. Fudagawa and A. Kawase, *Bunseki Kagaku*, **27**, 353 (1978).
[137] N. Fudagawa and A. Kawase, *Bunseki Kagaku*, **27**, 37 (1978).
[138] T. Tsushida and T. Takeo, *Nippon Shokuhin Kogyo Gakkaishi*, **27**, 585 (1980).
[139] N. Fudagawa and A. Kawase, *Bunseki Kagaku*, **29**, 6 (1980).
[140] W. Chorazy, K. Bliwert and M. Swoszowska, *Bromatol. Chem. Toksykol.*, **9**, 361 (1976).
[141] T. Tsushida and T. Takeo, *Agr. Biol. Chem.*, **43**, 1347 (1979).
[142] W. Lautenschlaeger and J. Maassen, *GIT Fachz. Lab.*, **23**, 176 (1979).
[143] P. Petrino, M. Cas and J. Estienne, *Ann. Falsif. Expert. Chim.*, **69**, 87 (1976).
[144] J. Ramírez-Muñoz, *Beckman Flame Notes*, **7**, 49 (1975).
[145] J. C. Meranger, *J. Assoc. Off. Anal. Chem.*, **58**, 1143 (1975).
[146] A. N. Sagredos, *Fette, Seifen, Anstrichm.*, **79**, 331 (1977).
[147] F. J. Langmyhr and J. Aamodt, *Anal. Chim. Acta*, **87**, 483 (1976).
[148] D. T. Gordon, *J. Assoc. Off. Anal. Chem.*, **61**, 715 (1978).
[149] H. J. Auslitz, *Arch. Lebensmittelhyg.*, **27**, 68 (1976).
[150] E. F. Dalton and A. J. Melanoski, *J. Assoc. Off. Anal. Chem.*, **52**, 1035 (1969).
[151] I. D. Pearce, R. R. Brooks and R. D. Reeves, *J. Assoc. Off. Anal. Chem.*, **59**, 655 (1976).
[152] R. P. Hullin, M. Kapel and J. A. Drinkall, *J. Food Technol.*, **4**, 235 (1969).
[153] H. Narasaki, J. L. Down and R. Ballah, *Analyst*, **102**, 537 (1977).

[154] G. T. C. Shum, H. C. Freeman and J. F. Uthe, *J. Assoc. Off. Anal. Chem.*, **60**, 1010 (1977).

[155] W. Oelschlaeger, S. Schmidt and L. Bestenlehner, *Landw. Forsch.*, **29**, 70 (1976).

[156] D. J. Hankinson, *J. Dairy Sci.*, **58**, 326 (1975).

[157] T. Ono and S. Odagiri, *Rakuno Kagaku No Kentyu*, **24**, A 133 (1975).

[158] K. E. Burke and C. H. Albright, *J. Assoc. Off. Agr. Chem.*, **53**, 531 (1970).

[159] S. W. Frey, *Atom. Absorp. Newslett.*, **3**, 127 (1964).

[160] S. W. Frey, W. G. DeWitt and B. R. Bellomy, *Am. Soc. Brewing Chemists Proc.*, 172 (1966).

[161] M. D. Garrido, C. Llaguno and J. Garrido, *Am. J. Enology Viticult.*, 44 (1971).

[162] G. L. Hill and A. Caputi, *Am. J. Enology Viticult.*, **20**, 227 (1969).

[163] J. W. Hwang, Ch. J. Mokeler and P. A. Ullucci, *Anal. Chem.*, **44**, 2018 (1972).

[164] K. T. Joseph, V. K. Panday, S. J. Raut and S. D. Soman, *Atom. Absorp. Newslett.*, 7, 25 (1968).

[165] R. K. Roschnik, *Analyst*, **98**, 596 (1973).

[166] D. H. Strunk and A. A. Andreasen, *Atom. Absorp. Newslett.*, **6**, 111 (1967).

[167] M. E. Varju, *Atom. Absorp. Newslett.*, **11**, 45 (1972).

[168] P. B. Zeeman and L. R. P. Butler, *Tegnikon*, **13**, 96 (1960).

[169] L. T. Black, *J. Am. Oil Chem. Soc.*, **47**, 313 (1970).

[170] R. A. Foss and D. M. Houston, *Atom. Absorp. Newslett.*, **8**, 82 (1969).

[171] J. M. Hall and C. Woodward, *Spectrosc. Lett.*, **2**, 113 (1969).

[172] A. Prévôt, *Atom. Absorp. Newslett.*, **5**, 13 (1966).

[173] W. J. Price, J. T. H. Ross and A. F. Clay, *Analyst*, **95**, 760 (1970).

[174] W. Slavin, *Atom. Absorp. Newslett.*, **4**, 330 (1965).

[175] J. Anderson, *Atom. Absorp. Newslett.*, **11**, 88 (1972).

[176] N. Antonacopoulos, *Chem. Mikrobiol. Technol. Lebensm.*, **3**, 8 (1974).

[177] A. Bouchard, *Atom. Absorp. Newslett.*, **12**, 115 (1973).

[178] B. Boppel, *Z. Anal. Chem.*, **268**, 114 (1974).

[179] A. G. Cameron and D. R. Hackett, *J. Sci. Food Agric.*, **21**, 535 (1970).

[180] D. J. David, *Analyst*, **89**, 747 (1964).

[181] D. J. David, *Spectrochim. Acta*, **20**, 1185 (1964).

[182] M. Heckman, *J. Assoc. Off. Agr. Chem.*, **50**, 45 (1967).

[183] M. Heckman, *J. Assoc. Off. Agr. Chem.*, **53**, 923 (1970).

[184] M. Heckman, *J. Assoc. Off. Agr. Chem.*, **54**, 666 (1971).

[185] W. Holak, *J. Assoc. Off. Agr. Chem.*, **54**, 1138 (1971).

[186] W. Holak, *Atom. Absorp. Newslett.*, **12**, 63 (1973).

[187] W. Holak, B. Krinitz and J. C. Williams, *J. Assoc. Off. Agr. Chem.*, **55**, 741 (1972).

[188] W. Hoover, J. R. Melton and P. A. Howard, *J. Assoc. Off. Agr. Chem.*, **54**, 860 (1971).

[189] M. T. Jeffus, J. S. Elkins and C. T. Kenner, *J. Assoc. Off. Agr. Chem.*, **53**, 1172 (1970).

[190] J. Jordan, *Atom. Absorp. Newslett.*, **7**, 48 (1968).

[191] K. Julshamn and O. R. Braekken, *Atom. Absorp. Newslett.*, **12**, 139 (1973).

[192] G. A. Knauer, *Analyst*, **95**, 476 (1970).

[193] E. N. Leonard, *Atom. Absorp. Newslett.*, **10**, 84 (1971).

[194] R. K. Munns and D. C. Holland, *J. Assoc. Off. Agr. Chem.*, **54**, 202 (1971).

[195] S. H. Omang, *Anal. Chim. Acta*, **63**, 247 (1973).

[196] E. G. Pappas and L. A. Rosenberg, *J. Assoc. Off. Agr. Chem.*, **49**, 792 (1966).

[197] H. E. Parker, *Atom. Absorp. Newslett.*, **2**, 23 (1963).

[198] R. K. Roschnik, *Analyst*, **98**, 596 (1973).

[199] G. Thilliez, *Chim. Anal. Paris*, **50**, 226 (1968).

[200] V. A. Thorpe, *J. Assoc. Off. Agr. Chem.*, **54**, 206 (1971).

[201] M. H. Thompson, *J. Assoc. Off. Agr. Chem.*, **52**, 55 (1969).

[202] D. J. Trent and W. Slavin, *Atom. Absorp. Newslett.*, **3**, 53 (1964).

[203] J. Tušl, *J. Assoc. Off. Agr. Chem.*, **53**, 1190 (1970).
[204] I. B. Brooks, G. A. Luster and D. G. Easterly, *Atom. Absorp. Newslett.*, **9**, 93 (1970).
[205] J. A. Fiorino, R. A. Moffitt, A. L. Woodson, R. J. Gajan, G. E. Huskey and R. G. Scholz, *J. Assoc. Off. Agr. Chem.*, **56**, 1246 (1973).
[206] P. Haelen, G. Cooper and C. Pampel, *Atom. Absorp. Newslett.*, **13**, 1 (1974).
[207] J. Lagathu and J. Desirant, *Rev. Franc. Corps Gras*, **19**, 169 (1972).
[208] D. C. Manning and F. J. Fernandez, *Atom. Absorp. Newslett.*, **9**, 65 (1970).
[209] J. T. M. Roos, *Spectrochim. Acta*, **24B**, 255 (1969).
[210] J. B. Willis, *Aust. J. Dairy Technol.*, 70 (1964).
[211] W. Lautenschlaeger and J. Maassen, *GIT Fachz. Lab.*, **23**, 176 (1979).
[212] M. T. Friend, C. A. Smith and D. Wishart, *Atom. Absorp. Newslett.*, **16**, 46 (1977).
[213] J. A. Fiorino, J. W. Jones and S. G. Capar, *Anal. Chem.*, **48**, 120 (1976).
[214] R. G. Lewis, *Residue Rev.*, **68**, 123 (1977).
[215] N. T. Crosby, *Analyst*, **102**, 225 (1977).
[216] M. Kurokawa, M. Kaneko, N. Nishiyama, S. Fukui and S. Kanno, *Eisei Kagaku*, **21**, 77 (1975).
[217] D. D. Seimer, R. K. Vitek, P. Koteel and W. C. Houser, *Anal. Lett.*, **10**, 357 (1977).
[218] W. Holak, *J. Assoc. Off. Anal. Chem.*, **63**, 485 (1980).
[219] S. Gherardi, G. Dall'Aglio and A. Versitano, *Ind. Conserve*, **50**, 284 (1975).
[220] J. Flanjak, *J. Assoc. Off. Anal. Chem.*, **61**, 1299 (1978).
[221] D. I. Rees, *J. Assoc. Public Anal.*, **16**, 71 (1978).
[222] W. H. Evans, F. J. Jackson and D. Dellar, *Analyst*, **104**, 16 (1979).
[223] H. Agemian and V. Cheam, *Anal. Chim. Acta*, **101**, 193 (1978).
[224] M. Ihnat and H. J. Miller, *J. Assoc. Off. Anal. Chem.*, **60**, 813 (1977).
[225] M. Ihnat and H. J. Miller, *J. Assoc. Off. Anal. Chem.*, **60**, 1414 (1977).
[226] H. Woidich and W. Pfannhauser, *Nährung*, **21**, 685 (1977).
[227] M. Palliere and G. Gernez, *Ann. Falsif. Expert. Chim.*, **69**, 821 (1976).
[228] P. Allenby, J. W. Robertson and F. C. Shenton, *J. Assoc. Public Anal.*, **15**, 61 (1977).
[229] Z. Grobenski and H. Schulze, *GIT Fachz. Lab.*, **24**, 1156 (1980).
[230] F. La Villa and F. Queraud, *Rev. Inst. Franc. Petrole Ann. Combust. Liquides*, **32**, 413 (1977).
[231] B. Welz, *Chim. Ind. (Milan)*, **59**, 771 (1977).
[232] International Union of Pure and Applied Chemistry, Applied Chemistry Division, Commission on Food Contaminations, *Pure Appl. Chem.*, **51**, 385 (1979).
[233] E. Florence and A. R. S. Audsley, *J. Sci. Food Agr.*, **29**, 429 (1978).
[234] D. D. Fetterolf and A. Syty, *J. Agr. Food Chem.*, **27**, 377 (1979).
[235] U. Mueller, E. Hauser, A. Kappeler, E. Merk, K. Steiner and H. Windemann, *Mitt. Geb. Lebensmittelunters. Hyg.*, **68**, 126 (1977).
[236] N. Oudart, C. Guichard and C. Delage, *J. Eur. Toxicol.*, **9**, 69 (1976).
[237] S. Gherardi, D. Bigliardi and G. Bellucci, *Ind. Conserve*, **51**, 273 (1976).
[238] M. Feinberg and C. Ducauze, *Anal. Chem.*, **52**, 207 (1980).
[239] H. Woidich and W. Pfannhauser, *Nährung*, **21**, 685 (1977).
[240] W. H. Evans, J. I. Read and B. E. Lucas, *Analyst*, **103**, 580 (1978).
[241] R. W. Dabeka, *Anal. Chem.*, **51**, 902 (1979).
[242] K. Ranfft, *Landwirtsch. Forsch.*, **31**, 135 (1975).
[243] E. V. Williams, *J. Food Technol.*, **13**, 367 (1978).
[244] G. Knezevic and K. Hueppe, *Dtsch. Lebensm.-Rundsch.*, **76**, 50 (1980).
[245] C. Delage, N. Oudart and C. Guichard, *Ann. Pharm. Franc.*, **34**, 315 (1976).
[246] I. Trzebska-Jeske and U. Rutkowska, *Rocz. Panstw. Zakl. Hig.*, **29**, 197 (1978).
[247] K.-T. Lee, Y.-F. Chang and S.-F. Tan, *Mikrochim. Acta*, 505 (1976 II).
[248] A. Kuentz, R. E. Simard, J. A. Zee and M. Desmarais, *Can. Inst. Food Sci. Technol. J.*, **9**, 147 (1976).

[249] W. H. Evans, D. Dellar, B. E. Lucas, F. J. Jackson and J. I. Read, *Analyst*, 105, 529 (1980).

[250] J. Maurer, *Z. Lebensm.-Unters. Forsch.*, 165, 1 (1977).

[251] F. J. Szydlowski, *Atom. Absorp. Newslett.*, 17, 65 (1978).

[252] U. Rutkowska and I. Trzebska-Jeske, *Rocz. Panstw. Zakl. Hig.*, 29, 33 (1978).

[253] K. Lee and F. M. Clydesdale, *J. Food Sci.*, 44, 549 (1979).

[254] J.-D. Aubort, A. Etournaud, Y. Jedrzejewska and A. Ramuz, *Mitt. Geb. Lebensmittel-unters. Hyg.*, 70, 416 (1979).

[255] International Union of Pure and Applied Chemistry, Applied Chemistry Division, Commission on Food Contaminants, *Pure Appl. Chem.*, 51, 2527 (1979).

[256] W. Oelschlaeger, S. Schmidt, L. Bestenlehner and W. Lautenschlaeger, *Landwirtsch. Forsch.*, 31, 336 (1978).

[257] F. J. Szydlowski and F. R. Vianzon, *Atom. Spectrosc.*, 1, 39 (1980).

[258] M. Ihnat, *J. Assoc. Off. Anal. Chem.*, 59, 911 (1976).

[259] G. Papanastasiu and K. Bronsch, *Z. Tierphysiol., Tierernaehr. Futtermittelkd.*, 40, 325 (1978).

[260] G. M. George, L. J. Frahm and J. P. McDonnell, *J. Assoc. Off. Anal. Chem.*, 60, 1054 (1977).

[261] P. N. Vijan and C. Y. Chan, *Anal. Chem.*, 48, 1788 (1976).

[262] A. W. Varnes and V. F. Gaylor, *Anal. Chim. Acta*, 101, 393 (1978).

[263] L. Capacho-Delgado and D. C. Manning, *Atom. Absorp. Newslett.*, 5, 1 (1966).

[264] C. L. Chakrabarti and G. Hall, *Spectrosc. Lett.*, 6, 385 (1973).

[265] S. H. Omang, *Anal. Chim. Acta*, 56, 470 (1971).

[266] E. J. Moore, O. I. Milner and J. R. Glass, *Microchem. J.*, 10, 148 (1966).

[267] I. Lang, G. Sebor, O. Weisser and V. Sychra, *Anal. Chim. Acta*, 88, 313 (1977).

[268] N. Mosescu, G. Kalmutchi and S. Badea, *Rev. Chim. (Bucharest)*, 28, 373 (1977).

[269] R. C. Barras and J. D. Helwig, *Proc. Am. Petrol. Inst.*, Section III, 43, 223 (1963).

[270] P. L. Grizzle, C. A. Wilson, E. P. Ferrero and H. J. Coleman, *U.S. Energy Res. Dev. Adm., Rept.*, No. BERC/RI-77/8, 1977.

[271] T. Tarui, T. Hasegawa and H. Tokairin, *Bunseki Kagaku*, 29, 423 (1980).

[272] T. Shimizu, Y. Shijo and K. Sakai, *Bunseki Kagaku*, 29, 685 (1980).

[273] D. C. Girvin, T. Hadeishi and J. P. Fox, *U.S. Energy Rept.*, LBL-8888, 1979.

[274] V. Sychra, O. Vyskočilová, I. Lang and O. Weisser, *Chem. Listy*, 73, 195 (1979).

[275] R. A. Mostyn and A. F. Cunningham, *J. Inst. Petrol.*, 53, 101 (1967).

[276] I. Lang, G. Sebor, V. Sychra, D. Kolihová and O. Weisser, *Anal. Chim. Acta*, 84, 299 (1976).

[277] N. M. Potter, *Anal. Chim. Acta*, 102, 201 (1978).

[278] T. T. Bartels and C. E. Wilson, *Atom. Absorp. Newslett.*, 8, 3 (1969).

[279] J. W. Owens and E. S. Gladney, *Atom. Absorp. Newslett.*, 14, 76 (1975).

[280] W. K. Robbins, J. H. Runnels and R. Merryfield, *Anal. Chem.*, 47, 2095 (1975).

[281] J. Scott and F. C. A. Killer, *Proc. Soc. Anal. Chem.*, 7, 18 (1970).

[282] B. Welz, *Werkstofftechik*, 4, 285 (1973).

[283] R. Karwowska, E. Bulska, K. A. Barakat and A. Hulanicki, *Chem. Anal. (Warsaw)*, 25, 1043 (1980).

[284] M. S. Vigler and V. F. Gaylor, *Appl. Spectrosc.*, 28, 342 (1974).

[285] J. W. Robinson, *Anal. Chim. Acta*, 24, 451 (1961).

[286] R. M. Dagnall and T. S. West, *Talanta*, 11, 1553 (1964).

[287] H. W. Wilson, *Anal. Chem.*, 38, 920 (1966).

[288] D. J. Trent, *Atom. Absorp. Newslett.*, 4, 348 (1965).

[289] M. Kashiki and S. Oshima, *Anal. Chim. Acta*, 55, 436 (1971).

[290] M. Kashiki and S. Oshima, *Bunseki Kagaku*, 20, 1398 (1971).

[291] M. Kashiki, S. Yamazoe and S. Oshima, *Anal. Chim. Acta*, 53, 95 (1971).

[292] M. Kashiki, S. Yamazoe, N. Ikeda and S. Oshima, *Anal. Lett.*, 7, 53 (1974).

[293] M. P. Bratzel, Jr. and C. L. Chakrabarti, *Anal. Chim. Acta*, 61, 25 (1972).

[294] W. K. Robbins, *Anal. Chim. Acta*, 65, 285 (1973).

[295] T. M. Vickrey, G. V. Harrison and G. J. Ramelow, *Atom. Spectrosc.*, 1, 116 (1980).

[296] B. Kolb, G. Kemmner, F. H. Schleser and E. Wiedeking, *Z. Anal. Chem.*, 221, 166 (1966).

[297] D. T. Coker, *Anal. Chem.*, 47, 386 (1975).

[298] D. A. Seger, *Anal. Lett.*, 7, 89 (1974).

[299] J. W. Robinson, L. E. Vidarreta, D. K. Wolcott, J. P. Goodbread and E. Kiesel, *Spectrosc. Lett.*, 8, 491 (1975).

[300] W. R. A. De Jonghe, D. Chakraborti and F. Adams, *Anal. Chim. Acta*, 115, 89 (1980).

[301] J. D. Messman and T. C. Rains, *Anal. Chem.*, 53, 1632 (1981).

[302] J. Koizumi, R. D. McLaughlin and T. Hadeishi, *Anal. Chem.*, 51, 387 (1979).

[303] American Society for Testing and Materials, D 3237 (1974).

[304] J. Meszaros, T. Mandy and J. Glencser, *Banyasz. Kohasz. Lapok, Koolaj Foldgaz*, 11, 121 (1978).

[305] V. Berenguer, J. L. Guinon and M. de la Guardia, *Z. Anal. Chem.*, 294, 416 (1979).

[306] M. Madec and F. La Villa, *Rev. Inst. Franc. Petrole, Ann. Combust. Liquides*, 31, 687 (1976).

[307] S. Slavin and W. Slavin, *Atom. Absorp. Newslett.*, 5, 106 (1966).

[308] E. A. Means and D. Ratcliff, *Atom. Absorp. Newslett.*, 4, 174 (1965).

[309] C. A. Waggoner and H. P. Dominique, *Materials Rept.*, 68-E, Defence Research Board of Canada 1968.

[310] S. Sprague and W. Slavin, *Atom. Absorp. Newslett.*, 2, 20 (1963).

[311] D. R. Jackson, C. Salama and R. Dunn, *Can. Spectrosc. J.*, 15, 17 (1970).

[312] S. Sprague and W. Slavin, *Atom. Absorp. Newslett.*, 4, 367 (1965).

[313] K. G. Brodie and J. P. Matoušek, *Anal. Chem.*, 43, 1557 (1971).

[314] F. S. Chuang and J. D. Winefordner, *Appl. Spectrosc.*, 28, 215 (1974).

[315] J. S. Dits, *Anal. Chim. Acta*, 130, 395 (1981).

[316] S. K. Kappor, I. D. Singh and P. L. Gupta, *Indian J. Technol.*, 15, 162 (1977).

[317] R. A. Mostyn and A. F. Cunningham, *Anal. Chem.*, 38, 121 (1966).

[318] G. E. Peterson and H. L. Kahn, *Atom. Absorp. Newslett.*, 9, 71 (1970).

[319] D. Trent and W. Slavin, *Atom. Absorp. Newslett.*, No. 19, 1 (March 1964).

[320] J. D. Kerber, *Appl. Spectrosc.*, 20, 212 (1966).

[321] E. A. Means and D. Ratcliffe, *Atom. Absorp. Newslett.*, 4, 174 (1975).

[322] S. T. Holding and J. J. Rowson, *Analyst*, 100, 465 (1975).

[323] Z. Wittmann, *Analyst*, 104, 156 (1979).

[324] V. Berenguer-Navarro and J. Hernandez-Mendez, *Quim. Anal.*, 31, 81 (1977).

[325] J. Hernandez-Mendez, L. Polo-Diez and A. Bernal-Melchor, *Anal. Chim. Acta*, 108, 39 (1979).

[326] G. R. Supp, *Atom. Absorp. Newslett.*, 11, 122 (1972).

[327] G. R. Supp, I. Gibbs and M. Juszli, *Atom. Absorp. Newslett.*, 12, 66 (1973).

[328] J. A. Burrows, J. C. Heerdt and J. B. Willis, *Anal. Chem.*, 37, 579 (1965).

[329] G. Norwitz and H. Gordon, *Talanta*, 20, 905 (1973).

[330] P. K. Hon, O. W. Lau and C. S. Mok, *Analyst*, 105, 919 (1980).

[331] S. T. Holding and P. H. D. Matthews, *Analyst*, 97, 189 (1972).

[332] V. A. Vilenkin, L. L. Kalinin and Y. V. Mikulin, *Khim. Tekhnol. Topl. Masel.*, 12, 54 (1975).

[333] R. H. Kriss and T. T. Bartels, *Atom. Absorp. Newslett.*, 9, 78 (1970).

[334] T. T. Bartels and M. P. Slater, *Atom. Absorp. Newslett.*, 9, 75 (1970).

[335] C. S. Saba and K. J. Eisentraut, *Anal. Chem.*, 49, 454 (1977).

[336] C. S. Saba and K. J. Eisentraut, *Anal. Chem.*, 49, 454 (1977).

[337] J. R. Brown, C. S. Saba, W. E. Rhine and K. J. Eisentraut, *Anal. Chem.*, **52**, 2365 (1980).

[338] C. S. Saba and K. J. Eisentraut, *Anal. Chem.*, **51**, 1927 (1979).

[339] R. E. Kauffman, C. S. Saba, W. E. Rhine and K. J. Eisentraut, *Anal. Chem.*, **54**, 975 (1982).

[340] M. Kulka and F. Umland, *Z. Anal. Chem.*, **288**, 273 (1977).

[341] W. Ross and F. Umland, *Talanta*, **26**, 727 (1979).

[342] S. H. Kägler, *Erdöl. Kohle Erdgas, Petrochemie*, **19**, 879 (1966).

[343] R. A. Mostyn and A. F. Cunningham, *J. Inst. Petrol.*, **53**, 101 (1967).

[344] F. Castellani, R. Riccioni, M. Gusteri, V. Bartocci and P. Cescon, *Atom. Absorp. Newslett.*, **16**, 57 (1977).

[345] J. C. Sheridan, E. P. K. Lau and B. Z. Senkowski, *Anal. Chem.*, **41**, 247 (1969).

[346] J. Chlebicki and W. Garnacz, *Chem. Anal. (Warsaw)*, **24**, 675 (1979).

[347] J. Chlebicki and W. Garncarz, *Tenside Detergents*, **17**, 13 (1980).

[348] J. Courtot-Coupez and A. Le Bihan, *Anal. Lett.*, **2**, 567 (1969).

[349] A. Le Bihan and J. Courtot-Coupez, *Analusis*, **6**, 339 (1978).

[350] S. Giacobetti, A. Lagana, B. M. Petronio and M. V. Russo, *Riv. Ital. Sostanze Grasse*, **55**, 176 (1978).

[351] S. Setsuda, S. Itoh, A. Utsunomiya and S. Naito, *Eisei Kagaku*, **25**, 199 (1979).

[352] S. Setsuda, S. Itoh, A. Utsunomiya and S. Naito, *Bull. Kanagawa P.H. Lab.*, **10**, 15 (1980).

[353] P. T. Crisp, J. M. Eckert and N. A. Gibson, *Anal. Chim. Acta*, **104**, 93 (1979).

[354] A. Le Bihan and J. Courtot-Coupez, *Bull. Soc. Chim. France*, 406 (1970).

[355] A. Le Bihan and J. Courtot-Coupez, *Analusis*, **2**, 695 (1974).

[356] A. Le Bihan and J. Courtot-Coupez, *Analusis*, **6**, 346 (1978).

[357] P. T. Crisp, J. M. Eckert and N. A. Gibson, *Anal. Chim. Acta*, **78**, 391 (1975).

[358] P. T. Crisp, J. M. Eckert, N. A. Gibson, G. F. Kirkbright and T. S. West, *Anal. Chim. Acta*, **87**, 97 (1976).

[359] T. Matsueda and M. Morimoto, *Bunseki Kagaku*, **29**, 769 (1980).

[360] T. Matsueda, *Bunseki Kagaku*, **30**, 375 (1981).

[361] W. N. Elliott and R. A. Mostyn, *Analyst*, **96**, 452 (1971).

[362] G. C. Toralballa, G. I. Spielholtz and R. J. Steinberg, *Mikrochim. Acta*, 484 (1972).

[363] C. Tonini, *Tinctoria*, **77**, 358 (1980).

[364] E. L. Henn, *Anal. Chim. Acta*, **73**, 273 (1974).

[365] F. J. Langmyhr, Y. Thomassen and A. Massoumi, *Anal. Chim. Acta*, **68**, 305 (1974).

[366] M. D. Danchev, N. I. Ekivina and V. P. Belyaev, *Zavodsk. Lab.*, **46**, 1110 (1980).

[367] I. Lewandowska, *Rocz. Panstw. Zakl. Hig.*, **29**, 295 (1978).

[368] S. Kojima, *Analyst*, **104**, 660 (1979).

[369] J. V. Simonian, *Atom. Absorp. Newslett.*, **7**, 63 (1968).

[370] M. Olivier, *Z. Anal. Chem.*, **248**, 145 (1969).

[371] M. Olivier, *Atom. Absorp. Newslett.*, **10**, 12 (1971).

[372] M. Olivier, *Z. Anal. Chem.*, **257**, 135 (1971).

[373] K. Kuga, *Bunseki Kagaku*, **29**, 342 (1980).

[374] C. Tonini, *Tinctoria*, **76**, 320 (1979).

[375] F. R. Hartley and A. S. Inglis, *Analyst*, **92**, 622 (1967).

[376] E. J. Parks, F. E. Brinckman and W. R. Blair, *J. Chromatog.*, **185**, 563 (1979).

[377] W. Slavin, *Atom. Absorp. Newslett.*, **4**, 192 (1965).

[378] D. Druckman, *Atom. Absorp. Newslett.*, **6**, 113 (1967).

[379] T. Matsuo, J. Shida and M. Motoki, *Bunseki Kagaku*, **18**, 521 (1969).

[380] C. Tonini, *Tinctoria*, **76**, 177 (1979).

[381] M. H. Farmer, *Atom. Absorp. Newslett.*, **6**, 121 (1967).

[382] I. G. Putov, S. A. Popova and A. G. Brashnarova, *Khim. Ind. (Sofia)*, **49**, 16 (1977).

[383] A. R. Matz, *Bull. Parenteral Drug Ass.*, **20**, 130 (1966).
[384] P. O. Bethge and R. Rådeström, *Svk. Papperstidn.*, **69**, 772 (1966).
[385] P. J. Simon, B. C. Giessen and T. R. Copeland, *Anal. Chem.*, **49**, 2285 (1977).
[386] J. M. Mendiola Ambrosio, A. Gonzalez Lopez and S. Arribas Jimeno, *Afinidad*, **37**, 251 (1980).
[387] J. M. Mendiola Ambrosio, A. Gonzalez Lopez and S. Arribas Jimeno, *Afinidad*, **37**, 39 (1980).
[388] J. M. Mendiola Ambrosio and A. Gonzalez Lopez, *Rev. Plast. Mod.*, **41**, 413 (1981).
[389] T. Korenaga, *Analyst*, **106**, 40 (1981).
[390] N. Rombach, R. Apel and F. Tschochner, *GIT Fachz. Lab.*, **24**, 1165 (1980).
[391] J. P. Price, *Atom. Absorp. Newslett.*, **11**, 1 (1972).
[392] J. M. Mendiola Ambrosio and A. Gonzalez Lopez, *Rev. Plast. Mod.*, **41**, 550 (1981).
[393] J. D. Kerber, *Atom. Absorp. Newslett.*, **10**, 104 (1971).
[394] J. D. Kerber, A. Koch and G. E. Peterson, *Atom. Absorp. Newslett.*, **12**, 104 (1973).
[395] O. Ant-Wuorinen and A. Visapää, *Paperi Ja Puu*, **48**, 649 (1966).
[396] C. M. Paralusz, *Appl. Spectrosc.*, **22**, 520 (1968).
[397] B. Searle, W. Chan, C. Jensen and B. Davidow, *Atom. Absorp. Newslett.*, **8**, 126 (1969).
[398] J. H. Barker, W. B. Chapman and A. J. Harrison, *J. Assoc. Pub. Anal.*, **2**, 89 (1964).
[399] E. Eng, *Skand. Tidskr. Faerg Lack*, **21**, 7 (1975).
[400] R. W. Scott, 166th ACS National Meeting. Chicago, Illinois, August (1973).
[401] E. L. Henn, *Atom. Absorp. Newslett.*, **12**, 109 (1973).
[402] E. L. Henn, *Paint Varn. Prod.*, **63**, 29 (1973).
[403] O. W. Lau and K. L. Li, *Analyst*, **100**, 430 (1975).
[404] D. G. Mitchell, K. M. Aldous and A. F. Ward, *Atom. Absorp. Newslett.*, **13**, 121 (1974).
[405] J. Dumanski and K. Sosin, 5th Polish Spectroanal. Conf., Wladislawow, Oct. 1976.
[406] O. Beniot and G. Gaiger, *Bull. Liaison Lab. Ponts et Chaussées*, **79**, 83 (1975).
[407] W. K. Porter, *J. Assoc. Off. Anal. Chem.*, **57**, 614 (1974).
[408] R. J. Noga, *Anal. Chem.*, **47**, 332 (1975).
[409] W. Holak, *Anal. Chim. Acta*, **74**, 216 (1975).
[410] P. Minkkinen, *Kem.-Kemi*, **3**, 282 (1976).
[411] C. W. Fuller, *Anal. Chim. Acta*, **62**, 261 (1972).
[412] J. Brandt, *Am. Paint J.*, **57**, 28 (1973).
[413] A. Lasarova and S. Arpadjan, *Z. Chem.*, **20**, 225 (1980).
[414] D. Kalihová, V. Sychra and N. Dudová, *Chem. Listy*, **72**, 1081 (1978).
[415] L. Moten, *J. Assoc. Off. Agr. Chem.*, **53**, 916 (1970).
[416] C. W. Fuller, Oil Col. Chem. Assoc. Meeting, Newcastle upon Tyne (March 1975).
[417] M. Kronstein, *Mod. Paint Coat.*, **67**, 57 (1977).
[418] S. S. Krishnan, *J. Forens. Sci.*, **19**, 789 (1974).
[419] R. L. Brunelle, C. M. Hoffman and K. B. Snow, *J. Assoc. Off. Anal. Chem.*, **53**, 270 (1970).
[420] A. L. Green and J. P. Sauve, *Atom. Absorp. Newslett.*, **11**, 93 (1972).
[421] S. S. Krishnan, *J. Forens. Sci.*, **19**, 351 (1974).
[422] C. M. Hoffman, *Identification News*, October 1968.
[423] M. L. Newburg, *J. Can. Soc. Forensic Sci.*, **13**, 19 (1980).
[424] G. D. Renshaw, C. A. Pounds and E. F. Pearson, *Atom. Absorp. Newslett.*, **12**, 55 (1973).
[425] A. B. Sakla and A. M. Shalaby, *Microchem. J.*, **24**, 168 (1979).
[426] J. Lede and J. Villermaux, *Anal. Chim. Acta*, **87**, 291 (1976).
[427] J. Courtot-Coupez and A. Le Bihan, *Anal. Lett.*, **2**, 567 (1969).

14

Laboratory procedures

The procedures given in this chapter are intended to give students and analysts first-hand experience with some typical applications of AAS to the analysis of organic materials. The 24 selected methods presented deal with the determination of elements, functional groups, compounds and metallic species in a wide range of materials.

14.1 ELEMENTS IN ORGANIC, INDUSTRIAL AND BIOLOGICAL MATERIALS

14.1.1 Determination of Arsenic in Organic Compounds [1]

Principle. Organoarsenic compounds are decomposed by wet oxidation, followed by treatment with magnesia mixture. The arsenate is precipitated as $MgNH_4AsO_4.6H_2O$ and the excess of magnesium is measured.

Reagents. Hydrogen peroxide (30%), perchloric acid (70%), hydrochloric acid ($0.1M$), ammonia solution (15%) and magnesia mixture (Mg 4 mg/ml) [2].

Compounds tested. 4-Hydroxy-3-nitrophenylarsonic acid, *o*-aminophenylarsonic acid, *p*-hydroxybenzene arsonic acid, *p*-toluene arsonic acid, 3,4-difluorophenyl arsonic acid, and *p*-toluene-3-nitroarsonic acid.

Procedure. Weigh accurately 3–10 mg of finely ground dried sample into a 25-ml Kjeldahl flask. Digest the sample with a mixture of 2 ml of 70% nitric acid and 1 ml of 30% hydrogen peroxide. Heat on a sand-bath at 250°C with continuous swirling. In about 5 min the solution begins to boil, and vigorous oxidation takes place soon thereafter. Heat to dryness, then add 0.1 ml of 70% perchloric acid. The solution should become perfectly clear. If any charred material remains, repeat the digestion treatment, heating until the digest is clear. Complete digestion usually requires 12–15 min. After digestion, cool the contents of the flask, and dissolve the residue with 3 ml of $0.1M$ hydrochloric acid. Add 0.5 ml of magnesia mixture and cool. Dilute to 10 ml with 15% ammonia solution added slowly and with continuous stirring. Allow to stand for 30 min, and centrifuge at 4000 rpm for 5 min. Pipette 0.25 ml of the supernatant liquid into a 50-ml standard flask and dilute to volume with $0.1M$ hydrochloric acid. Aspirate the solution into a hydrogen-air flame and measure the absorbance of Mg at 285.2 nm. Run a blank and calculate the arsenic contents (1 ppm Mg = 3.08 ppm As).

Accuracy and precision. The apparent recovery range is 98.9-103.5% with mean of 100.7% and standard deviation 0.4%.

14.1.2 Determination of Phosphorus in Organic Compounds [3]

Principle. Organophosphorus compounds are decomposed by combustion in an oxygen-filled flask, followed by reaction of the phosphate with molybdic acid to form molybdophosphoric acid which is extracted into MIBK and atomized in the flame.

Reagents. Ammonium molybdate solution (10%), perchloric acid (60%), $0.5M$ sodium hydroxide, saturated aqueous bromine solution.

Compounds tested. Triphenylphosphine, pyridoxal phosphate, histamine diphosphate.

Procedure. Weigh accurately 2-5 mg of the sample (= 0.1-0.5 mg of P) onto an L-shaped piece of Whatman No. 42 filter paper (2.5 × 2.5 cm), then fold the paper and clamp it in the platinum gauze of the combustion unit (use a 500-ml combustion flask). Add to the flask 5 ml of $0.5M$ sodium hydroxide and 5 ml of saturated aqueous bromine solution, and flush the flask with pure oxygen. Ignite the fuse of the filter paper and introduce the stopper into the neck of the flask. After combustion is complete, shake the flask well for 5 min and allow it to stand for 30 min with occasional shaking. Remove the stopper, rinse the platinum wire and gauze with 5 ml of demineralized water, add 1 ml of concentrated hydrochloric acid, heat gently on a boiling water-bath to expel all bromine and transfer to a 100-ml standard flask. Dilute to the mark with demineralized water and mix [4]. To a 5-ml aliquot of the solution, add 5 ml of 60% perchloric acid, mix and dilute to 40 ml with demineralized water. Add 3 ml of 10% ammonium molybdate solution, extract with three 10-ml portions of MIBK and wash the extract with demineralized water. Aspirate the organic extract into an air-acetylene flame and measure the absorbance of the molybdenum at 313.3 nm. Run a blank.

Accuracy and precision. The average recovery is 98% and standard deviation 2%.

14.1.3 Determination of Chlorine in Organic Compounds [5,6]

Principle. Amine hydrochlorides and quaternary ammonium chlorides are treated with silver nitrate, followed by measurement of the excess of silver. Compounds containing organically bound chlorine are decomposed in an oxygen-filled flask prior to precipitation with silver.

Reagents. Silver working standard solution (4-5 ppm Ag). Chloride precipitating solution, containing 7 mg of silver nitrate per ml.

Compounds tested. Apomorphine.HCl, betazol.HCl, bethanechol chloride, ephedrine.HCl, isoproterenol.HCl, methamphetamine.HCl, nalorphine.HCl, phenylephrine.HCl, pralidoxime chloride, procaine.HCl, tetracycline.HCl and poly(vinyl chloride).

Procedure. Transfer a 2-ml aliquot of a solution of amine hydrochloride or quaternary ammonium chloride (containing 600–700 μg of chloride per ml) to a 10-ml centrifuge tube. Add exactly 1 ml of silver nitrate solution, stand the tube in an ice-bath for 10 min and centrifuge for 10 min. Transfer 1 ml of the supernatant liquid to a 250-ml standard flask, add 12.5 ml of concentrated nitric acid and dilute to the mark. Aspirate into an air–acetylene flame and measure the absorbance of silver at 328.1 nm. Run a blank.

For organically bound chlorine, e.g. in poly(vinyl chloride), combust a 50-mg sample in an oxygen-filled flask containing 50 ml of 0.1M sodium hydroxide, 0.5 ml of 30% H_2O_2 and 70 ml of demineralized water. After the combustion, transfer the absorption solution to a 1-litre standard flask and add 2 ml of nitric acid (2+1). Add 8.5 ml of 0.05M silver nitrate, shake and dilute to the mark with demineralized water. Centrifuge a 10-ml aliquot of the solution for 10 min at 2700 rpm and aspirate the supernatant liquid into the flame. Run a blank.

Accuracy and precision. The average recovery ranges between 98.9 and 103.7% and the standard deviation is 2%.

14.1.4 Determination of Lead in Plastic Containers [7]

Principle. Trace levels of lead in polymers and in plastic containers can be measured by direct electrothermal atomization of the solid samples in a graphite tube furnace.

Reagents. Lead nitrate stock standard solution (1.0 mg/ml) in 1M nitric acid.

Samples tested. Polyethylene, polypropylene, poly(vinyl chloride) and polystyrene.

Procedure. Wash the container thoroughly with distilled water, dry it in air at room temperature and cut it into small pieces. Weigh accurately up to 5 mg of sample, rinse well with distilled water and transfer (without drying) to the graphite tube atomizer. Dry at 130°C for 90 sec, ash at 800°C for 60 sec and atomize at 1600°C for 2 sec (ramp rate 400°C/sec). Measure the absorbance of Pb at 217.0 nm. Inject a 5-μl volume of a standard lead nitrate solution, containing approximately the same amount of lead as the sample, before and after each measurement.

Accuracy and precision. The mean relative standard deviation is 15% at the 0.03–1 ppm level of lead.

14.1.5 Determination of Lead in Gasoline [8]

Principle. The gasoline sample is diluted with MIBK and the alkyl-lead compounds are stabilized by reaction with iodine and a quaternary ammonium salt before atomization in the flame.

Reagents. Aliquat 336-MIBK solution (1% v/v), 3% iodine solution in toluene. 'Lead-free' gasoline (gasoline containing less than 5 mg of Pb per gallon). Lead standard solutions (0.02, 0.05 and 0.10 g/gallon), prepared by dilution of 2.0, 5.0 and 10.0 ml of a 1.0 g/gallon solution to 100 ml with MIBK containing 5.0 ml of 1% Aliquat 336.

Limit of detection. Pb 2.5–25 mg/l.

Procedure. Transfer 5.0 ml of gasoline sample to a 50-ml standard flask containing 30 ml of MIBK. Add 0.10 ml of iodine–toluene solution and allow the mixture to react for about 1 min. Add 5.0 ml of 1% Aliquat 336-MIBK solution and mix. Dilute to the mark with MIBK and mix. Aspirate the sample solution and working standard into an air–acetylene flame and measure the absorbance of Pb at 283.3 nm.

Accuracy and precision. The average recovery is 98% and the standard deviation is 2%.

14.1.6 Determination of Tetra-alkyl-lead in Biological Materials [9]

Principle. Tetra-alkyl-lead compounds are extracted by shaking with a mixture of benzene and EDTA solution, then digested with acid; the resultant Pb^{2+} is measured by electrothermal atomization AAS.

Reagents. Tetraethyl- and tetramethyl-lead standards, 10.0 μg/ml and 1.0 μg/ml. Lead nitrate standard solution (Pb 1000 μg/ml) in $1M$ nitric acid. Scintillation grade benzene. EDTA, 0.4% solution, pH 6–7.

Limit of detection. 10 ng of lead per litre.

Procedure. Homogenize 5 g of fish tissue and place it in a 50-ml glass centrifuge tube fitted with a screw cap. Add 10 ml of benzene and 10 ml of EDTA solution and shake the tube for 10 min. Centrifuge at 2200 rpm for 30 min to separate the two phases. Transfer a 3.0-ml portion of the benzene layer to a 50-ml calibrated Folin-Wu digestion tube and acidify with 3 ml of concentrated nitric acid. Evaporate the benzene layer under a stream of highly pure nitrogen. Digest the residue for at least 2 hr on an aluminium block at 80-90°C or until evolution of large amounts of NO_2 has ceased. Dilute the sample to 10 ml with demineralized water and shake it with 2 ml of hexane. Discard the hexane layer, inject a portion of the aqueous phase into a graphite tube furnace and measure the absorbance of lead at 283.3 nm. Use the standard addition technique and the following conditions: drying at 100°C (40 sec), charring at 500°C (60 sec), atomization at 2000°C (5 sec).

Accuracy and precision. The relative standard deviation is 5% at the ng/g level.

14.1.7 Determination of Cadmium and Lead in Human Whole Blood [10]

Principle. Blood samples are diluted with Triton X-100 as a solubilizing agent and ammonium phosphate as a mixture modifier, and are then electro-thermally atomized.

Reagents. Standard Cd and Pb stock solutions (1000 mg/ml), diluted aₔ required. Diammonium hydrogen phosphate (50 g/l.) and Triton X-100 (5 ml/l.).

Limit of detection. The normal levels of Cd (0.6–11 μg/l.) and Pb (80–140 μg/l.) can be measured.

Procedure. Draw a fresh human blood sample from the antecubital vein with a disposable stainless-steel needle. Transfer it to a 10-ml glass tube previously cleaned and tested for Pb and Cd. Shake well, then transfer 0.20 ml of the blood to a precleaned 1-ml Pyrex glass standard flask, and add 0.1 ml each of the Triton X-100 and ammonium phosphate solutions. Complete to the mark with demineralized water, shake the flask for 20 sec and inject 10 μl of the solution into a pyrocoated graphite tube. Dry at 100°C (50 sec), ash at 500°C (Cd) and 700°C (Pb) for 30 sec and atomize at 2200°C for 10 sec (Cd) and 2300°C for 8 sec (Pb). Use nitrogen purge gas in the interrupt mode, at a flow-rate of 10.8 l./hr. Compare with a calibration graph prepared from fresh aqueous standard lead and cadmium solutions containing the same concentrations of Triton X-100 and ammonium phosphate.

Accuracy and precision. The average recoveries of Cd and Pb are 100.8 ± 4.3% and 98.7 ± 3.9%, respectively.

14.1.8 Determination of Trace Elements in Human Hair [11]

Principle. The concentrations of Cu, Fe, Mg and Zn in the hair are determined by prewashing, wet digestion with HNO_3–$HClO_4$ mixture, and atomization in the flame.

Reagents. Standard stock solutions of Cu, Fe, Mg and Zn (1000 μg/ml), nitric acid, perchloric acid, and non-ionic surfactant (e.g. 7X O-Matic).

Limit of detection. As little as 9 μg of Cu and Fe, 16 μg of Mg and 140 μg Zn per g of dry hair can be accurately measured.

Procedure. Collect ~ 0.5 g of hair sample, cut it into 1-cm lengths and thoroughly mix to ensure homogeneity. Transfer the hair sample to a 500-ml polyethylene bottle and add 150 ml of a 1% non-ionic surfactant solution containing negligible levels of the analyte element. Shake the bottle for 30 min at room temperature. After the washing, transfer the hair to a polyethylene filter crucible and wash it with 1 litre of demineralized water to remove the surfactant. Dry the sample overnight at 110°C. Weigh out 0.3–0.8 g of the dry sample, add 6 ml of concentrated nitric acid and leave to digest at room temperature. Warm the acid digest, add 1 ml of 70% perchloric acid and continue the digestion on a hot-plate at ~ 200°C until dense white fumes of perchloric acid are evolved. Transfer the digest to a 5-ml standard flask and dilute to the mark with demineralized water. Further dilution may be required. Shake the solution and aspirate it into an air–acetylene flame.

Accuracy and precision. The recoveries of Cu, Fe, Mg and Zn in hair samples are 92.5–102%, 94.5–107.5%, 95.8–100.8% and 99.1–102.1%, respectively.

14.2 ORGANIC FUNCTIONAL GROUPS

14.2.1 Determination of 1,2-Diols [12]

Principle. 1,2-Diols undergo oxidative cleavage by reaction with periodate. The excess of periodate is precipitated with lead nitrate, isolated, dissolved in $1M$ nitric acid and analysed for lead.

$$-\underset{\underset{OH}{|}}{\overset{|}{C}}——\underset{\underset{OH}{|}}{\overset{|}{C}}- + IO_4^- \longrightarrow 2 \,\,>\!\!C{=}O + H_2O + IO_3^-$$

$$2IO_4^- + Pb^{2+} \longrightarrow Pb(IO_4)_2$$

Reagents. Potassium periodate ($0.02M$) in water; lead nitrate (Pb 1000 $\mu g/ml$) in $1M$ nitric acid.

Compounds tested. 1,2-Ethanediol, 1,2-propanediol, 1-phenyl-1,2-ethanediol, DL-β-3,4-dihydroxyphenylalanine.

Procedure. Transfer an aliquot of sample solution containing 0.8–8.0 μmole of the diol to a 15-cm test-tube. Add 1.0 ml of $0.020M$ potassium periodate, mix well and leave to stand for 2–3 min. Add 1.0 ml of $0.04M$ lead nitrate and shake the tube for 30 sec to precipitate lead periodate (do not leave for more than 2 min, as lead iodate and periodate in aqueous solution will be adsorbed on the test-tube). Filter the contents of the test-tube and rinse the tube 3 times with approximately 6-ml portions of demineralized water. Dissolve the lead periodate precipitate with two 10.0-ml portions of $1M$ nitric acid, collect the lead solution in a 50-ml standard flask and dilute to the mark with demineralized water. Shake and aspirate into an air–acetylene flame. Measure the absorption of lead at 283.3 nm and run a blank.

Accuracy and precision. The relative deviation is 2.6–6.9% over the concentration range 1.5–4.0 $\mu g/ml$. The average recovery is 98.5%.

14.2.2 Determination of Aldehydes [13]

Principle. The aldehyde group is quantitatively oxidized by Tollen's reagent. The reduced silver is separated from excess of reagent and dissolved in nitric acid, then the resultant solution is analysed for silver.

$$RCHO + 2Ag(NH_3)_2^+ + 2OH^- \longrightarrow 2Ag + RCOONH_4 + 3NH_3 + H_2O$$

Reagents. Tollen's reagent is prepared by the addition of exactly 1.00 ml of $3M$ sodium hydroxide to 5.00 ml of $0.50M$ silver nitrate in a 50-ml beaker. The contents are mixed and sufficient aqueous ammonia solution (1+1) is added dropwise to dissolve the Ag_2O ($\sim$ 2 ml). The reagent must be prepared fresh and used immediately, any unused reagent being discarded within 4 hr (otherwise there is danger of explosive decomposition products being formed). Stock silver nitrate solution is prepared by dissolving 0.20–0.35 g of the salt (accurately weighed) in 250 ml of nitric acid (1+1).

Compounds tested. Benzaldehyde and its *p*- and *m*-nitro, *m*-cyano, *p*-chloro, *m*-methoxy and 3,5-dimethoxy derivatives, butyraldehyde, propaldehyde and formaldehyde.

Procedure. Transfer an aliquot of the aldehyde solution (containing 0.25–4.0 μmole) to a 6-inch glass tube. Add an equal volume of Tollen's reagent under minimum lighting conditions. Mix thoroughly, place the tube in a light-tight container and shake this on a mechanical agitator for 30–135 min. Filter through a fine-porosity fritted-glass funnel. Rinse the tube with two 5-ml portions of aqueous ammonia (1+1) to dissolve any silver oxide present, followed by two 5-ml portions of water. Place a clean 125-ml filter flask below the funnel and add 6 ml of nitric acid (1+1) to the test-tube to dissolve the adhering silver mirror. Rinse the tube with two 5-ml portions of water and transfer the filtrate to a 50-ml standard flask. Aspirate the solution into an air–acetylene flame, measure the absorption of silver at 328.1 nm, run a blank under identical conditions and compare with a calibration graph prepared from the stock silver nitrate solution.

Accuracy and precision. Relative standard deviations of 1.2–5.8% are obtained over the 1–4 μmole range.

14.2.3 Determination of Aliphatic and Aromatic Primary Amines [14]

Principle. Primary amines react with *p*-nitrosalicylaldehyde to give the corresponding Schiff's base, which upon reaction with copper(II) gives a water-insoluble copper complex.

Reagents. A copper complexing agent is prepared by mixing 3.372 g of triethanolamine, 0.205 g of 5-nitrosalicylaldehyde, 1 ml of 50% acetaldehyde solution and 5 ml of 4% copper sulphate and diluting to a total volume of 50 ml with water.

Compounds tested. Methylamine, isopropylamine, n-hexylamine, cyclo-hexylamine, hexamethylenediamine, triethylenetriamine, aniline, *o*-, *m*- and *p*-toluidine, *o*-, *m*- and *p*-aminophenol, *o*- and *m*-anisidine, *m*- and *p*-phenylene-diamine, 3,3'-dimethylbenzidine, *p*-chloroaniline, procaine, α,β-naphthylamine.

Procedure. Transfer an aliquot of the amine solution (containing 40 μg–10 mg) to a 15-cm test-tube. Add 1.0 ml of the complexing reagent solution and dilute to 10 ml with water. Let stand for 1 hr, then filter off the copper complex. Transfer 1.0 ml of the filtrate to a 25-ml standard flask and dilute to the mark with demineralized water. Aspirate the solution into an air–acetylene flame. Alternatively, dissolve the precipitate in 10 ml of 12.6M nitric acid, dilute accurately to 100 ml with demineralized water and aspirate into the flame. Measure the absorption of copper at 324.7 nm and run a blank.

Accuracy and precision. The average recovery ranges between 96 and 104% and the standard deviation is 2%.

14.2.4 Determination of Aliphatic Secondary Amines [15]

Principle. Reaction of secondary amines with carbon disulphide gives dialkyldithiocarbamic acids. Upon addition of nickel, water-insoluble nickel dialkyldithiocarbamates are formed. The precipitate is isolated, digested in a 1:1 mixture of nitric acid and hydrochloric acid, and the solution is analysed for nickel.

$$2R_2NH + 2CS_2 + NI^{2+} \longrightarrow (R_2NCS_2)_2Ni + 2H^+$$

Reagents. The ammoniacal nickel reagent is prepared by dissolving 200 g of ammonium acetate and 5 g of $NiCl_2.6H_2O$ in 300 ml of water, adding 200 ml of 50% sodium hydroxide solution and 200 ml of aqueous ammonia solution (s.g. 0.90) and diluting to 1 litre with demineralized water. Nickel di-n-butyl-dithiocarbamate is prepared, crystallized from acetone, dried *in vacuo* at room temperature and used for preparation of the standard calibration graph.

Compounds tested. Diethylamine *N*-ethyl-n-butylamine, di-n-butylamine, di-n-hexylamine, di-n-octylamine, *N*-butyl-n-dodecylamine, di-n-dedecylamine and piperidine.

Procedure. Transfer an aliquot of the aliphatic secondary amine solution (containing 1–4 μmole) to a 6-inch test-tube. Add 10 ml of ammoniacal nickel reagent and approximately 0.025–0.05 ml of carbon disulphide. Mix the contents of the test-tube thoroughly and place it in a water-bath at 30–35°C for 60–75 min. Filter off the precipitate on a medium-porosity fritted-glass funnel. Place a clean flask below the filter funnel and rinse the tube with 1 ml of warm benzene–acetone (1:1) mixture to ensure complete transfer of the adhering precipitate. Rinse the funnel with 2 portions of warm benzene–acetone mixture. Place the flask in a boiling water-bath to evaporate the solvent. Add 4 ml of concentrated nitric–hydrochloric acid mixture (1:1) and dissolve the precipitate by heating at 100°C for 15–20 min. Transfer the contents of the flask to a

10.0-ml standard flask, dilute to the mark, aspirate into an air–acetylene flame and measure the absorbance of nickel at 232.0 nm. Run a blank.

Accuracy and precision. The average recovery ranges from 97.4 to 100.7%, with a relative standard deviation of 1.1–3.9% for the range 1–3 μmole.

14.2.5 Determination of Nitrates and Nitramines [16]

Principle. Nitrates and nitramines undergo reduction with cadmium metal and 0.1M hydrochloric acid with the release of 4 equivalents of cadmium ions per mole of group.

$$2R\text{-}ONO_2 + 4Cd + 8H^+ \longrightarrow N_2O + 4Cd^{2+} + 2ROH + 3H_2O$$

Reagents. Cadmium metal of purity not less than 99.5%, washed successively with 6M hydrochloric acid, water and alcohol, and then dried. Stock standard cadmium nitrate solution (Cd/ml 0.1 mg). EDTA solution (0.1%) and 0.1M hydrochloric acid.

Compounds tested. Pentaerythritol tetranitrate, urea nitrate and guanidine nitrate.

Procedure. Transfer a portion of nitrate sample containing 0.2–2 mg of nitrate-nitrogen to a 100-ml conical flask having a ground-glass neck and a side-arm with gas-bubbling tube. Add 10 ml of 0.1M hydrochloric acid and attach a water condenser. Heat on a sand-bath, while passing carbon dioxide ($\sim$ 50 bubbles/min) through the side-arm. When the solution starts boiling, introduce 50–100 mg of cadmium metal turnings. Continue boiling for 10–15 min under a carbon dioxide atmosphere. Cool the solution and transfer it to a 25-ml standard flask, making up to the mark with demineralized water. Transfer a 1.00-ml aliquot of the test solution to a 100-ml standard flask, add 1 ml of 1% EDTA solution, dilute to the mark with 0.05M hydrochloric acid and mix. Aspirate into an air–acetylene flame and measure the absorbance at 228.8 nm. Compare with a calibration graph prepared by treating 0.50–3.00 ml of the 0.1-mg/ml cadmium solution in the same way (1 mg of Cd $\equiv$ 62.3 μg of nitrate-nitrogen).

Accuracy and precision. The average recovery is 98% and the mean standard deviation is 1% at the 0.2–2 mg nitrate-nitrogen level.

14.2.6 Determination of Sulphonamides [17]

Principle. Sulphonamides quantitatively react at pH 8 with silver or copper ions to form the insoluble metal sulphonamide. The precipitate is isolated and the excess of metal ion is measured.

$$R\text{-}SO_2NH\text{-}R + Ag^+ \longrightarrow RSO_2NAgR + H^+$$

$$2R\text{-}SO_2NHR + Cu^{2+} \longrightarrow (RSO_2NR)_2Cu + 2H^+$$

Reagents. Stock standard silver (5 mg/ml) and copper (2 mg/ml) nitrate solutions in 0.02M nitric acid, Thymol Blue indicator (0.1% in 96% ethanol) and 0.1M sodium hydroxide.

Compounds tested. Sulphadiazine, sulphadimidine, sulphamerazine, sulphathiazole, sulphamethoxypyridazine, sulphamethoxazole, sulphamethoxine and sulphapyridine.

Procedure. Accurately weigh 1 g of pulverized dried pure sulphonamide powder, transfer it to a 100-ml standard flask, add 10 ml of demineralized water and 5 drops of 0.1% Thymol Blue indicator. Add 0.1M sodium hydroxide dropwise until the solution is blue. Shake the flask until the sample has dissolved completely, and dilute to volume with demineralized water. Transfer a 10.0-ml aliquot of the sample solution to a 100-ml standard flask. Add 10.0 ml of stock silver nitrate (5 mg/ml) or copper nitrate (2 mg/ml) solution, shake well, and let stand for 10 min. Dilute to volume with demineralized water, shake, and centrifuge. Transfer 5.0 ml of the supernatant liquid to a 100-ml standard flask, dilute to volume with 0.02M nitric acid, and mix. Aspirate into an air–acetylene flame and measure the absorbance at 328.1 for silver or 324.8 nm for copper. Run a blank experiment and compare with the calibration graphs prepared by transferring aliquots of the stock silver and copper solutions (0.10–1.00 ml) to 100-ml standard flasks and diluting to volume with 0.02M nitric acid followed by aspiration into the flame as described above.

Accuracy and precision. The average recovery is 99.4% and the mean standard deviation 0.6%.

14.2.7 Determination of Thiols [18]

Principle. Thiols react with alcoholic silver nitrate to form a silver mercaptide precipitate. The precipitate is separated from excess of reagent and dissolved in nitric acid, then the solution is analysed for silver.

$$R\text{-}SH + Ag^+ \longrightarrow RSAg + H^+$$

Reagents. Silver nitrate reagent at a concentration of 10000 ppm in 2-propanol–water azeotrope. Aldehyde-free 2-propanol is prepared by refluxing with 10% sodium hydroxide solution.

Compounds tested. Benzenethiol, butanethiol, pentanethiol, hexanethiol, heptanethiol, *o*-, *m*- and *p*-toluenethiol, 2,3-dimercaptopropionic acid and thioacetic acid.

Procedure. Transfer a 1.0-ml aliquot of the thiol solution (0.8–20 μmole/ml) in 2-propanol–water azeotrope to a 100-ml beaker. Add 1–2 ml of silver nitrate reagent solution with an Eppendorf pipette. Add 2 ml of distilled water, stir and allow to stand at room temperature for 20–30 min. Transfer the precipitate to a fine-porosity sintered-glass funnel, and filter. Wash the reaction vessel with less than 10 ml of the wash solution (5–10% v/v aqueous ammonia), transferring the washings into the funnel, and wash the precipitate with demineralized water. Place a clean 25-ml flask below the funnel, add 20 ml of hot concentrated nitric acid to the precipitate and let stand for 5 min before applying suction. Rinse with 20-30 ml of demineralized water and transfer the filtrate quantitatively to a suitable standard flask, dilute with water and measure the absorbance of silver at 328.1 nm, using an air–acetylene flame. Run a blank and compare with standards containing 1–6 ppm silver.

Accuracy and precision. The relative deviations are 0.8–3.5% in the 0.8–20 μmole/ml range and the average recovery is 96.7–103%.

14.2.8 Determination of Sulphides [19]

Principle. Organic sulphides are oxidized to the corresponding sulphoxides with aqueous sodium metaperiodate. The iodate ions released can be quantitatively precipitated by the addition of silver nitrate in nitric acid. The precipitate is isolated, dissolved in aqueous ammonia and atomized in the flame.

$$R\text{-}S\text{-}R + IO_4^- \longrightarrow R\overset{\overset{\textstyle O}{\uparrow}}{\text{-}S\text{-}}R + IO_3^-$$

Reagents. Sodium metaperiodate solution (15 μmole/ml). Acid silver nitrate reagent prepared by mixing equal volumes of a nitric acid solution (3+1) with $2M$ silver nitrate.

Compounds tested. Butyl sulphide, isobutyl sulphide, hexyl sulphide, phenyl sulphide, L-(–)-methionine, DL-methionine and potassium penicillin G.

Procedure. Transfer a 1.0-ml aliquot of the sulphide test solution (containing 100–500 μg) to a 15-cm test-tube. Add 1.0 ml of sodium metaperiodate reagent. Place the tube on a mechanical shaker and shake overnight ($\sim$ 15 hr). Add 1 ml of silver nitrate reagent and shake again for an additional 10 min to flocculate the AgIO$_3$ precipitate. Place the test-tube in a freezer at -15°C for 30–60 min to promote further coagulation and suppress the solubility of AgIO$_3$. Remove the test-tube from the freezer and transfer its contents to a fine-porosity sintered-glass filter funnel with 3 ml of 1:1 acetone–water mixture at 0°C. Wash the precipitate three times with the ice-cold 1:1 acetone–water mixture. Rinse the stem of the funnel and fit the filter in a clean suction flask. Add 5 ml of concentrated aqueous ammonia solution to the test-tube, then transfer to the filter to dissolve the precipitate; wash the funnel with three 15-ml portions of demineralized water and transfer the solution to a 100-ml standard flask. Dilute to the mark with demineralized water and measure the absorbance of silver at 328.1 nm, using an air–acetylene flame. Run a blank.

Accuracy and precision. The precision of the method is 2–4% over the range 1–5 μmole/ml and the average recovery is 96–103%.

14.3 ORGANIC COMPOUNDS

14.3.1 Determination of Strychnine [20]

Principle. Strychnine reacts with molybdophosphoric acid to form a strychnine molybdophosphate precipitate, the molybdate content of which can be stripped with a basic buffer and measured.

$$PMo_{12}O_4^{3-} + 3(C_{21}H_{22}N_2O_2H)^+ \longrightarrow (C_{21}H_{22}N_2O_2H)_3.PMo_{12}O_{40}$$

Reagents. Sodium molybdate dihydrate solution (6%), sulphuric acid (7.2N), potassium dihydrogen phosphate solution (P 100 μg/ml), standard strychnine sulphate solution (500 μg/ml) and MIBK.

Limit of detection. 0.7 ppm.

Procedure. Transfer 4 ml of 6% sodium molybdate solution to a 60-ml separating funnel. Add 1.0 ml of standard phosphate solution and 1.0 ml of 7.2N sulphuric acid. Mix and wait 5 min for complete formation of molybdophosphoric acid. Add the strychnine solution (containing up to 1.0 mg of strychnine) and dilute to approximately 20 ml. Swirl to mix and wait 10 min for complete formation of the strychnine molybdophosphate precipitate. Add 4.0 ml of 10% sodium citrate solution, swirl to mix and wait 3-5 min. Extract the complex with 10 ml of MIBK. Shake for 1 min, allow the phases to separate and remove and discard the lower aqueous phase. Wash the organic extract three times with 10-ml portions of 1.0N sulphuric acid. For each wash, shake the separating funnel with dilute acid for 1 min, allow the phases to separate, then remove and discard the lower phase. After the final wash, rinse the tip of the separating funnel with distilled water to remove any traces of excess of molybdate. Aspirate the MIBK extract into an air-acetylene flame, measure the absorbance of molybdenum at 313 nm and run a blank.

Accuracy and precision. The standard deviation is 2.5%.

14.3.2 Determination of Biuret in Urea [21,22]

Principle. Biuret in mixed fertilizers and urea is determined by reaction with copper in basic medium to form the soluble biuret-copper complex. The excess of insoluble copper reagent is isolated and the complexed copper in solution is measured.

$$2H_2NCOCNHCONH_2 + Cu^{2+} \longrightarrow$$

Reagents. Copper sulphate solution (15 g/l.), buffer solution (24.6 g of potassium hydroxide and 30 g potassium chloride in 1 litre of water), starch solution (1 g of soluble starch in 200 ml of boiling water containing 1 g of oxalic acid) and biuret standard solution (0.4 g in 1 litre of water).

Compounds tested. Biuret, biuret in mixed fertilizers, biuret in urea.

Procedure. Transfer aliquots of biuret standard solution containing 4, 8 and 12 mg of biuret to 100-ml standard flasks. Dilute each to about 30 ml with demineralized water and add 25 ml of 95% ethanol. Stir, add 2 ml of the starch solution, 10 ml of copper sulphate solution and 20 ml of the buffer solution. Dilute to the mark with water, mix and let stand for 10 min. Filter 50 ml through a dry 150-ml medium-porosity fritted glass funnel into a dry flask. Transfer a 25-ml aliquot of filtrate to a 250-ml standard flask, acidify with 5 ml of $1M$ hydrochloric acid and dilute to volume with water. Aspirate the solution into an air–acetylene flame and measure the absorbance of copper at 324.8 nm. From the concentration of copper obtained, calculate the factor relating mg of copper found to mg of biuret. For determination of biuret in urea, weigh a sample containing < 10 mg of biuret, dissolve it in water, transfer to a 100-ml standard flask and complete as above.

Accuracy and precision. The average recovery is 97–100% and the mean standard deviation is 1.3%.

14.3.3 Determination of Chloramphenicol in Pharmaceutical Preparations [23]

Principle. Chloramphenicol in various pharmaceutical preparations is reduced with cadmium metal, 6 equivalents of cadmium ions being released per mole of chloramphenicol, and measured in the flame at 228.8 nm.

Reagents. Cadmium metal (purity 99.5%), stock cadmium chloride solution (Cd 0.1 mg/ml) in $0.05M$ hydrochloric acid.

Limit of detection. Down to 3 mg of chloramphenicol or its esters.

Procedure. Weigh accurately 1 g of the pulverized dried chloramphenicol powder or its ester, dissolve it in the least amount of ethanol needed, transfer to a 100-ml standard flask and dilute to the mark with ethanol. For capsules, weigh the contents of 20 capsules in a small dish, mix the powder and weigh out a portion equivalent to 2 capsules. Dissolve in the least amount of ethanol, filter into a 100-ml standard flask and dilute to the mark with ethanol. For suppositories, dissolve 4 suppositories in the least amount of ethanol, homogenize, transfer to a 100-ml standard flask and make up to the mark with ethanol. For suspensions, shake well and transfer 20 ml to a 150-ml stoppered conical flask. Add 50 ml of ethanol, shake for 2 min, let stand for 10 min, filter into a 100-ml standard flask, and wash the precipitate and dilute to the mark with ethanol. For injections, dissolve the contents of 5 vials in the least amount of water, transfer quantitatively to a 50-ml standard flask and dilute to the mark with demineralized water. Mix, transfer a 10-ml aliquot to a 100-ml standard flask and make up to the mark with water. Eye and ear drops are used directly without further dilution.

Transfer a 1- or 2-ml aliquot of the sample solution to a 100-ml conical flask with a ground-glass neck and a side-arm with bubbler. Add 10 ml of $0.05M$ hydrochloric acid and attach a water condenser. Heat on a sand-bath while passing carbon dioxide ($\sim$ 50 bubbles/min) through the side-arm. When the solution starts boiling introduce 50–100 mg of cadmium metal turnings, previously washed with $6M$ hydrochloric acid and thoroughly with doubly distilled water, and continue boiling for 15-20 min in a carbon dioxide atmosphere. Cool, transfer the reaction solution to a 50-ml standard flask and make up to the mark with doubly distilled water. Pure powders or pharmaceutical preparations containing chloramphenicol palmitate, stearate and succinate should be hydrolysed before reduction with cadmium. Transfer an aliquot containing 10-15 mg of the ester to a test-tube (20 × 2 cm), add 1 ml of $1M$ alcoholic potassium hydroxide solution and shake the mixture for 2 min, then transfer it to the reaction vessel, add 10 ml of $0.15M$ hydrochloric acid and complete the reaction as above.

Transfer a 1.00 ml aliquot of the test solution to a 100-ml standard flask, dilute to the mark with $0.05M$ hydrochloric acid and mix. Aspirate into an air-acetylene flame and measure the absorbance at 228.8 nm. Compare with a calibration graph prepared by treating 0.50-3.00 ml portions of standard cadmium stock solution (0.1 mg/ml) in the same way (1 mg of Cd ≡ 0.959 mg of chloramphenicol, 1.667 mg of chloramphenicol palmitate, 1.749 mg of chloramphenicol stearate, 1.330 mg of chloramphenicol succinate).

Accuracy and precision. The average recovery is 98%, the standard deviation being 2%.

14.3.4 Determination of Vitamin B_1 [24]

Principle. Vitamin B_1 undergoes a desulphurization reaction with potassium plumbite. After the precipitation of lead sulphide is completed, the remaining lead is measured.

Reagents. Potassium plumbite ($0.02M$) prepared by dissolving 6.624 g of lead nitrate in 1 litre of $0.2M$ potassium hydroxide.

Limit of detection. 1 mg/ml.

Procedure. Transfer 5-10 mg of vitamin B_1 to the bottom of a Pyrex test-tube (10 × 1 cm). Add 3 potassium hydroxide pellets (∼ 0.2-0.3 g) and place the tube in a 250-280°C sand-bath for 5-8 min. Cool to room temperature, add 5 ml of potassium plumbite stock solution to the alkaline reaction product. Shake, place in a boiling water-bath for 2 min, add 5 ml of $0.2M$ disodium EDTA, and cool. Transfer to a 25-ml standard flask, filter and dilute to volume with demineralized water. Transfer 5.0 ml of the filtrate to a 100-ml standard flask and dilute to volume with $0.01M$ nitric acid. Shake, aspirate into an acetylene-air flame, and measure the absorbance of lead at 217 nm. Prepare a blank and subtract the absorbance of the blank from those of the samples and standards. Compare with a calibration graph prepared by diluting 10.0 ml of potassium plumbite stock solution to 100.0 ml with water. Transfer portions of dilute plumbite solution (Pb 0.1 mg/ml) ranging from 1.0 to 10.0 ml to separate 100-ml standard flasks, add 1 ml of $0.2M$ EDTA solution to each, and dilute to volume with $0.01M$ nitric acid. Mix, aspirate into the flame, and measure the absorbance at 217 nm.

Accuracy and precision. The average recovery is 99.1% and the mean standard deviation 0.8%.

14.3.5 Determination of Ethambutol [25]

Principle. Ethambutol reacts with copper(II) in basic medium to form a copper chelate extractable into MIBK.

$$C_2H_5-\underset{\underset{CH_2}{|}}{\underset{|}{\overset{|}{CH}}-CH_2OH} \quad + \quad Cu^{2+} \quad \longrightarrow \quad \text{[chelate]} \quad + \quad 2H^+$$

Reagents. Copper sulphate solution (2%), sodium hydroxide solution (0.1M), sodium chloride and MIBK.

Limit of detection. 50 μg/ml.

Procedure. Transfer 1.0 ml of sample solution containing up to 400 μg of ethambutol to a stoppered 10-ml test-tube. Add 0.4 ml of 0.1M sodium hydroxide, 0.2 ml of 2% copper sulphate solution and 0.4 g of solid sodium chloride. Dilute to about 6 ml and add 10 ml of MIBK. Shake, leave to stand, then aspirate the organic layer into an air–acetylene flame and measure the absorbance of copper at 324.7 nm. Run a blank and compare with a calibration graph similarly prepared with 0.30, 0.50, 0.70 and 1.00 ml of standard ethambutol solution (400 μg/ml).

Accuracy and precision. The recovery is 98%, the standard deviation 2%.

14.3.6 Determination of Phthalic Acid [26,27]

Principle. Phthalic acid is extracted as an ion-pair complex of the biphthalate ion with bis(neocuproine)copper(I).

$$C_6H_4(COOH)(COO^-) + [Cu(neocuproine)_2]^+ \longrightarrow$$

$$[Cu(neocuproine)_2]^+ [C_6H_5(COOH)COO]^-$$

Reagents. Copper sulphate solution ($10^{-2}M$), phosphate buffer of pH 8 (equal volumes of 0.25M KH$_2$PO$_4$ and 0.25M Na$_2$HPO$_4$), $2 \times 10^{-3}M$ neocuproine in MIBK, 5% hydroxylamine sulphate solution and standard 0.01M phthalic acid.

Limit of detection. $4 \times 10^{-6}M$.

Procedure. Transfer 1.0 ml of copper sulphate solution, 2 ml of 5% hydroxylamine sulphate solution, 5 ml of the phosphate buffer solution and 5.0 ml of $10^{-4}M$ standard phthalic acid solution to a 100-ml separating funnel, in that order. Dilute the mixture to 25 ml with demineralized water and add 10 ml of $2 \times 10^{-3}M$ neocuproine in MIBK. Shake the funnel for 2 min. Leave to stand for

20 min and separate the organic layer. Dry the organic layer with 1 g of anhydrous sodium sulphate. Measure the absorbance of the copper in the organic layer at 324.7 nm, using an air–acetylene flame. Run a blank.

Accuracy and precision. The relative standard deviation is 1% and the recovery is almost quantitative.

14.3.7 Determination of Anionic Surfactants [28,29]

Principle. Anionic surfactants (LAS^-) are extracted into chloroform as an ion-association compound with the bis(ethylenediamine)copper(II) cation.

$$[Cu(en)_2(H_2O)_2]^{2+} + 2LAS^- \longrightarrow [Cu(en)_2(LAS)_2]$$

Reagents. Bis(ethylenediamine)copper(II) reagent prepared by dissolving 124.6 g of copper sulphate pentahydrate and 99.2 g of ammonium sulphate in warm water, followed by addition of 90.2 g ($\sim$ 100 ml) of ethylenediamine and dilution to 1 litre with demineralized water. Standard reference anionic surfactant (e.g. linear alkyl sulphonic acid of mean molecular weight 318) solution (750 μg/l.).

Compound tested. Linear alkyl sulphonic acid in fresh water and sea-water samples.

Procedure. Place a 750-ml water sample containing not more than 50 μg of LAS per litre in a 1-litre separating funnel. Adjust the pH to 5–9 and add 25 ml of bis(ethylenediamine)copper(II) reagent. Add 20 ml of chloroform, shake the funnel for 1 min and let stand until the phases separate. Transfer about 13 ml of the chloroform layer to a 15-ml graduated centrifuge tube. Add 1 ml of demineralized water to prevent evaporation of chloroform, stopper the tube and centrifuge it at 2000 rpm for 30 min. Transfer 10 ml of the organic extract to a 25-ml standard flask and make up to the mark with $0.1M$ nitric acid. Shake the flask for 30 sec, let the phases separate and aspirate the aqueous layer into an air–acetylene flame for copper measurement at 324.7 nm. Alternatively inject a 50-μl aliquot of the chloroform extract into a graphite tube furnace. Use the temperature programme: 100°C for 30 sec (evaporation), 950°C for 30 sec (ashing), 2500°C for 10 sec (atomization) and 3000°C for 10 sec (burning off). Run a blank on 750 ml of demineralized water.

Accuracy and precision. The mean recovery of LAS from sea-water ranges from 80 to 98%, the standard deviation being 0.01–0.13 μg/ml, and the limit of detection 0.03 μg/ml.

14.3.8 Determination of Non-Ionic Surfactants [30]

Principle. Non-ionic surfactants are extracted into 1,2-dichlorobenzene as a neutral adduct with potassium tetrathiocyanatozincate(II).

Reagents. Standard (1500 mg/l.) reference non-ionic surfactant such as Triton X-100 containing an average of approximately 10 ethoxy units; this solution is diluted as required. Zinc thiocyanate prepared by dissolving 116 g of zinc sulphate heptahydrate, 312 g of potassium thiocyanate and 40 g of potassium acetate in hot water and diluting to 2 litres with demineralized water, and extracting the reagent with three 50-ml portions of 1,2-dichlorobenzene previously purified by passage through a 20-cm column of activated alumina.

Limit of detection. The limit of detection is 0.03 mg/l. (as Triton X-100).

Procedure. Transfer a 150-ml aliquot of the water sample, containing not more than 2 mg of non-ionic surfactant per litre to a 500-ml separating funnel fitted with a Teflon stopcock. Adjust the pH to 6–8, add 50 ml of zinc thiocyanate reagent and 20.0 ml of 1,2-dichlorobenzene, shake the funnel for 5 min and allow it to stand until the phases separate. Run about 13 ml of the organic phase into a dry 15-ml graduated centrifuge tube. Centrifuge at 2500 rpm for 30 min at room temperature. Transfer a 10.0-ml aliquot of the clarified extract into a 25-ml standard flask and add 10 ml of $0.1M$ hydrochloric acid. Stopper the flask, shake it vigorously for 2 min and allow the phases to separate. Aspirate the aqueous layer directly into an air–acetylene flame and measure the absorbance of the zinc at 213.9 nm. Compare with a calibration graph made with standard Triton X-100. Run a blank with 150 ml of distilled water.

Accuracy and precision. The average recovery ranges from 98 to 103% for water samples containing > 0.5 mg of Triton X-100 per litre.

REFERENCES

[1] S. S. M. Hassan and M. H. Eldesouki, *Z. Anal. Chem.,* **259**, 346 (1972).
[2] A. I. Vogel, *Quantitative Inorganic Analysis*, p. 435, Longmans, London (1957).
[3] Y. Kidani, H. Takemura and H. Koike, *Bunseki Kagaku,* **23**, 212 (1974).
[4] S. S. M. Hassan and M. H. Eldesouki, *Mikrochim. Acta,* 261 (1981II).
[5] R. V. Smith and M. A. Nessen, *Microchem. J.,* **17**, 638 (1972).
[6] E. D. Truscott, *Anal. Chem.,* **42**, 1657 (1970).
[7] P. Girgis-Takla and I. Chroneos, *Analyst,* **103**, 122 (1978).
[8] American Society for Testing and Materials, D 3237-74 (1975).
[9] G. R. Sirota and J. F. Uthe, *Anal. Chem.,* **49**, 823 (1977).
[10] K. S. Subramanian and J. C. Meranger, *Clin. Chem.,* **27**, 1866 (1981).
[11] W. W. Harrison J. P. Yurachek and C. A. Benson, *Clin. Chim. Acta,* **23**, 83 (1969).
[12] B. Tan, P. Melius and M. V. Kilgore, *Anal. Chem.,* **52**, 602 (1980).
[13] T. Mitsui and T. Kojima, *Bunseki Kagaku,* **26**, 182 (1977).
[14] T. Mitsui and Y. Fujimura, *Bunseki Kagaku,* **23**, 1309 (1974).
[15] P. J. Oles and S. Siggia, *Anal. Chem.,* **45**, 2150 (1973).
[16] S. S. M. Hassan, *Talanta,* **28**, 89 (1981).
[17] S. S. M. Hassan and M. H. Eldesouki, *J. Assoc. Off. Anal. Chem.,* **64**, 1158 (1981).
[18] J. S. Marhevka and S. Siggia, *Anal. Chem.,* **51**, 1259 (1979).
[19] R. P. D'Alonzo, A. P. Carpenter, Jr., S. Siggia and P. C. Uden, *Anal. Chem.,* **50**, 326 (1978).
[20] S. J. Simon and D. F. Boltz, *Microchem. J.,* **20**, 468 (1975).
[21] T. C. Woodis, Jr., B. Hunter and F. J. Johnson, *J. Assoc. Off. Anal. Chem.,* **59**, 22 (1976).
[22] L. F. Corominas, *J. Assoc. Off. Anal. Chem.,* **60**, 1214 (1977).
[23] S. S. M. Hassan and M. H. Eldesouki, *Talanta,* **26**, 531 (1979).

[24] S. S. M. Hassan, M. T. Zaki and M. H. Eldesouki, *J. Assoc. Off. Anal. Chem.*, **62**, 315 (1979).

[25] A. V. Kovatsis and M. A. Tsougas, *Arzneimittel-Forschung Drug Research*, **28**, 248 (1978).

[26] T. Kumamaru, Y. Hayashi, N. Okamoto, E. Tao and Y. Yamamoto, *Anal. Chim. Acta*, **35**, 524 (1966).

[27] T. Kumamaru, *Anal. Chim. Acta*, **43**, 19 (1968).

[28] P. T. Crisp, J. M. Eckert, N. A. Gibson, G. F. Kirkbright and T. S. West, *Anal. Chim. Acta*, **87**, 97 (1976).

[29] P. T. Crisp, J. M. Eckert and N. A. Gibson, *Anal. Chim. Acta*, **78**, 391 (1975).

[30] P. T. Crisp, J. M. Eckert and N. A. Gibson, *Anal. Chim. Acta*, **104**, 93 (1979).

Appendix

Manufacturers of atomic-absorption spectrometers and related equipment*

Baird-Atomic Ltd., Springwood Industrial Estate, Braintree, Essex CM7 7YL, U.K.

Baird-Atomic Inc., 125 Middlesex Turnpike, Bedford, Mass. 01730, U.S.A.

Beckman Instruments GmbH, 8 Munich 40, Frankfurter Ring 115, German Federal Republic.

Beckman-RIIC Ltd., Cressex Industrial Estate, High Wycombe, Bucks. HP12 3NR, U.K.

Cathodeon Ltd., Nuffield, Cambridge CB4 1TF, U.K.

Erdmann & Grün KG Feinmechanik & Optik, Solmsestr. 90, Postfach 1580, D-6330 Wetzlar, W. Germany.

GCA/MC Pherson Instrument, 530 Main Street, Acton, Mass. 01720, U.S.A.

Hitachi Ltd., Nissei Sangyo Co. Ltd., Mori 17th Building, 26-5 Toranomon, 1-Chome, Minato-Ku, Tokyo, Japan.

Instrumentation Laboratory Inc., 68 Jonspin Road, Wilmington, Mass. 01887, U.S.A.

Instrumentation Laboratory (U.K.) Ltd., Station House, Stamford New Road, Altrincham, Cheshire WA14 1BR, U.K.

Jarrell-Ash Division, Fisher Scientific Co., 590 Lincoln Street, Waltham, Mass. 02154, U.S.A.

S. & J. Juniper & Co., 7 Potter Street, Harlow, Essex, U.K.

Nissei Sangyo Instruments Inc., 392 Potorero Avenue, Sunnyvale, Calif. 94086, U.S.A.

Nissei Sangyo GmbH, 4 Düsseldorf, Am Wehrhahn 41, German Federal Republic.

Perkin-Elmer Corp., Main Avenue, Norwalk, Conn. 06856, U.S.A.

Perkin-Elmer Ltd., Post Office Lane, Beaconsfield, Bucks. HP9 1QA, U.K.

Perkin-Elmer & Co. GmbH, Postfach 1120, D-7770 Überlingen, German Federal Republic.

Philips Science & Industry, Pye Unicam, York Street, Cambridge CB12 2PX, U.K.

---

* A survey of the general features of the commercially available instruments is given by J. A. C. Broekaert, *Spectrochim. Acta*, **36B**, 931 (1981).

Pye Unicam Ltd., York Street, Cambridge CB1 2PX, U.K.

Rank-Hilger, Westwood Industrial Estate, Ramsgate Road, Margate, Kent CT9 4JL, U.K.

Shimadzu-Seisakusho Ltd., 14-5 Uchikanda, 1-Chome, Chiyoda-Ku, Tokyo 101, Japan.

Varian Techtron Pty, 679 Springvale Road, Mulgrave, Victoria 3170, Australia.

Varian Associates Ltd., Instrument Group, 28 Manor Road, Walton on Thames, Surrey, U.K.

Varian Instrument Division, 611 Hansen Way, Palo Alto, Calif. 94303, U.S.A.

VEB Carl Zeiss Jena, Carl-Zeiss Str. 1, 69 Jena, German Democratic Republic.

Westinghouse Electric Corporation, I & G Tube Division, Westinghouse Circle, Horseheads, New York 14845, U.S.A.

Westinghouse Electric International, S.A., 1 Curfew Yard, Thames Street, Windsor, Berkshire, U.K.

Index

C